Berger   Automating with STEP 7 in LAD and FBD

Automating with STEP 7 in LAD and FBD

Programmable Controllers
SIMATIC S7-300/400

by Hans Berger

Publicis MCD Verlag

Die Deutsche Bibliothek – CIP-Cataloguing-in-Publication-Data

A catalogue record for this publication is available from Die Deutsche Bibliothek

The programming examples concentrate on describing the LAD and FBD functions and providing SIMATIC S7 users with programming tips for solving specific tasks with this controller.

The programming examples given in the book do not pretend to be complete solutions or to be executable on future STEP 7 releases or S7-300/400 versions. Additional care must be taken in order to comply with the relevant safety regulations.

The author and publisher have taken great care with all texts and illustrations in this book. Nevertheless, errors can never be completely avoided. The publisher and the author accept no liability, regardless of legal basis, for any demage resulting from the use of the programming examples

ISBN 3-89578-131-2

Editor: Siemens Aktiengesellschaft, Berlin and Munich
Publisher: Publicis MCD Verlag, Erlangen and Munich
© 2000 by Publicis MCD Werbeagentur GmbH, Munich
This publication and all parts thereof are protected by copyright. All rights reserved.
Any use of it outside the strict provisions of the copyright law without the consent of the publisher is forbidden and will incur penalties. This applies particularly to reproduction, translation, microfilming or other processing, and to storage or processing in electronic systems. It also applies to the use of extracts from the text.

Printed in Germany

Preface

The new SIMATIC automation system unites all the subsystems of an automation solution under uniform system architecture into a homogeneous whole from the field level right up to process control. This is achieved with integrated configuring and programming, data management and communications with programmable controllers (SIMATIC S7), automation computers (SIMATIC M7) and control systems (SIMATIC C7). With the programmable controllers, three series cover the entire area of process and production automation: S7-200 as compact controllers ("micro PLCs"), S7-300 and S7-400 as modularly expandable controllers for the low-end and high-end performance ranges.

STEP 7, a further development of STEP 5, is the programming software for the new SIMATIC. Microsoft Windows 95/98 or Microsoft Windows NT has been chosen as the operating system, to take advantage of the familiar user interface of standard PCs (windows, mouse operation).

For block programming STEP 7 provides programming languages that comply with DIN EN 6.1131-3: STL (statement list; an Assembler-like language), LAD (ladder logic; a representation similar to relay logic diagrams), FBD (function block diagram) and the SCL optional package (a Pascal-like high-level language). Several optional packages supplement these languages: S7-GRAPH (sequential control), S7-HiGraph (programming with state-transition diagrams) and CFC (connecting blocks; similar to function block diagram). The various methods of representation allow every user to select the suitable control function description. This broad adaptability in representing the control task to be solved significantly simplifies working with STEP 7.

This book contains the description of the LAD and FBD programming languages for S7-300/400. In the first section, the book introduces the S7-300/400 automation system and explains the basic handling of STEP 7. The next section addresses first-time users or users changing from relay contactor controls; the "Basic Functions" of a binary control are described here. The digital functions explain how digital values are combined; for example, basic calculations, comparisons, data type conversion. With LAD or FBD, you can control program processing (program flow) and design structured programs. As well as a cyclically processed main program, you can also incorporate event-driven program sections as well as influence the behavior of the controller at startup and in the event of errors/faults.

The book concludes with a general overview of the system functions and the function set for LAD and FBD.

The contents of this book describe Version 5.0 of the STEP 7 programming software with Service Pack 2.

Erlangen, January 2000   Hans Berger

The Contents of the Book at a Glance

Overview of the
S7-300/400 programmable
logic controller

PLC functions
comparable to a contactor
control system

Handling numbers and
digital operands

| Introduction | Basic functions | Digital functions |

1 SIMATIC S7-300/400 Programmable Controller

Structure of the Programmable Controller (Hardware Components of S7-300/400)

Memory Areas;

Distributed I/O (PROFIBUS DP);

Communications (Subnets);

Modul Addresses;

Addresses Areas

2 STEP 7 Programming Software

Editing Projects;

Configuring Stations;

Configuring the Network;

Symbol Editor;

LAD/FBD Program Editor;

Online Mode;
Testing LAD and FBD Programs

3 SIMATIC S7 Program

Program Processing;

Block Types;

Programming Code Blocks ans Data Blocks;

Addressing Variables, Constant Representation, Data Types Description

4 Binary Logic Operations

AND, OR and Exclusive OR Functions;

Nesting Functions

5 Memory Functions

Assign, Set and Reset;
Midline Outputs;
Edge Evaluation;

Example of a Conveyor Belt Control System

6 Move Functions

Load and Transfer Functions;

MOVE Box;

System Functions for Data Transfer

7 Timers

Start SIMATIC Timers with Five Different Characteristics, Resetting and Scanning

IEC Timer Functions

8 Counters

SIMATIC Counters;

Count up, Count down, Set, Reset und Scan Counters;

IEC Counter Functions

9 Comparison Functions

Comparison According to Data Types INT, DINT and REAL

10 Arithmetic Functions

Four-function Math with INT, DINT and REAL numbers;

11 Mathematical Functions

Trigonometric Functions;

Arc Functions;

Squaring, Square-root Extraction, Exponentiation, Logarithms

12 Conversion Functions

Data Type Conversion,

Complement Formation

13 Shift Functions

Shifting and Rotating

14 Word Logic

Processing a
AND, OR and Exclusive OR Word Logic Operation

Controlling program execution, block functions

Processing the user program

Supplements to LAD and FBD; block libraries, Function overviews

Program Flow Control

15 Status Bits

Binary Flags,
Digital Flags;

Setting and Evaluating the Status Bits;

EN/ENO Mechanism

16 Jump Functions

Unconditional Jump;
Jump if RLO = "1";
Jump if RLO = "0"

17 Master Control Relay

MCR Dependency,
MCR Area,
MCR Zone

18 Block Functions

Block Call,
Block End;

Temporary and Static Local Data, Local Instances;

Addressing Data Operands
Opening a Data Block

19 Block Parameters

Formal Parameters,
Actual Parameters;

Declarations and Assignments, "Parameter Passing"

Program Processing

20 Main Program

Program Structure;

Scan Cycle Control
(Response Time,
Start Information,
Background Scanning);

Program Functions;

Communications via Distributed I/O and Global Data; SFC and SFB Communications

21 Interrupt Handling

Hardware Interrupts;
Watchdog Interrupts;
Time-of-Day Interrupts;
Time-Delay Interrupts;
Multiprocessor Interrupt;

Handling Interrupt Events

22 Restart Characteristics

Cold Restart, Warm Restart, Complete Restart;

STOP, HOLD, Memory Reset;

Parameterizing Modules

23 Error Handling

Synchronous Errors;
Asynchronous Errors;

System Diagnostics

Appendix

24 Supplments to Graphic Programming

Block Protection
KNOW_HOW_PROTECT;

Indirect Addressing,
Pointers: General Remarks;

Brief Description of the "Message Frame Example"

25 Block Libraries

Organization Blocks;

System Function Blocks;

IEC Function Blocks;

S5-S7 Converting Blocks;

TI-S7 Converting Blocks;

PID Control Blocks;

Communication Blocks

26 Function Set LAD

Basic Functions;

Digital Functions;

Program Flow Control

27 Function Set FBD

Basic Functions;

Digital Functions;

Program Flow Control

The Contents of the Diskette at a Glance

The present book provides many figures representing the use of the LAD and FBD programming languages. You can find all the program sections shown in the book on the diskette accompanying the book in two libraries LAD_Book and FBD_Book. When dearchived with the Retrieve function, these libraries occupy approximately 2 MB (dependent on the PG/PC file system used).

The libraries LAD_Book and FBD__Book contain eight programs that are essentially illustrations of the graphical representation. Two extensive examples show the programming of functions, function blocks and local instances (Conveyor Example) and the handling of data (Message Frame Example). All the examples contain symbols and comments

Library LAD_Book

Data Types	
Examples of Definition and Application	
FB 101	Elementary Data Types
FB 102	Complex Data Types
FB 103	Parameter Types

Basic Functions	
LAD Representation Examples	
FB 104	Chapter 4: Seroes and Parallel Circuits
FB 105	Chapter 5: Memory Functions
FB 106	Chapter 6: Move Functions
FB 107	Chapter 7: Timer Functions
FB 108	Chapter 8: Counter Functions

Digital Functions	
LAD Representation Examples	
FB 109	Chapter 9: Comparison Functions
FB 110	Chapter 10: Arithmetic Functions
FB 111	Chapter 11: Math Functions
FB 112	Chapter 12: Conversion Functions
FB 113	Chapter 13: Shift Functions
FB 114	Chapter 14: Word Logic

Program Flow Control	
LAD Representation Examples	
FB 115	Chapter 15: Status Bits
FB 116	Chapter 16: Jump Functions
FB 117	Chapter 17: Master Control Relay
FB 118	Chapter 18: Block Functions
FB 119	Chapter 19: Block Parameters

Program Processing	
Examples of SFC Calls	
FB 120	Chapter 20: Main Program
FB 121	Chapter 21: Interrupt Processing
FB 122	Chapter 22: Start-up Characteristics
FB 123	Chapter 23: Error Handling

Conveyor Example	
Examples of Basic Functions and Local Instances	
FC 11	Belt Control
FC 12	Conter Control
FB 20	Feed
FB 21	Conveyor Belt
FB 22	Parts Counter

Message Frame Example	
Data Handling Examples	
UDT 51	Data Structure for the Frame Header
UDT 52	Data Structure for a Message
FB 51	Generate Message Frame
FB 52	Store Message Frame
FC 51	Time-of-day Check
FC 52	Copy Data Area with indirect Addressing

General Examples	
FC 41	Range Monitor
FC 42	Limit Value Detection
FC 43	Compount Interest Calculation
FC 44	Doubleword-wise Edge Evaluation

The libraries are supplied in archived form. Before you can start working with them, you must dearchive the libraries. Select the FILE → DEARCHIVE menu item in the SIMATIC Manager and follow the instructions (see also the README.TXT file on the disk).

To try the programs out, set up a project corresponding to your hardware configuration and then copy the program, including the symbol table from the library to the project. Now you can call the example programs, adapt them for your own purposes and test them online.

Library FBD_Book

Data Types Examples of Definition and Application	**Program Processing** Examples of SFC Calls
FB 101 Elementary Data Types FB 102 Complex Data Types FB 103 Parameter Types	FB 120 Chapter 20: Main Program FB 121 Chapter 21: Interrupt Processing FB 122 Chapter 22: Start-up Characteristics FB 123 Chapter 23: Error Handling
Basic Functions FBD Representation Examples	**Conveyor Example** Examples of Basic Functions and Local Instances
FB 104 Chapter 4: Seroes and Parallel Circuits FB 105 Chapter 5: Memory Functions FB 106 Chapter 6: Move Functions FB 107 Chapter 7: Timer Functions FB 108 Chapter 8: Counter Functions	FC 11 Belt Control FC 12 Conter Control FB 20 Feed FB 21 Conveyor Belt FB 22 Parts Counter
Digital Functions FBD Representation Examples	**Message Frame Example** Data Handling Examples
FB 109 Chapter 9: Comparison Functions FB 110 Chapter 10: Arithmetic Functions FB 111 Chapter 11: Math Functions FB 112 Chapter 12: Conversion Functions FB 113 Chapter 13: Shift Functions FB 114 Chapter 14: Word Logic	UDT 51 Data Structure for the Frame Header UDT 52 Data Structure for a Message FB 51 Generate Message Frame FB 52 Store Message Frame FC 51 Time-of-day Check FC 52 Copy Data Area with indirect Addressing
Program Flow Control FBD Representation Examples	**General Examples**
FB 115 Chapter 15: Status Bits FB 116 Chapter 16: Jump Functions FB 117 Chapter 17: Master Control Relay FB 118 Chapter 18: Block Functions FB 119 Chapter 19: Block Parameters	FC 41 Range Monitor FC 42 Limit Value Detection FC 43 Compount Interest Calculation FC 44 Doubleword-wise Edge Evaluation

Automating with STEP 7

Automating with STEP 7

This double page shows the basic procedure for using the STEP 7 programming software.

Start the SIMATIC Manager and set up a new project or open an existing project. All the data for an automation task are stored in the form of objects in a project. When you set up a project, you create containers for the accumulated data by setting up the required stations with at least the CPUs; then the containers for the user programs are also created. You can also create a program container direct in the project.

In the next steps, you configure the hardware and, if applicable, the communications connections. Following this, you create and test the program.

The order for creating the automation data is not fixed. Only the following general regulation applies: if you want to process objects (data), they must exist; if you want to insert objects, the relevant containers must be available.

You can interrupt processing in a project at any time and continue again from any location the next time you start the SIMATIC Manager.

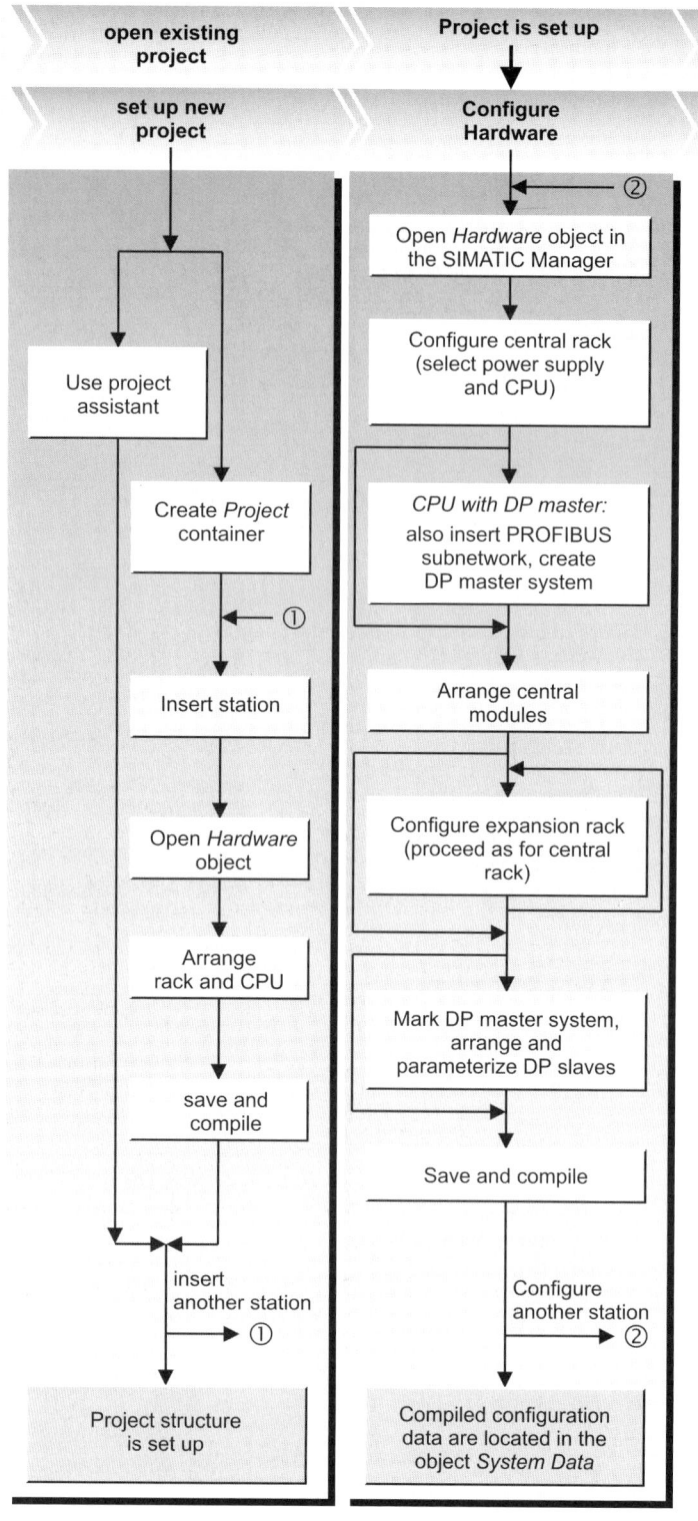

10

Automating with STEP 7

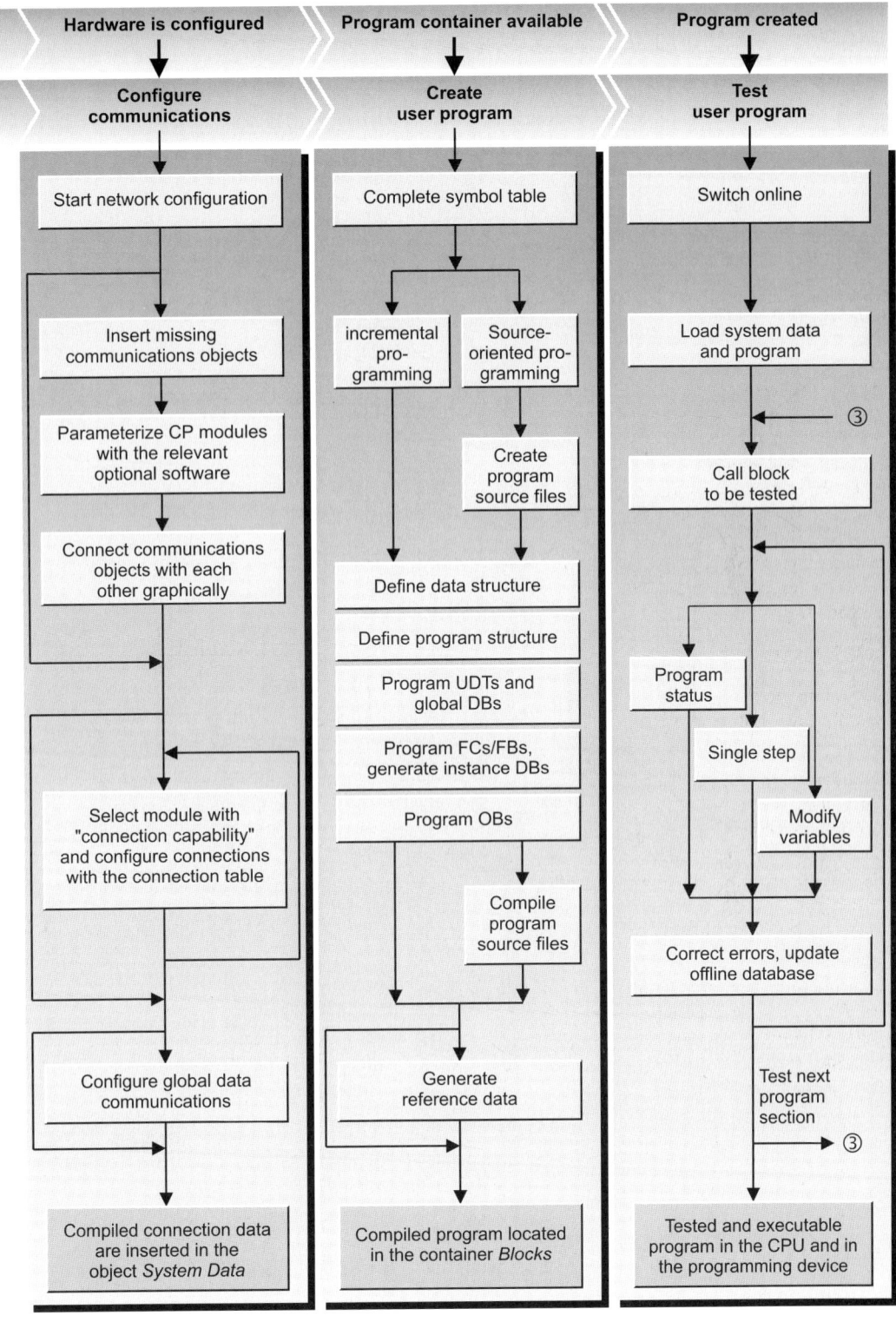

Contents

Indroduction		**19**
1	**SIMATIC S7-300/400 Programmable Controller**	**20**
1.1	Structure of the Programmable Controller	20
1.1.1	Components	20
1.1.2	S7-300 Station	20
1.1.3	S7-400 Station	22
1.1.4	CPU Memory Areas	23
1.1.5	Memory Card	24
1.1.6	System Memory	24
1.2	Distributed I/O	25
1.2.1	DP Master System	25
1.2.2	DP Master	25
1.2.3	DP Slaves	25
1.2.4	Connection to PROFIBUS-PA	27
1.2.5	Connection to AS-Interface	27
1.2.6	Connection to a Serial Interface	28
1.3	Communications	28
1.3.1	Introduction	29
1.3.2	Subnets	31
1.3.3	Communications Services	32
1.3.4	Connections	33
1.4	Module Addresses	33
1.4.1	Signal Path	33
1.4.2	Slot Address	34
1.4.3	Module Start Address	35
1.4.4	Diagnostics Address	35
1.4.5	Address for Bus Nodes	35
1.5	Address Areas	36
1.5.1	User Data Area	36
1.5.2	Process Image	37
1.5.3	Bit Memory	38

2	**STEP 7 Programming Software**	**39**
2.1	STEP 7 Basic Package	39
2.1.1	Installation	39
2.1.2	Authorization	39
2.1.3	SIMATIC Manager	40
2.1.4	Projects and Libraries	41
2.1.5	Online Help	43
2.2	Editing Projects	43
2.2.1	Creating Projects	43
2.2.2	Managing, Rearranging and Archiving	45
2.2.3	Project Versions	45
2.3	Configuring Stations	46
2.3.1	Arranging Modules	48
2.3.2	Addressing Modules	48
2.3.3	Parameterizing Modules	48
2.3.4	Networking Modules with MPI	49
2.4	Configuring the Network	49
2.4.1	Configuring the Network View	51
2.4.2	Configuring a DP Master System with the Network Configuration	51
2.4.3	Configuring Connections	52
2.4.4	Loading the Connection Data	55
2.5	Creating the S7 Program	55
2.5.1	Introduction	55
2.5.2	Symbol Table	56
2.5.3	Program Editor	57
2.5.4	Updating or generating source files	58
2.5.5	Address Priority	60
2.5.6	Reference Data	61
2.6	Online Mode	62
2.6.1	Connecting a PLC	62
2.6.2	Protecting the User Program	63
2.6.3	CPU Information	63
2.6.4	Loading the User Program into the CPU	64
2.6.5	Block Handling	64

2.7	Testing the Program	65
2.7.1	Diagnosing the Hardware	66
2.7.2	Determining the Cause of a STOP	66
2.7.3	Monitoring and Modifying Variables	66
2.7.4	Forcing Variables	68
2.7.5	Enabling Peripheral Outputs	69
2.7.6	LAD/FBD Program Status	69
3	**SIMATIC S7 Program**	**71**
3.1	Program Processing	71
3.1.1	Program Processing Methods	71
3.1.2	Priority Classes	72
3.1.3	Specifications for Program Processing	73
3.2	Blocks	74
3.2.1	Block Types	74
3.2.2	Block Structure	75
3.2.3	Block Properties	76
3.3	Programming Code Blocks	78
3.3.1	Generating blocks	78
3.3.2	Editing LAD Elements	82
3.3.3	Editing FBD Elements	84
3.4	Programming Data Blocks	86
3.4.1	Creating Blocks	86
3.4.2	Types of Data Blocks	86
3.4.3	Block Window	86
3.5	Variables, Constants and Data Types	87
3.5.1	General Remarks Concerning Variables	87
3.5.2	Addressing Variables	88
3.5.3	Overview of Data Types	90
3.5.4	Elementary Data Types	91
3.5.5	Complex Data Types	96
3.5.6	Parameter Types	99
3.5.7	User Data Types	99

	Basic Functions	**101**
4	**Binary Logic Operations**	**102**
4.1	Series and Parallel Circuits (LAD)	102
4.1.1	NO Contact and NC Contact	102
4.1.2	Series Circuits	103
4.1.3	Parallel Circuits	103
4.1.4	Combinations of Binary Logic Operations	104
4.1.5	Negating the Result of the Logic Operation	104
4.2	Binary Logic Operations (FBD)	106
4.2.1	Elementary Binary Logic Operations	106
4.2.2	Combinations of Binary Logic Operations	109
4.2.3	Negating the Result of the Logic Operation	110
4.3	Taking Account of the Sensor Type	110
5	**Memory Functions**	**113**
5.1	LAD Coils	113
5.1.1	Single Coil	113
5.1.2	Set and Reset Coil	113
5.1.3	Memory Box	115
5.2	FBD Boxes	117
5.2.1	Assign	117
5.2.2	Set and Reset Box	118
5.2.3	Memory Box	118
5.3	Midline Outputs	120
5.3.1	Midline Outputs in LAD	121
5.3.2	Midline Outputs in FBD	123
5.4	Edge Evaluation	123
5.4.1	How Edge Evaluation Works	123
5.4.2	Edge Evaluation in LAD	125
5.4.3	Edge Evaluation in FBD	125
5.5	Binary Scaler	126
5.5.1	Solution in LAD	126
5.5.2	Solution in FBD	127
5.6	Example of a Conveyor Control System	128

6	**Move Functions**	**132**	8.5	IEC Counters	154
6.1	General	132	8.5.1	Up Counter SFB 0 CTU	154
6.2	MOVE Box	133	8.5.2	Down Counter SFB 1 CTD	155
6.2.1	Processing the MOVE Box	133	8.5.3	Up/down Counter SFB 2 CTUD	155
6.2.2	Moving Operands	134	8.6	Parts Counter Example	155
6.2.3	Moving Constants	135			
6.3	System Functions for Data Transfer	136	**Digital Functions**		**160**
6.3.1	ANY Pointer	136	**9**	**Comparison Functions**	**161**
6.3.2	Copy Data Area	136	9.1	Processing a ComparisonFunction	161
6.3.3	Uninterruptible Copying of a Data Area	137	9.2	Description of the Comparison Functions	163
6.3.4	Fill Data Area	137	**10**	**Arithmetic Functions**	**165**
7	**Timers**	**139**	10.1	Processing an Arithmetic Function	165
7.1	Programming a Timer	139			
7.1.1	General Representation of a Timer	139	10.2	Calculating with Data Type INT	167
7.1.2	Starting a Timer	140	10.3	Calculating with Data Type DINT	168
7.1.3	Specifying the Duration of Time	141	10.4	Calculating with Data Type REAL	168
7.1.4	Resetting A Timer	142	**11**	**Mathematical Functions**	**170**
7.1.5	Checking a Timer	142	11.1	Processing a Mathematical Function	170
7.1.6	Sequence of Timer Operations	142			
7.1.7	Timer Box in a Rung (LAD)	143	11.2	Trigonometric Functions	172
7.1.8	Timer Box in a Logic Circuit (FBD)	143	11.3	Arc Functions	172
7.2	Pulse Timer	143	11.4	Miscellaneous Mathematical Functions	172
7.3	Extended Pulse Timer	144			
7.4	On-Delay Timer	144	**12**	**Conversion Functions**	**175**
7.5	Retentive On-Delay Timer	146	12.1	Processing a Conversion Function	175
7.6	Off-Delay Timer	147	12.2	Conversion of INT and DINT Numbers	177
7.7	IEC Timers	147			
7.7.1	Pulse Timer SFB 3 TP	148	12.3	Conversion of BCD Numbers	178
7.7.2	On-Delay Timer SFB 4 TON	148	12.4	Conversion of REAL Numbers	178
7.7.3	Off-Delay Timer SFB 5 TOF	149	12.5	Miscellaneous Conversion Functions	180
8	**Counters**	**150**	**13**	**Shift Functions**	**181**
8.1	Programming a Counter	150	13.1	Processing a Shift Function	181
8.2	Setting and Resetting Counters	153	13.1.1	Representation	181
8.3	Counting	153	13.2	Shift	183
8.4	Checking a Counter	154	13.3	Rotate	184

14	Word Logic	185
14.1	Processing a Word Logic Operation.	185
14.2	Description of the Word Logic Operations	187

Program Flow Control 188

15	Status Bits	189
15.1	Description of the Status Bits . .	189
15.2	Setting the Status Bits	190
15.3	Evaluating the Status Bits	192
15.4	Using the Binary Result	193
15.4.1	Setting the Binary Result BR . .	193
15.4.2	Main Rung, EN/ENO Mechanism	193
15.4.3	ENO in the Case of User-written Blocks 194	

16	Jump Functions.	195
16.1	Processing a Jump Function . . .	195
16.2	Unconditional Jump	196
16.3	Jump if RLO = "1".	197
16.4	Jump if RLO = "0"	197

17	Master Control Relay	198
17.1	MCR Dependency	198
17.2	MCR Area	199
17.3	MCR Zone	200
17.4	Setting and Resetting I/O Bits . .	202

18	Block Functions.	203
18.1	Block Functions for Code Blocks	203
18.1.1	Block Calls: General	204
18.1.2	Call Box	205
18.1.3	CALL Coil/Box	206
18.1.4	Block End Function	207
18.1.5	Temporary Local Data	207
18.1.6	Static Local Data	209
18.2	Block Functions for Data Blocks	212
18.2.1	Two Data Block Registers	212
18.2.2	Accessing Data Operands	212
18.2.3	Opening a Data Block	214
18.2.4	Special Points in Data Addressing	214

18.3	System Functions for Data Blocks	216
18.3.1	Creating a Data Block.	216
18.3.2	Deleting a Data Block.	216
18.3.3	Testing a Data Block	216

19	Block Parameters	218
19.1	Block Parameters in General . .	218
19.1.1	Defining the Block Parameters .	218
19.1.2	Processing the Block Parameters	219
19.1.3	Declaration of the Block Parameters	219
19.1.4	Declaration of the Function Value	220
19.1.5	Initializing Block Parameters . .	220
19.2	Formal Parameters	220
19.3	Actual Parameters.	222
19.4	"Forwarding" Block Parameters	225
19.5	Examples	225
19.5.1	Conveyor Belt Example.	225
19.5.2	Parts Counter Example	226
19.5.3	Feed Example.	228

Program Processing 233

20	Main Program	234
20.1	Program Organization.	234
20.1.1	Program Structure.	234
20.1.2	Program Organization.	235
20.2	Scan Cycle Control	236
20.2.1	Process Image Updating	236
20.2.2	Scan Cycle Monitoring Time . .	237
20.2.3	Minimum Scan Cycle Time, Background Scanning.	238
20.2.4	Response Time	239
20.2.5	Start Information	240
20.3	Program Functions	242
20.3.1	Real-Time Clock	242
20.3.2	Read System Clock	242
20.3.3	Run-Time Meter	242
20.3.4	Compressing CPU Memory. . .	243
20.3.5	Waiting and Stopping.	243
20.3.6	Multiprocessing Mode	244

Contents

20.4	Communication via Distributed I/O.	245
20.4.1	Addressing Distributed I/O . . .	245
20.4.2	Configuring Distributed I/O . . .	248
20.4.3	System Functions for Distributed I/O.	256
20.5	Global Data Communication . .	259
20.5.1	Fundamentals	259
20.5.2	Configuring GD communication	261
20.5.3	System Functions for GD Communication.	263
20.6	SFC Communication	263
20.6.1	Station-Internal SFC Communication	263
20.6.2	System Functions for Data Interchange Within a Station . .	264
20.6.3	Station-External SFC Communication	266
20.6.4	System Functions for Station-External SFC Communication. .	266
20.7	SFB Communication	269
20.7.1	Fundamentals	269
20.7.2	Two-Way Data Exchange	270
20.7.3	One-Way Data Exchange	272
20.7.4	Transferring Print Data	273
20.7.5	Control Functions	274
20.7.6	Monitoring Functions	274
21	**Interrupt Handling. . . . . . .**	**277**
21.1	General Remarks	277
21.2	Hardware Interrupts	278
21.2.1	Generating a Hardware Interrupt	279
21.2.2	Servicing Hardware Interrupts .	279
21.2.3	Configuring Hardware Interrupts with STEP 7	280
21.3	Watchdog Interrupts.	280
21.3.1	Handling Watchdog Interrupts .	280
21.3.2	Configuring Watchdog Interrupts with STEP 7	281
21.4	Time-of-Day Interrupts.	282
21.4.1	Handling Time-of-Day Interrupts	282
21.4.2	Configuring Time-of-Day Interrupts with STEP 7	283
21.4.3	System Functions for Time-of-Day Interrupts.	283
21.5	Time-Delay Interrupts	284
21.5.1	Handling Time-Delay Interrupts .	284
21.5.2	Configuring Time-Delay Interrupts with STEP 7	285
21.5.3	System Functions for Time-Delay Interrupts	285
21.6	Multiprocessor Interrupt	286
21.7	Handling Interrupts.	287
22	**Restart Characteristics . . . . .**	**289**
22.1	General Remarks	289
22.1.1	Operating Modes	289
22.1.2	HOLD Mode	290
22.1.3	Disabling the Output Modules . .	290
22.1.4	Restart Organization Blocks . . .	290
22.2	Power-Up.	291
22.2.1	STOP Mode	291
22.2.2	Memory Reset	291
22.2.3	Retentivity	291
22.2.4	Restart Parameterization	292
22.3	Types of Restart	293
22.3.1	START-UP Mode	293
22.3.2	Cold Restart	293
22.3.3	Complete Restart	295
22.3.4	Warm Restart.	295
22.4	Ascertaining a Module Address .	296
22.5	Parameterizing Modules	298
23	**Error Handling . . . .**	**301**
23.1	Synchronous Errors.	301
23.2	Synchronous Error Handling . . .	302
23.2.1	Error Filters.	302
23.2.2	Masking Synchronous Errors. . .	304
23.2.3	Unmasking Synchronous Errors .	304
23.2.4	Reading the Error Register	305
23.2.5	Entering a Substitute Value . . .	305
23.3	Asynchronous Errors	305

23.4	System Diagnostics 307
23.4.1	Diagnostic Events and Diagnostic Buffer 307
23.4.2	Writing User Entries in the Diagnostic Buffer 308
23.4.3	Evaluating Diagnostic Interrupts 308
23.4.4	Reading the System Status List . 309

Appendix 310

24	**Supplements to Graphic Programming 311**
24.1	Block Protection 311
24.2	Indirect Addressing 312
24.2.1	Pointers: General Remarks. . . . 312
24.2.2	Area Pointer 312
24.2.3	DB Pointer 312
24.2.4	ANY Pointer 314
24.2.5	"Variable" ANY Pointer 314
24.3	Brief Description of the "Message Frame Example" . . . 315

25	**Block Libraries 318**
25.1	Organization Blocks 318
25.2	System Function Blocks 319
25.3	IEC Function Blocks 321
25.4	S5-S7 Converting Blocks 321
25.5	TI-S7 Converting Blocks 323
25.6	PID Control Blocks 323
25.7	Communication Blocks 323

26	**Function Set LAD 324**
26.1	Basic Functions 324
26.2	Digital Functions 325
26.3	Program Flow Control 327

27	**Function Set FBD 328**
27.1	Basic Functions 328
27.2	Digital Functions 329
27.3	Program Flow Control 331

Index 332

Abbreviations 338

The author and publisher are always grateful to hear your responses to the contents of the book.
Publicis MCD Verlag
Postfach 3240
D-91052 Erlangen
Federal Republic of Germany
Fax: ++49 9131/72 78 38
E-mail: publishing-books@publicis-mcd.de

Indroduction

This portion of the book provides an overview of the SIMATIC S7-300/400.

The S7-300/400 programmable controller is of modular design. The modules with which it is configured can be central (in the vicinity of the CPU) or distributed without any special settings or parameter assignments having to be made. In SIMATIC S7 systems, distributed I/O is an integral part of the system. The CPU, with its various memory areas, forms the hardware basis for processing of the user programs. A load memory contains the complete user program: the parts of the program relevant to its execution at any given time are in a work memory whose short access times are the prerequisite for fast program processing.

STEP 7 is the programming software for S7-300/400 and the automation tool is the SIMATIC Manager. The SIMATIC Manager is a Windows 95/98/NT application and contains all functions needed to set up a project. When necessary, the SIMATIC Manager starts additional tools, for example to configure stations, initialize modules, and to write and test programs.

You formulate your automation solution in the STEP 7 programming languages. The SIMATIC S7 program is structured, that is to say, it consists of blocks with defined functions that are composed of networks or rungs. Different priority classes allow a graduated interruptibility of the user program currently executing. STEP 7 works with variables of various data types starting with binary variables (data type BOOL) through digital variables (e.g. data type INT or REAL for computing tasks) up to complex data types such as arrays or structures (combinations of variables of different types to form a single variable).

The first chapter contains an overview of the hardware in an S7-300/400 programmable controller, and the second chapter contains an overview of the STEP 7 programming software.

The basis for the description is the function scope for STEP 7 Version 5.0.

Chapter 3, "The SIMATIC S7 Program" serves as an introduction to the most important elements of an S7 program and shows the programming of individual blocks in the programming languages LAD and FBD. The functions and operations of LAD and FBD are then described in the subsequent chapters of the book. All the descriptions are explained using brief examples.

1 **SIMATIC S7-300/400 Programmable Controller**

   Structure of the programmable controller; distributed I/O; communications; module addresses; operand areas

2 **STEP 7 Programming Software**

   SIMATIC Manager; processing a project; configuring a station; configuring a network; writing programs (symbol table, program editor); switching online; testing programs

3 **SIMATIC S7 Program**

   Program processing with priority classes; program blocks; addressing variables; programming blocks with LAD and FBD; variables and constants; data types (overview)

1 SIMATIC S7-300/400 Programmable Controller

1.1 Structure of the Programmable Controller

1.1.1 Components

The SIMATIC S7-300/400 is a modular programmable controller comprising the following components:

- Racks
  Accommodate the modules and connect them to each other

- Power supply (PS);
  Provides the internal supply voltages

- Central processing unit (CPU)
  Stores and processes the user program

- Interface modules (IMs);
  Connect the racks to one another

- Signal modules (SMs);
  Adapt the signals from the system to the internal signal level or control actuators via digital and analog signals

- Function modules (FMs);
  Execute complex or time-critical processes independently of the CPU

- Communications processors (CPs)
  Establish the connection to subsidiary networks (subnets)

- Subnets
  Connect programmable controllers to each other or to other devices

A programmable controller (or station) may consist of several racks, which are linked to one another via bus cables. The power supply, CPU and I/O modules (SMs, FMs and CPs) are plugged into the central rack. If there is not enough room in the central rack for the I/O modules or if you want some or all I/O modules to be separate from the central rack, expansion racks are available which are connected to the central rack via interface modules (Figure 1.1).

It is also possible to connect distributed I/O to a station (see Section 1.2 "Distributed I/O").

The racks connect the modules with two buses: the I/O bus (or P bus) and the communication bus (or K bus). The I/O bus is designed for high-speed exchange of input and output signals, the communication bus for the exchange of large amounts of data. The communication bus connects the CPU and the programming device interface (MPI) with function modules and communications processors.

1.1.2 S7-300 Station

Centralized configuration

In an S7-300 controller, as many as 8 I/O modules can be plugged into the central rack. Should this single-tier configuration prove insufficient, you have two options for controllers equipped with a CPU 314 or a more advanced CPU:

- Either choose a two-tier configuration (with IM 365 up to 1 meter between racks)

- or choose a configuration of up to four tiers (with IM 360 and IM 361 up to 10 meters between racks)

You can operate a maximum of 8 modules in a rack. The number of modules may be limited by the maximum permissible current per rack, which is 1.2 A (0.8 A for CPU 312 IFM).

The modules are linked to one another via a backplane bus, which combines the functions of the P and K buses.

Local bus segment

A special feature regarding configuration is the use of the FM 356 application module from the M7-300 family of automation computers. An FM 356 is able to "split" a module's backplane bus and to take over control of the remaining

1.1 Structure of the Programmable Controller

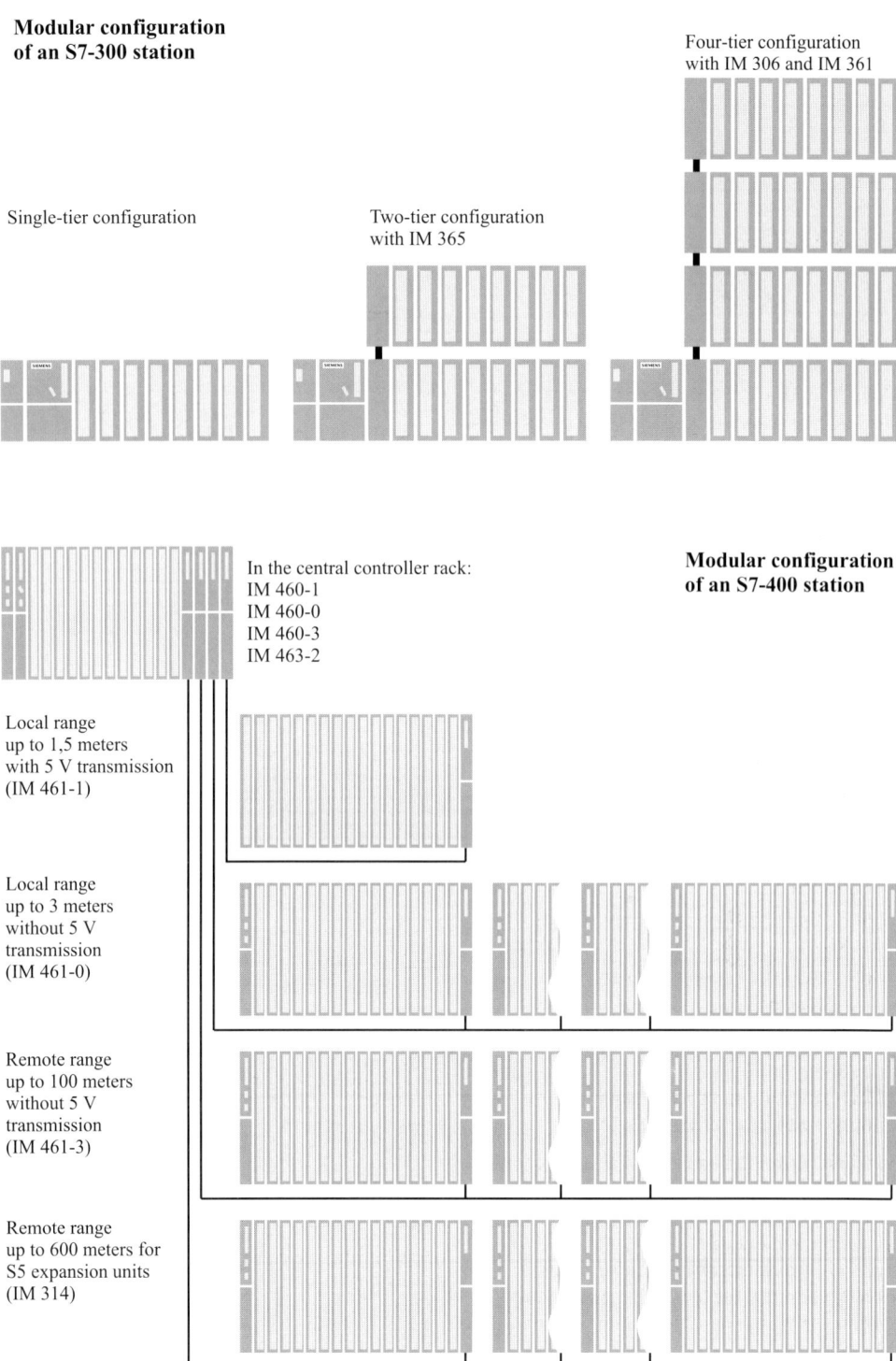

Figure 1.1 Hardware Configuration for S7-300/400

modules in the split-off "local bus segment" itself. The limitations mentioned above regarding the number of modules and the power consumption also apply in this case.

SIMATIC Outdoor

There are SIMATIC S7-300 modules available for use in harsh environments. The have an extended temperature range from –25 to +60°C, enhanced vibration and shock immunity in accordance with IEC 68 Part 2-6 and they fulfill the requirements for humidity, condensation and freezing in accordance with IEC 721-3-3 Class 3 K5 as well as the requirements for rolling stock in accordance with EN 50155 (available soon). All other technical specifications are identical to those of the standard modules.

1.1.3 S7-400 Station

Centralized configuration

In S7-400s, there are central racks with 18 or 9 slots (UR1 or UR2); the power supply module and the CPU also occupy slots, sometimes even two or more per module. The IM 460-1 and IM 461-1 interface modules make it possible to have one expansion rack per interface up to 1.5 meters from the central rack, including the 5 V supply voltage. In addition, as many as four expansion racks can be operated up to 3 meters away using IM 360-0 and IM 361-0 interface modules. And finally, IM 360-3 and IM 361-3 interface modules can be used to operate as many as four expansion racks at a distance of up to 100 meters away.

A maximum of 21 expansion racks can be connected to a central rack. To distinguish between racks, you set the number of the rack on the coding switch of the receiving IM.

The backplane bus consists of a parallel P bus and a serial K bus. Expansion racks ER1 and ER2, with 18 or 9 slots, are designed for "simple" signal modules which generate no hardware interrupts, do not have to be supplied with 24 V voltage via the P bus, require no back-up voltage, and have no K bus connection. The K bus is in racks UR1, UR2 and CR2 either when these racks are used as central racks or expansion racks with the numbers 1 to 6.

Segmented rack

A special feature is the segmented rack CR2. The rack can accommodate two CPUs with a shared power supply while keeping them functionally separate. The two CPUs can exchange data with one another via the K bus, but have completely separate P buses for their own signal modules.

Multiprocessor mode

In an S7-400, as many as four specially designed CPUs in a suitable rack can take part in multiprocessor mode. Each module in this station is assigned to only one CPU, both with its address and its interrupts. See Section 20.3.4 "Multiprocessor Mode" and 21.6 "Multiprocessor Interrupt" for more details.

Connecting SIMATIC S5 modules

The IM 463-2 interface module allows you to connect S5 expansion units (EG 183U, EG 185U, EG 186U as well as ER 701-2 and ER 701-3) to an S7-400, and also allows centralized expansion of the expansion units. An IM 314 in the S5 expansion unit handles the link. You can operate all analog and digital modules allowed in these expansion units. An S7-400 can accommodate as many as four IM 463-2 interface modules; as many as four S5 expansion units can be connected in a distributed configuration to each of an IM 463-2's interfaces.

Software redundancy

Using SIMATIC S7-300/400 standard components, you can establish a software-based redundant system with a master station controlling the process and a standby station assuming control in the event of the master failing.

Fault tolerance through software redundancy is suitable for slow processes because transfer to the standby station can require several seconds depending on the configuration of the programmable controllers. The process signals are "frozen" during this time. The standby station then continues operation with the data last valid in the master station.

Redundancy of the input/output modules is implemented with distributed I/O (ET 200M with IM 153-3 interface module for redundant

1.1 Structure of the Programmable Controller

PROFIBUS-DP). The optional "Software Redundancy" software is available for configuring.

Fault-tolerant SIMATIC S7-400H

The SIMATIC S7-400H is a fault-tolerant programmable controller with redundant configuration comprising two central racks, each with an H CPU and a synchronization module for data comparison via fiber optic cable. Both controllers operate in "hot standby" mode; in the event of a fault, the intact controller assumes operation alone via automatic bumpless transfer.

The I/O can have normal availability (single-channel, single-sided configuration) or enhanced availability (single-channel switched configuration with ET 200M). Communication is carried out over a simple or a redundant bus.

The user program is the same as that for a non-redundant controller; the redundancy function is handled exclusively by the hardware and is invisible to the user. The optional "S7-400H" software is required for configuring.

1.1.4 CPU Memory Areas

User memory

Figure 1.2 shows the CPU memory areas important for your program. The user program itself is in two areas, namely load memory and work memory.

Load memory can be integrated in the CPU or it can be a plug-in memory card. The entire user program, including configuration data, is in load memory.

Work memory is designed in the form of high-speed RAM fully integrated in the CPU. Work memory contains the relevant portions of the user program; these are essentially the program code and the user data. "Relevant" is a characteristic of the existing objects and does not mean that a particular code block will necessarily be called and executed.

The programming device transfers the entire user program including the configuration data to load memory. The operating system of the CPU then copies the "relevant" program code and the user data to work memory. When load-

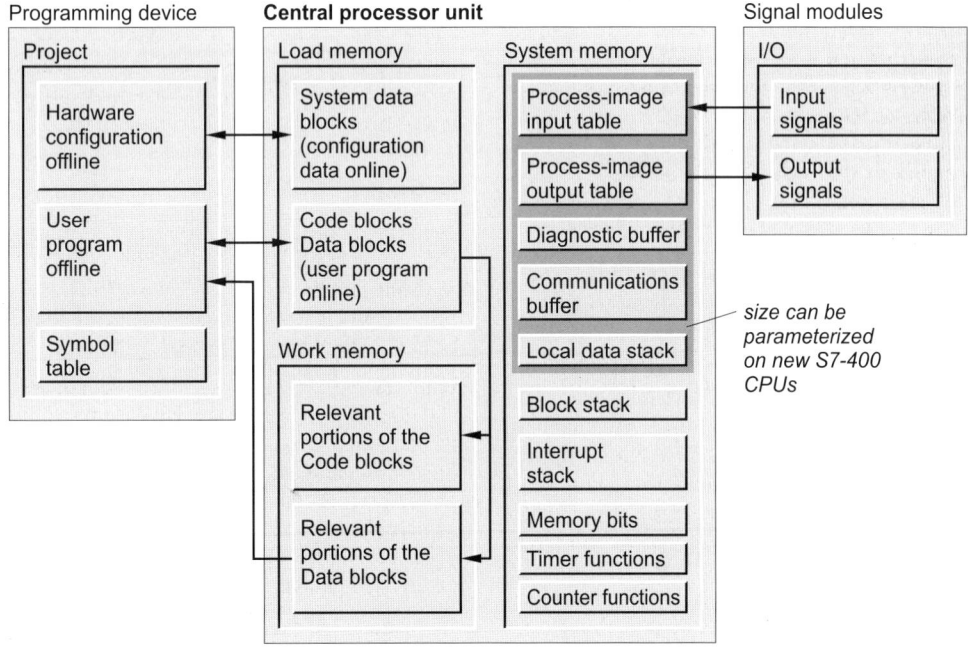

Figure 1.2 CPU Memory Areas

ing the program back to the programming device, the blocks are fetched from load memory, supplemented by the current values of the data addresses from work memory (please refer to Section 2.6.4 "Loading the User Program into the CPU" and 2.6.5 "Block Handling" for more information).

If load memory consists of RAM, a backup battery is required in order to keep the user program retentive. Where load memory is implemented as integrated EEPROM or as a plug-in flash EPROM memory card, the CPU can be operated without battery backup.

Load memory in the CPUs 3xxIFM comprises a RAM and an EEPROM component. You transfer and test the program in RAM, and you can then store the tested program in the integral EEPROM per menu command, safe from power failure.

S7-300 CPUs (with the exception of CPU 318) have an integral RAM load memory that can take the entire program. You can use a flash EPROM memory card as data carrier or as a power failure-proof memory medium for the user program.

On the S7-300, current values from parts of the user memory (data blocks) and the system memory (memory bits, timers and counters) can be stored in a non-volatile form. In this way, you can retain your data without a backup battery in the event of a power failure.

The integral RAM load memory on S7-400 CPUs is designed for small programs or for modifying individual blocks. If the full program is larger than the integral load memory, you require a RAM memory card for testing. You can use a flash EPROM memory card as a data medium or as a power failure-proof memory medium.

On new S7-400 CPUs, the work memory can be expanded with plug-in modules.

1.1.5 Memory Card

There are two types of memory card: RAM cards and flash EPROM cards.

If you want to expand load memory only, use a RAM card (e.g. on S7-400 CPUs). A RAM card allows you to modify the entire user program online. RAM memory cards lose their contents when unplugged.

If you want to protect your user program, including configuration data and module parameters, against power failure, use a flash EPROM card. In this case, load the entire program offline onto the flash EPROM card with the card plugged into the programming device. With the relevant CPUs, you can also load the program online with the memory card plugged into the CPU.

1.1.6 System Memory

System memory contains the addresses (variables) that you access in your program. The addresses are combined into areas (address areas) containing a CPU-specific number of addresses. Addresses may be, for example, inputs used to scan the signal states of momentary-contact switches and limit switches, and outputs that you can use to control contactors and lamps. The system memory on a CPU contains the following address areas:

▷ Inputs (I)
Inputs are an image ("process image") of the digital input modules.

▷ Outputs (Q)
Outputs are an image ("process image") of the digital output modules.

▷ Bit memory (M)
Stores of information accessible throughout the whole program.

▷ Timers (T)
Timers are locations used to implement waiting and monitoring times.

▷ Counters (C)
Counters are software-level locations, which can be used for up and down counting.

▷ Temporary local data (L)
Locations used as dynamic intermediate buffers during block processing. The temporary local data are located in the L stack, which the CPU occupies dynamically during program execution.

The letters enclosed in parentheses represent the abbreviations to be used for the different addresses when writing programs. You may also assign a symbol to each variable and then use the symbol in place of the address identifier.

The system memory also contains buffers for communication jobs and system messages (diagnostics buffer). The size of these data buffers, as well as the size of the process image and the L stack, are parameterizable on the new S7-400 CPUs.

1.2 Distributed I/O

PROFIBUS-DP provides a standardized interface for transferring predominantly binary process data between an "interface module" in the (central) programmable controller and the field devices. This "interface module" is called the DP master and the field devices are the DP slaves. Distributed I/O refers to modules connected via PROFIBUS-DP to a PROFIBUS master module. PROFIBUS-DP complies with EN 50170 and is a vendor-independent standard for connecting DP standard slaves.

For more information about PROFIBUS-DP please refer to Section 1.3.2 "Subnets".

The DP master and all the slaves it controls form a DP master system. There can be up to 32 stations in one segment and up to 127 stations in the entire network. A DP master can control a number of DP slaves specific to itself. You can also connect programming devices to the PROFIBUS-DP network as well as, for example, devices for human machine interface, ET 200 devices or SIMATIC S5 DP slaves.

1.2.1 DP Master System

Mono master system

PROFIBUS-DP is usually operated as a "mono master system", that is, one DP master controls several DP slaves. The DP master is the only master on the bus, with the exception of a temporarily available programming device (diagnostics and service device). The DP master and the DP slaves assigned to it form a DP master system (Figure 1.3).

Multi master system

You can also install several DP master systems on one PROFIBUS subnet (multi master system). However, this reduces the response time in individual cases because when a DP master has initialized "its" DP slaves, the access rights fall to the next DP master that in turn initializes "its" DP slaves, etc.

Several DP master systems per station

You can reduce the response time if a DP master system contains only a few DP slaves. Since it is possible to operate several DP masters in one S7 station, you can distribute the DP slaves of a station over several DP master systems. In multiprocessor mode, every CPU has its own DP master systems.

1.2.2 DP Master

The DP master is the active node on the PROFIBUS network. It exchanges cyclic data with "its" DP slaves. A DP master can be

▷ A CPU with integral DP master interface or plug-in interface submodule (e.g. CPU 315-2DP, CPU 417)

▷ An interface module in conjunction with a CPU (e.g. IM 467)

▷ A CP in conjunction with a CPU (e.g. CP 342-5, CP 443-5)

There are "Class 1 masters" for data exchange in process operation and "Class 2 masters" for service and diagnostics (e.g. a programming device).

1.2.3 DP Slaves

The DP slaves are the passive nodes on PROFIBUS. In SIMATIC S7, a distinction is made between

▷ Compact DP slaves
They behave like a single module towards the DP master

▷ Modular DP slaves
They comprise several modules (submodules)

▷ Intelligent DP slaves
They contain a control program that controls the lower-level (own) modules

1 SIMATIC S7-300/400 Programmable Controller

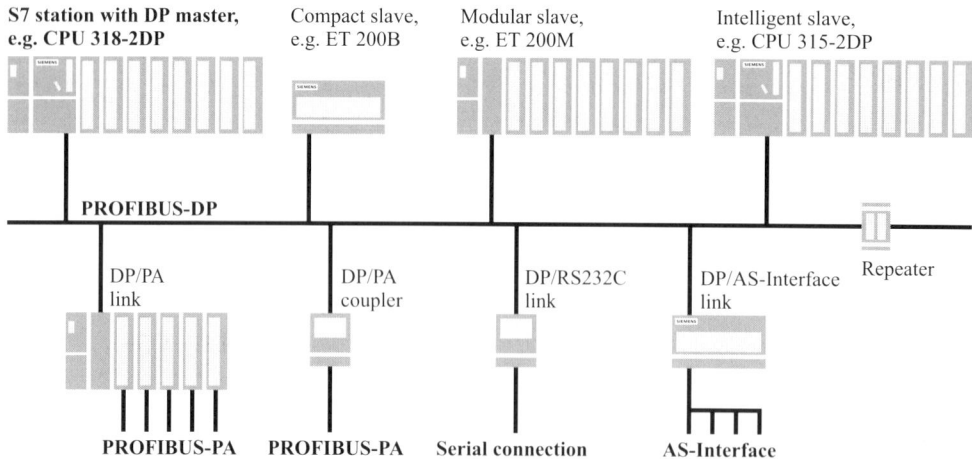

Figure 1.3 Components of a PROFIBUS-DP Master System

Compact PROFIBUS DP slaves

Examples of DP slaves include the ET 200B (version with digital input/output modules or analog input/output modules; degree of protection IP 20; max. data transfer rate 12 Mbits/s), the ET 200C (rugged construction IP 66/67; different variants with digital inputs/outputs and analog inputs/outputs; data transfer rate 1.5 Mbits/s or 12 Mbits/s) and the ET 200L-SC (discrete modularity with freely combinable digital input/output modules and analog input/output modules; degree of protection IP 20; data transfer rate 1.5 Mbits/s). The bus gateways such as DP/AS-i link behave like a compact slave on PROFIBUS.

Modular PROFIBUS-DP slaves

The ET 200M is an example of a modular DP slave. The design corresponds to an S7-300 station with DIN rail, power supply, IM 153 interface module instead of the CPU and with up to 8 signal modules (SMs) or function modules (FMs). The data transfer rate is 9.6 kbits/s to 12 Mbits/s.

The ET 200M can also be designed with *active bus modules* if the DP master is an S7-400 station. This means that the S7-300 input/output modules can be plugged in and removed during operation under power. Operation of the remaining modules continues. The modules no longer have to be plugged in without gaps.

The ET 200M can be used with the IM 153-3 interface module as a slave in a *redundant bus*. The IM 153-3 has two connections, one for the DP master in the master station and one for the DP master in the standby station.

Intelligent PROFIBUS-DP slaves

Examples of intelligent DP slaves are an S7-300 station in which is operated a CPU with DP interface that can be switched to slave mode (e.g. CPU 315-2DP) or an S7-300 station with a CP 342-5 in slave mode.

The ET 200X with the basic module BM 147/CPU can be operated as an intelligent DP slave. It comprises the basic module and up to 7 expansion modules. As basic modules, you can use "passive" basic modules with digital inputs or outputs or you can use the "intelligent" basic module BM 147/CPU capable of executing a STEP 7 user program. Expansion modules are available with digital input/output modules, analog input/output modules and as load feeders (switching and protection of any three-phase a.c. loads to 5.5 kW at 400 V AC). The basic modules operate with data transfer rates of 9.6 Kbits/s to 12 Mbits/s.

1.2.4 Connection to PROFIBUS-PA

PROFIBUS-PA

PROFIBUS-PA (Process Automation) is a bus system for process engineering in intrinsically-safe areas (Ex area Zone 1) e.g. in the chemical industry, as well as in non-intrinsically-safe areas such as the food and drinks industry.

The protocol for PROFIBUS-PA is based on the EN 50170 standard, Volume 2 (PROFIBUS-DP), and the transmission method is based on IEC 1158-2.

There are two possible methods of connecting PROFIBUS-DP to PROFIBUS-PA:

▷ DP/PA interface, if the PROFIBUS-DP network can be operated at 45.45 kbits/s

▷ DP/PA link that converts the data transfer rates of PROFIBUS-DP to the transfer rate of PROFIBUS-PA.

DP/PA interface

The DP/PA interface enables connection of PA field devices to PROFIBUS-DP. On PROFIBUS-DP, the DP/PA interface is a DP slave operated at 45.45 kbits/s. Up to 31 PA field devices can be connected to one DP/PA interface. These field devices form a PROFIBUS-PA segment with a data transfer rate of 31.25 kbits/s. Taken together, all PROFIBUS-PA segments form a shared PROFIBUS-PA bus system.

The DP/PA interface is available in two variants: a non-Ex version with up to 400 mA output current and an Ex version with up to 100 mA output current.

DP/PA link

The DP/PA link enables connection of PA field devices to PROFIBUS-DP at a data transfer rate of 9.6 kbits/s to 12 Mbits/s. A DP/PA link consists of an IM 157 interface module and up to 5 DP/PA interfaces linked together via SIMATIC S7 bus connectors. It takes the bus system comprising all the PROFIBUS-PA segments and maps it to a PROFIBUS-DP slave. You can connect up to 31 PA field devices per DP/PA link.

SIMATIC PDM

SIMATIC PDM (Process Device Manager, previously SIPROM) is a vendor-independent tool for parameterizing, startup and diagnostics of intelligent field devices with PROFIBUS-PA or HART functionality. The DDL (Device Description Language) is available for parameterizing HART transducers (Highway Addressable Remote Transducers).

1.2.5 Connection to AS-Interface

Actuator-sensor interface

The actuator-sensor interface (AS-i) is a networking system for the lowest process level in automation systems. An AS-i master controls up to 31 AS-i slaves over a 2-wire AS-i line that transmits both the control signals and the supply voltage. AS-i slaves can be actuators or sensors with bus capability or AS-i modules to which up to 8 binary ("normal") actuators or sensors can be connected.

An AS-i segment can be up to 100 m long; a segment can be extended by up to 2 times 100 m with a repeater (AS-i slaves and AS-i power supplies at both ends) or with an extender (AS-i slaves and AS-i power supply only on the line directed to the master).

AS-i master

The AS-i master updates its data and the data of all connected AS-i slaves in up to 5 ms. You can connect the AS-i bus direct to SIMATIC S7 with the CP 342-2, or to PROFIBUS-DP with a DP/AS-Interface link (Figure 1.4).

The AS-i master CP 342-2 can be used in an S7 300 station or in an ET 200M station. It supports two operating modes:

In *standard mode*, the CP 342-2 behaves like an input/output module. It occupies 16 input bytes and 16 output bytes in the analog address area (from 128 on). The AS-i slaves are parameterized with default data stored in the CP.

In *extended mode*, the full functionality of the AS-i master specification is available. If the supplied FC block is used, master calls can be made from the user program in addition to

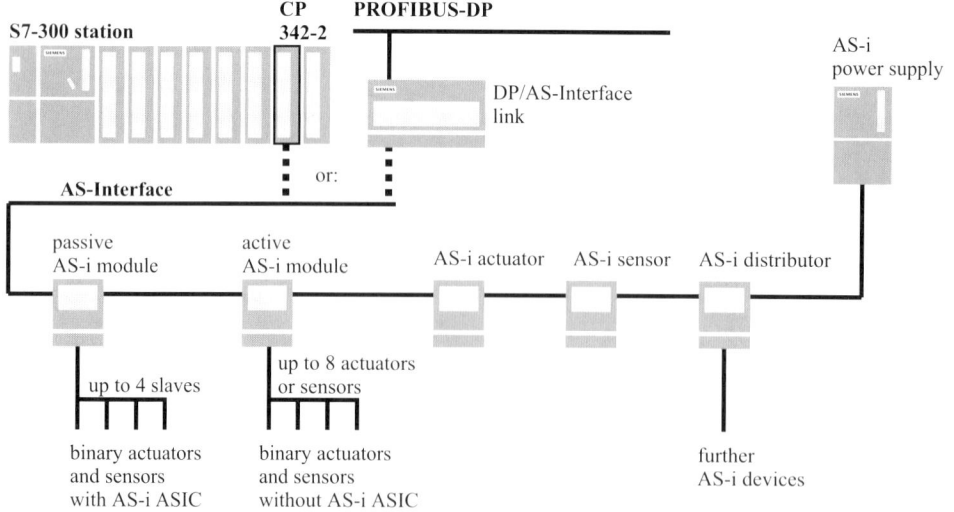

Figure 1.4 Connecting the AS-i Bus System to SIMATIC S7

standard mode (transfer of parameters during operation, testing of the setpoint/actual configuration, test and diagnostics).

A DP/AS-Interface link enables connection of AS-i actuators and AS-i sensors to PROFIBUS-DP. On PROFIBUS-DP, the link is a modular DP slave, and on the AS-Interface, it is an AS-i master that can control up to 31 AS-i slaves. With the maximum 31 AS-i slaves, a DP/AS-Interface link occupies 16 input bytes and 16 output bytes. The data transfer rate can be up to 12 Mbits/s.

The DP/AS-Interface link is available in two versions: the rugged DP/AS-Interface link 65 with degree of protection IP 66/67 and the DP/AS-Interface link 20 with degree of protection IP 20, which can be set with an additional command interface so that the input and output range can each be increased to 20 bytes.

1.2.6 Connection to a Serial Interface

The PROFIBUS-DP/RS 232C link is a converter between an RS 232C (V.24) interface and PROFIBUS-DP. Devices with an RS 232C interface can be connected to PROFIBUS-DP with the DP/RS 232C link. The DP/RS 232C link supports the 3964R and free ASCII protocol procedures.

The PROFIBUS-DP/RS 232C link is connected to the device via a point-to-point connection. Conversion to the PROFIBUS-DP protocol takes place in the PROFIBUS-DP/RS 232C link. The data are transferred consistently in both directions. Up to 224 bytes of user data are transmitted per frame.

The data transfer rate on PROFIBUS-DP can be up to 12 Mbits/s; RS 232C can be operated at up to 38.4 kbits/s with no parity, even or odd parity, 8 data bits and 1 stop bit.

1.3 Communications

Communications – data exchange between programmable modules – is an integral component of SIMATIC S7. Almost all communications functions are handled via the operating system. You can exchange data without any additional hardware and with just one connecting cable between the two CPUs. If you use CP modules, you can achieve powerful network links and the facility of linking to non-Siemens systems.

SIMATIC NET is the umbrella term for SIMATIC communications. It represents information exchange between programmable controllers and between programmable controllers and hu-

1.3 Communications

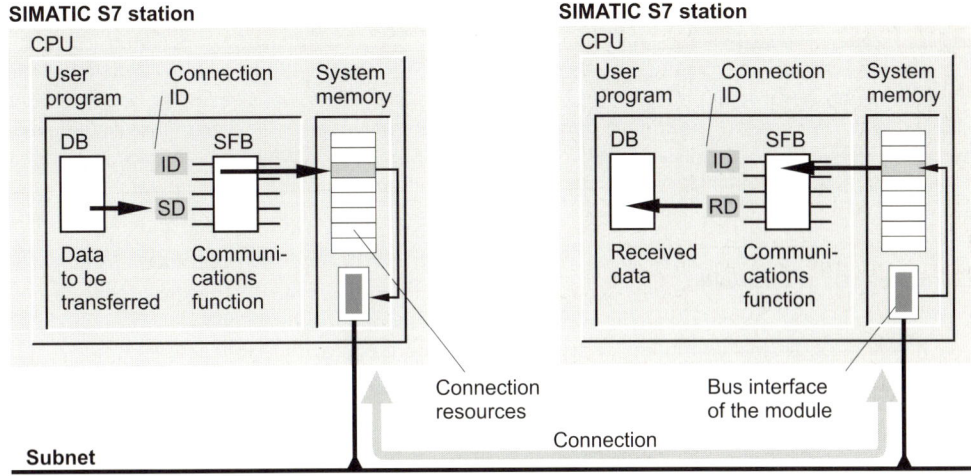

Figure 1.5 Data Exchange Between Two SIMATIC S7 Stations

man machine interface devices. There are various communications paths available depending on performance requirements.

1.3.1 Introduction

Figure 1.5 shows the most significant communications objects. You have SIMATIC stations or non-Siemens devices between which you want to exchange data. You require modules with communications capability here. With SIMATIC S7, all CPUs have an MPI interface over which they can handle communications.

In addition, there are communications processors (CPs) available that enable data exchange at higher throughput rates and with different protocols. You must link these modules via networks. A network is the hardware connection between communication nodes.

Data is exchanged via a "connection" in accordance with a specific execution plan ("communications service") which is based, among other things, on a specific coordination procedure ("protocol"). S7 connection is the standard between S7 modules with communications capability, for example.

Network

A network is a connection between several devices for the purpose of communication. It comprises one or more identical or different subnets linked together.

Subnet

In a subnet, all the communications nodes are linked via a hardware connection with uniform physical characteristics and transmission parameters, such as the data transfer rate, and they exchange data via a shared transmission procedure. SIMATIC recognizes MPI, PROFIBUS, Industrial Ethernet and point-to-point connection (PTP) as subnets.

Communications service

A communications service determines how the data are exchanged between communications nodes and how the data are to be handled. It is based on a protocol that describes, amongst other things, the coordination procedure between the communications nodes. SIMATIC has the following services: S7 functions, global data communication, PROFIBUS-DP, PROFIBUS-FMS, PROFIBUS-FDL (SDA), ISO transport and ISO-on-TCP.

Connection

A connection defines the communications relationships between two communications nodes. It is the logical assignment of two nodes for the execution of a specific communications service and also contains special characteristics such as the type of connection (dynamic, static) and how it is established. SIMATIC has the connection types S7 connection, point-to-point connection, FMS and FDL connection, ISO transport connection, ISO-on-TCP and UDP connection.

Communications functions

The communications functions are the user program's interface to the communications service. For SIMATIC S7-internal communications, the communications functions are integrated in the operating system of the CPU and they are called via system blocks. Loadable blocks are available for communication with non-Siemens devices via communications processors.

Overview of communications objects

Table 1.1 shows the relationships between subnets, modules with communications capability and communications services.

Table 1.1 Communications Objects

Subnet	Modules	Communications Service	Configuring, Interface
MPI	All CPUs	Global data communications	GD table
		Station-external SFC communications	SFC calls
		SFB communications (only S7-400 active)	Connection table, FB calls
PROFIBUS	CPUs with DP master	PROFIBUS-DP (master, also slave possible)	Hardware configuration, inputs/outputs, SFC calls
		Station-internal SFC communications	SFC calls
	IM 467	PROFIBUS-DP (master or slave)	Hardware configuration, inputs/outputs, SFC calls
		Station-internal SFC communications	SFC calls
	CP 342-5 CP 443-5 Extended	PROFIBUS-DFL PROFIBUS-DP (master or slave)	NCM, connection table, SEND/RECEIVE
		Station-internal SFC communications	SFC calls
		SFB communications (only S7-400 active)	Connection table, SFB calls
	CP 343-5 CP 443-5 Basic	PROFIBUS-FMS PROFIBUS-FDL	NCM, connection table, FMS interface, SEND/RECEIVE
		Station-internal SFC communications	SFC calls
		SFB communications (only S7-400 active)	Connection table, SFB calls
Industrial Ethernet	CP 343-1 CP 443-1	ISO transport	NCM, connection table, SEND/RECEIVE
		SFB communications (only S7-400 active)	Connection table, SFB calls
	CP 343-1 TCP CP 443-1 TCP	ISO-on-TCP transport	NCM, connection table, SEND/RECEIVE
		SFB communications (only S7-400 active)	Connection table, SFB calls

NCM is the configuring software for the CP; NCM is available for PROFIBUS and for Industrial Ethernet.

1.3.2 Subnets

Subnets are communications paths with the same physical characteristics and the same communications procedure. Subnets are the central objects for communication in the SIMATIC Manager.

The subnets differ in their performance capability:

▷ MPI
Low-cost method of networking a few SIMATIC devices with small data volumes.

▷ PROFIBUS
High-speed exchange of small and mid-range volumes of data, used primarily with distributed I/O.

▷ Industrial Ethernet
Communications between computers and programmable controllers for high-speed exchange of large volumes of data.

▷ Point-to-point (PTP)
Serial link between two communications partners with special protocols.

From STEP 7 V5, you can use a programming device to reach SIMATIC S7 stations via subnets, for the purposes of, say, programming or parameterizing. The gateways between the subnets must be located in an S7 station with "routing capability".

MPI

Every CPU has an "interface with multipoint capability" (multipoint interface, or MPI). It enables establishment of subnets in which CPUs, human machine interface devices and programming devices can exchange data with each other. Data exchange is handled via a Siemens proprietary protocol.

As transmission medium, MPI uses either a shielded twisted-pair cable or a glass or plastic fiber-optic cable. The cable length in a bus segment can be up to 50 m. This can be increased by inserting RS485 repeaters (up to 1100 m) or optical link modules (up to > 100 km). The data transfer rate is usually 187.5 kbits/s.

The maximum number of nodes is 32. Each node has access to the bus for a specific length of time and may send data frames. After this time, it passes the access rights to the next node ("token passing" access procedure).

Over the MPI network, you can exchange data between CPUs with global data communications, station-external SFC communications or SFB communications. No additional modules are required.

PROFIBUS

PROFIBUS stands for "Process Fieldbus" and is a vendor-independent standard complying with EN 50170 for the networking of field devices.

A shielded twisted-pair cable or a glass or plastic fiber-optic cable is used as the transmission medium. The cable length in a bus segment depends on the data transfer rate; it is 100 m at the fastest transfer rate (12 Mbits/s) and 1000 m at the slowest (9.6 kbits/s). Network range can be increased with repeaters or optical link modules.

The maximum number of nodes is 127; a distinction is made between active and passive nodes. An active node receives access rights to the bus for a specific length of time and may send data frames. After this time, it passes the access rights to the next node ("token passing" access procedure). If passive nodes (slaves) are assigned to an active node (master), the master executes data exchange with the slaves assigned to it while it is in possession of the access rights. A passive node does not receive access rights.

You implement connection of distributed I/O via a PROFIBUS network; the relevant PROFIBUS-DP communications service is implicit. You can use either CPUs with integral or plug-in DP master, or the relevant CPs. You can also operate station-internal SFC communications or SFB communications via this network.

You can transfer data with PROFIBUS-FMS and PROFIBUS-FDL using the relevant CPs. There are loadable blocks (FMS interface or SEND/RECEIVE interface) available as the interface to the user program.

Industrial Ethernet

Industrial Ethernet is the subnet for connecting computers and programmable controllers, with the focus on the industrial area, defined by the international standard IEEE 802.3.

The electrical physical connection is a double-shielded coaxial cable or an industrial twisted pair, and the optical connection is a glass fiber-optic cable. On an electrical network, the range is 1.5 km and on an optical network it is 4.5 km. The data transfer rate is set at 10 Mbits/s.

More than 1000 nodes can be networked with Industrial Ethernet. Before accessing, each node checks to see if another node is currently transmitting. If this is the case, the node waits for a random time before attempting another access (CSMA/CD access procedure). All nodes have equal access rights.

You can also exchange data with SFB communications via Industrial Ethernet and you can use the S7 functions. You require the relevant CPs for Industrial Ethernet and you can then also establish ISO transport connections or ISO-on-TCP connections and control with the SEND/RECEIVE interface.

Point-to-point connection

A point-to-point connection (PTP) enables data exchange via a serial link. A point-to-point connection is handled by the SIMATIC Manager as a subnet and configured similarly.

The transmission medium is an electrical cable with interface-dependent assignment. RS 232C (V.24), 20 mA (TTY) and RS 422/485 are available as interfaces. The data transfer rate is in the range 300 bits/s to 19.2 kbits/s with a 20 mA interface or 76.8 kbits/s with RS 232C and RS 422/485. The cable length depends on the physical interface and the data transfer rate; it is 10 m with RS 232C, 1000 m with a 20 mA interface at 9.6 kbits/s and 1200 with RS 422/485 at 19.2 kbits/s.

AS-Interface

The AS-Interface (actuator/sensor interface, AS-i) networks the appropriately designed binary sensors and actuators in accordance with the AS-Interface specification IEC TG 178. The AS-Interface does not appear in the SIMATIC Manager as a subnet; only the AS-I master is configured with the hardware configuration or with the network configuration.

The transmission medium is an unshielded twisted-pair cable that supplies the actuators and sensors with both data and power (power supply required). Network range can be up to 300 m with repeaters. The data transfer rate is set at 167 kbits/s.

A master controls up to 31 slaves through cyclic scanning and so guarantees a defined response time.

1.3.3 Communications Services

Data exchange over the subnets is controlled by different communications services – depending on the connection selected. These services are used with the focus on the following:

▷ Programming device (PG) functions: test, startup and service functions; used by a programming device, for example, to execute the function "monitor variables" or "read diagnostics buffer" or to load user programs.

▷ HMI functions: human machine interface functions; used by connected OPs, for example, to read or write variables.

▷ SFB communications: an event-driven service for exchanging larger volumes of data. It is started with SFB calls in the user program, with modify and monitor functions; static, configured connections.

▷ SFC communications: an event-driven service for exchanging up to 76 bytes per transmission. It is started with SFC calls in the user program; dynamic, non-configured connections.

The S7 functions can be executed via the MPI, PROFIBUS and Industrial Ethernet subnets.

Global data communications enables exchange of small volumes of data between several CPUs without additional programming overhead in the user program. Transfer can be cyclic or event-driven.

Global data communications is a broadcast procedure; data reception is not acknowledged. Communication status is reported.

Global data communications is only possible over the MPI bus or the K bus.

With **PROFIBUS-DP**, data is exchanged between the master and the slaves over distributed

I/O. Communication is transparent and standardized in accordance with EN 50170 Volume 2. SIMATIC S7 slaves and non-SIMATIC standard slaves can be accessed via a PROFIBUS subnet with this service.

PROFIBUS FMS (Fieldbus Message Specification) provides services for transferring structured variables (FMS variables) in accordance with EN 50170 Volume 2. Communication takes place exclusively with static connections over a PROFIBUS subnet.

PROFIBUS-FDL (Fieldbus Data Link) transfers data with the SDA (Send Data with Acknowledge) function, standardized in accordance with EN 50170 Volume 2. Communication takes place with static connections. With PROFIBUS-FDL, data can be exchanged, for example, with a SIMATIC S5 controller via a PROFIBUS subnet.

The communications service ISO transport enables data transmission in accordance with ISO 8073 Class 4. Communication takes place over static connections. With ISO transport, data can be exchanged, for example, with a SIMATIC S5 controller via Industrial Ethernet.

The communications service ISO-on-TCP corresponds to the TCP/IP standard with the RFC 1006 extension. Communication takes place with static connections via Industrial Ethernet.

1.3.4 Connections

A connection is either dynamic or static depending on the communications service selected. Dynamic connections are not configured; their buildup or cleardown is event-driven ("Communications via non-configured connections"). There can only ever be one non-configured connection to a communications partner.

Static connections are configured in the connection table; they are built up at startup and remain throughout the entire program execution ("communications via configured connections"). Several connections can be established in parallel to one communications partner. You use a "Connection type" to select the desired communications service in the network configuration (see Section 2.4 "Configuring the Network").

You do not need to configure connections with the network configuration for global data communications and PROFIBUS-DP or for SFC communications in the case of S7 functions. You define the communications partners for global data communications in the global data table; in the case of PROFIBUS-DP and SFC communications, you define the partners via the node addresses.

Connection resources

Each connection requires connection resources on the participating communications partner for the end point of the connection or the transition point in a CP module. If, for example, S7 functions are executed via the MPI interface of the CPU, a connection is assigned in the CPU; the same functions via the MPI interface of the CP occupy one connection in the CP and one connection in the CPU.

Each CPU has a specific number of possible connections. One connection is reserved for a programming device and one connection for an OP (these cannot be used for any other purpose).

Connection resources are also required temporarily for the "non-configured connections" in SFC communications.

1.4 Module Addresses

1.4.1 Signal Path

When you wire your machine or plant, you determine which signals are connected where on the programmable controller (Figure 1.6).

An input signal, for example the signal from momentary-contact switch +HP01-S10, the one for "Switch motor on", is run to an input module, where it is connected to a specific terminal. This terminal has an "address" called the I/O address (for instance byte 5, bit 2).

Before every program execution start, the CPU then automatically copies the signal to the process input image, where it is then accessed as an "input" address (I 5.2, for example). The expression "I 5.2" is the absolute address.

1 SIMATIC S7-300/400 Programmable Controller

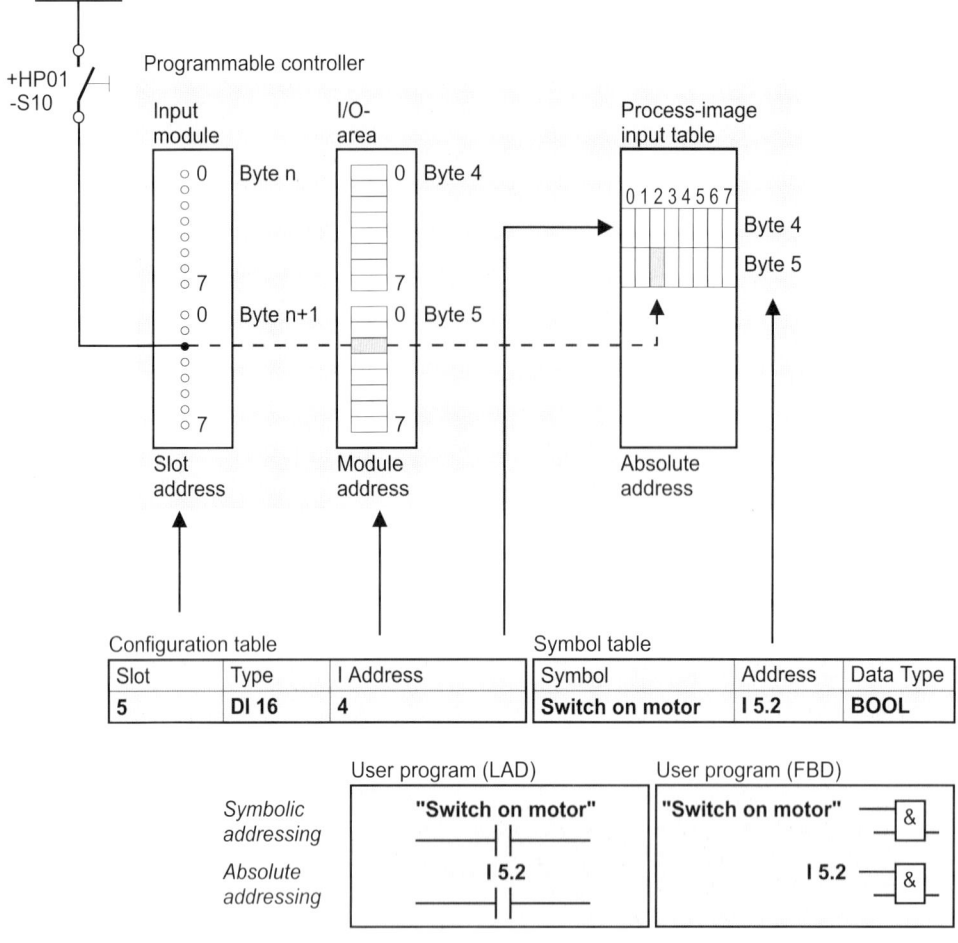

Figure 1.6
Correlation between Module Address, Absolute Address and Symbolic Address
(Path of a Signal from Sensor to Scanning in the Program)

You can now give this input a name by assigning an alphanumeric symbol corresponding to this input signal (such as "Switch motor on") to the absolute address in the symbol table. The expression "Switch motor on" is the symbolic address.

1.4.2 Slot Address

Every slot has a fixed address in the programmable controller (an S7 station). This slot address consists of the number of the mounting rack and the number of the slot. A module is uniquely described using the slot address ("geographical address").

If the module contains interface cards, each of these cards is also assigned a submodule address. In this way, each binary and analog signal and each serial connection in the system has its own unique address.

Correspondingly, distributed I/O modules also have a "geographical address". In this case, the number of the DP master system and the station number replace the rack number.

You use STEP 7's "Hardware Configuration" tool to plan the hardware configuration of an S7 station as per the physical location of the modules. This tool also makes it possible to set the module start addresses and parameterize the modules (see Section 2.3 "Configuring Stations").

1.4.3 Module Start Address

In addition to the slot address, which defines the slot, each module has a start address, which defines the location in the logical address space (I/O address space). The I/O address space begins at address 0 and ends at a CPU-specific upper limit.

The module start address determines how the input and output signals are addressed (accessed) by the program. In the case of digital modules, the individual signals (bits) are bundled into groups of eight called "bytes". There are modules with one, two or four bytes. These bytes have the relative addresses 0, 1, 2 and 3; addressing of the bytes begins at the module start address. Example: In the case of a digital module with four bytes and the start address 8, the individual bytes are accessed by addresses 8, 9, 10 and 11. In the case of analog modules, the individual analog signals (voltages, currents) are called "channels", each of which occupies two bytes. Analog modules are available, depending on design, with 2, 4, 8 and 16 channels, corresponding to 4, 8, 16 or 32 bytes of address area.

On power-up (if there is no setpoint configuration), the CPU defaults to a slot-oriented module start address which depends on the module type, the slot, and the rack. This module start address corresponds to (relative) byte 0. You can view this address in the configuration table.

On S7-3xx with integral DP interface, S7-318 and S7-400 systems, you can change this address. You have the option of assigning the start addresses of the modules within the permissible address space. You also have the option of assigning different start addresses for inputs and outputs on a hybrid digital or analog module. FMs and CPs normally occupy the same start address for inputs and outputs.

Like centralized modules, distributed I/O modules (stations) reserve a specific number of bytes in the I/O address space. The addresses of the centralized modules and those of the distributed I/O must not overlap.

Appropriately equipped DP slaves can be parameterized so that a specific number of bytes constitute consistent (logically associated) data for data transfers. These slaves display an I/O address of only one byte, via which they are addressed with the system functions SFC 14 DPRD_DAT and SFC 15 DPWR_DAT.

The digital modules are usually arranged according to address in the process image so that their signal states can be automatically updated and they can be accessed with the address areas "Input" and "Output". Analog modules, FMs and CPs receive an address that is not in the process image.

1.4.4 Diagnostics Address

Appropriately equipped modules can supply diagnostics data that you can evaluate in your program. If centralized modules have a user data address (module start address), you access the module via this address when reading the diagnostics data. If the modules have no user data address (e.g. power supplies), or if they are part of the distributed I/O, there is a diagnostics address for this purpose.

The diagnostics address is always an address in the peripheral input range and always has a length of one byte.

STEP 7 automatically assigns the diagnostics address counting down from the highest possible I/O address. You can change the diagnostics address with the Hardware Configuration function.

The diagnostics data can only be read with special system functions; accessing this address with load statements has no effect (see also Section 20.4.1 "Addressing Distributed I/O").

1.4.5 Address for Bus Nodes

Node address, station number

Every DP station (e.g. DP master, DP slave, programming device) on PROFIBUS has an additional node address with which it can be uniquely addressed on the bus.

MPI address

Modules that are nodes on an MPI network (CPUs, FMs and CPs) also have an MPI address. This address is decisive for the link to programming devices, human machine interface devices and for global data communications.

Please note that with older versions of the S7-300 CPUs, the FMs and CPs operated in the same station receive an address derived from the MPI address of the CPU.

1.5 Address Areas

The address areas available in every programmable controller are

▷ the peripheral inputs and outputs

▷ the process input image and the process output image

▷ the bit memory area

▷ the timer and counter functions (see Chapter 7 "Timer Functions" and 8 "Counter Functions")

▷ the L stack (see Section 18.1.5 "Temporary Local Data")

To this are added the code and data blocks with the block-local variables, depending on the user program.

1.5.1 User Data Area

In SIMATIC S7, each module can have two address areas: a user data area, which can be directly addressed with Load and Transfer statements, and a system data area for transferring data records.

When modules are accessed, it makes no difference whether they are in racks with centralized configuration or used as distributed I/O. All modules occupy the same (logical) address space.

A module's user data properties depend on the e type. In the case of signal modules, they are either digital or analog input/output signals, and in the case of function modules and communications processors, they might, for example, be control or status information. The volume of user data is module-specific. There are modules that occupy one, two, four or more bytes in this area. Addressing always begins at relative byte 0. The address of byte 0 is the module start address; it is stipulated in the configuration table.

The user data represent the I/O address area, subdivided, depending on the direction of transfer, into peripheral inputs (PIs) and peripheral outputs (PQs). If the user data are in the area of the process images, the CPU automatically handles the transfers when updating the process images.

Peripheral inputs

You use the peripheral input (PI) address area when you read from the user data area on input modules. Part of the PI address area leads to the process image. This part always begins at I/O address 0; the length of the area is CPU-specific.

With a Direct I/O Read operation, you can access the modules whose interfaces do not lead to the process input image (for instance analog input modules). The signal states of modules that lead to the process input image can also be read with a Direct Read operation. The momentary signal states of the input bits are then scanned. Please note that this signal state may differ from the relevant inputs in the process image since the process input image is updated at the beginning of the program scan.

Peripheral inputs may occupy the same absolute addresses as peripheral outputs.

Peripheral outputs

You use the peripheral output (PQ) address area when you write values to the user data area on an output module. Part of the PQ address area leads to the process image. This part always begins at I/O address 0; the length of the area is CPU-specific.

With a Direct I/O Write operation, you can access modules whose interfaces do not lead to the process output image (such as analog output modules). The signal states of modules controlled by the process output image can also be directly affected. The signal states of the output bits then change immediately. Please note that a Direct I/O Write operation also updates the signal states of the relevant modules in the process output image! Thus, there is no difference between the process output image and the signal states on the output modules.

Peripheral outputs can reserve the same absolute addresses as peripheral inputs.

1.5.2 Process Image

The process image contains the image of the digital input and digital output modules, and is thus subdivided into process input image and process output image. The process input image is accessed via the address area for inputs (I), the process output image via the address area for outputs (Q). As a rule, the machine or process is controlled via the inputs and outputs.

The process image can be subdivided into subsidiary process images that can be updated either automatically or via the user program. Please refer to Section 20.2.1 "Updating the Process Image" for more details.

On the S7-300 CPUs and, from 10/98, also on S7-400 CPUs, you can use the addresses of the process image not occupied by modules as additional memory area similar to the bit memory area. This applies both for the process input image and the process output image.

On suitably equipped CPUs, say, the CPU 417, the size of the process image can be parameterized. If you enlarge the process image, you reduce the size of the work memory accordingly. Following a change to the size of the process image, the CPU executes initialization of the work memory, with the same effect as a cold restart.

Inputs

An input is an image of the corresponding bit on a digital input module. Scanning an input is the same as scanning the bit on the module itself. Prior to program execution in every program cycle, the CPU's operating system copies the signal state from the module to the process input image.

The use of a process input image has many advantages:

▷ Inputs can be scanned and linked bit by bit (I/O bits cannot be directly addressed).

▷ Scanning an input is much faster than accessing an input module (for example, you avoid the transient recovery time on the I/O bus, and the system memory response times are shorter than the module's response times). The program is therefore executed that much more quickly.

▷ The signal state of an input is the same throughout the entire program cycle (there is data consistency throughout a program cycle). When a bit on an input module changes, the change in the signal state is transferred to the input at the start of the next program cycle.

▷ Inputs can also be set and reset because they are located in random access memory. Digital input modules can only be read. Inputs can be set during debugging or startup to simulate sensor states, thus simplifying program testing.

These advantages are offset by an increased program response time (please also refer to Section 20.2.4 "Response Time").

Outputs

An output is an image of the corresponding bit on a digital output module. Setting an output is the same as setting the bit on the output module itself. The CPU's operating system copies the signal state from the process output image to the module.

The use of a process output image has many advantages:

▷ Outputs can be set and reset bit by bit (direct addressing of I/O bits is not possible).

▷ Setting an output is much faster than accessing an output module (for example, you avoid the transient recovery time on the I/O bus, and the system memory response times are shorter than the module response times). The program is therefore executed that much more quickly.

▷ A multiple signal state change at an output during a program cycle does not affect the bit on the output module. It is the signal state of the output at the end of the program cycle that is transferred to the module.

▷ Outputs can also be scanned because they are located in random access memory. While it is possible to write to digital output modules, it is not possible to read them. The scanning and linking of the outputs makes additional storage of the output bit to be scanned unnecessary.

These advantages are offset by an increased program response time. Section 20.2.4, "Response Time" describes how a programmable controller's response time comes about.

1.5.3 Bit Memory

The area called bit memory holds what could be regarded as the controller's "auxiliary contactors". Bit memory is used primarily for storing binary signal states. The bits in this area can be treated as outputs, but are not "externalized". Bit memory is located in the CPU's system memory area, and is therefore available at all times. The number of bits in bit memory is CPU-specific.

Bit memory is used to store intermediate results that are valid beyond block boundaries and are processed in more than one block. Besides the data in global data blocks, the following are also available for storing intermediate results:

▷ Temporary local data, which are available in all blocks but valid for the current block call only, and

▷ Static local data, which are available only in function blocks but valid over multiple block calls.

Retentive bit memory

Part of bit memory may be designated "retentive", which means that the bits in that part of bit memory retain their signal states even under off-circuit conditions. Retentivity always begins with memory byte 0 and ends at the designated location. Retentivity is set when the CPU is parameterized. Please refer to Section 22.2.3 "Retentivity" for additional information.

Clock memory

Many procedures in the controller require a periodic signal. Such a signal can be implemented using timers (clock pulse generator), watchdog interrupts (time-controlled program execution), or simply by using clock memory.

Clock memory consists of bits whose signal states change periodically with a mark-to-space ratio of 1:1. The bits are combined into a byte, and correspond to fixed frequencies (Figure 1.7). You specify the number of clock memory bits when you parameterize the CPU. Please note that the updating of clock memory is asynchronous to execution of the main program.

CLock memory byte

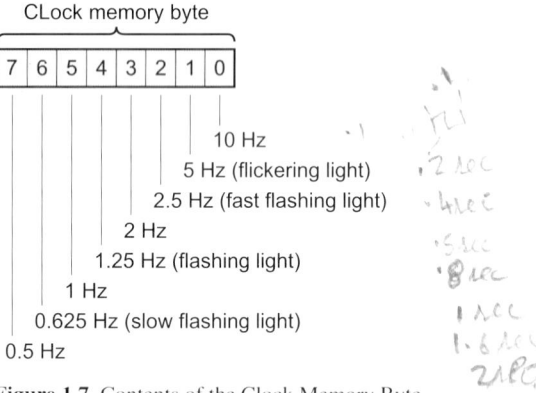

| 7 | 6 | 5 | 4 | 3 | 2 | 1 | 0 |

10 Hz
5 Hz (flickering light)
2.5 Hz (fast flashing light)
2 Hz
1.25 Hz (flashing light)
1 Hz
0.625 Hz (slow flashing light)
0.5 Hz

Figure 1.7 Contents of the Clock Memory Byte

2 STEP 7 Programming Software

2.1 STEP 7 Basic Package

This chapter describes the STEP 7 basic package, Version 5.0. While the first chapter presented an overview of the properties of the programmable controller, this chapter tells you how to set these properties.

The basic package contains the statement list (STL), ladder logic (LAD) and function block diagram (FBD) programming languages. In addition to the basic package, option packages such as S7-SCL (Structured Control Language), S7-GRAPH (sequence planning) and S7-HiGraph (state-transition diagram) are also available.

2.1.1 Installation

STEP 7 V5 is a 32-bit application requiring Microsoft Windows 95/98 or Microsoft Windows NT as the operating system. To work with the STEP 7 software under Windows 95/98, you required a programming device (PG) or a PC with an 80486 processor or higher and at least 16 MB of RAM, with 32 MB recommended. For Windows NT, you require a Pentium processor and at least 32 MB of RAM; you must have administration authorization to install STEP 7 under Windows NT.

STEP 7 V5 occupies approximately 105 MB per language (for example, English) on the hard disk. A swap-out file is also needed. This file is approximately 128 MB minus main memory; for example, if the main memory configuration is 16 MB, the swap-out file would comprise about 112 MB. You should reserve around 50 MB for your user data, and between 10 and 20 MB for each additional option package. The memory requirements may increase for certain operations, such as copying a project. If there is insufficient space for the swap-out file, errors such as program crashes may occur.

Windows 95/98/NT's SETUP program, which is on the first diskette, is used for installation. On the programming device, STEP 7 is already factory-installed.

In addition to STEP 7 V5, the CD also includes the authorization program (see below), the NCM programs for configuring CPs and the STEP 7 electronic manuals with Acrobat Reader V3.01.

An MPI interface is needed for the online connection. The programming devices have the multipoint interface already built in, but PCs must be retrofitted with an MPI module. If you want to use PC memory cards, you will need a prommer.

STEP 7 V5 has multi-user capability, that is, a project that is stored, say, on a central server can be edited simultaneously from several workstations. You make the necessary settings in the Windows Control Panel with the "SIMATIC Workstation" program. In the dialog box that appears, you can parameterize the workstation as a single-user system or a multi-user system with the protocols used.

2.1.2 Authorization

An authorization (right of use) is required to operate STEP 7. The authorization is supplied on diskette. Following installation of STEP 7, you will be prompted for your authorization if the hard disk does not already contain an authorization. You can also provide your authorization later.

You may also transfer the authorization to another device by copying it back to the (original) authorization diskette, then transferring it to the new device.

Should you lose your authorization for some reason, such as a hard disk defect, you can use the emergency license, which is also on the authorization diskette and is good for a limited time only, until you are able to obtain a replacement authorization.

2.1.3 SIMATIC Manager

The SIMATIC Manager is the main tool in STEP 7; you will find its icon in Windows.

SIMATIC Manager

The SIMATIC Manager is started by double-clicking on its icon.

When first started, the project wizard is displayed. This can be used for simple creation of new projects. You can deactivate it with the check box "Display Wizard on starting the SIMATIC Manager" since it can also be called, if required, via the menu command FILE → 'NEW PROJECT' WIZARD.

Programming begins with opening or creating a 'project'. The example projects supplied are a good basis for familiarization.

When you open project ZEn01_09_S7_ZEBRA with FILE → OPEN, you will see the split project window: on the left is the structure of the open object (the object hierarchy), and on the right is the selected object (Figure 2.1). Clicking on the box containing a plus sign in the left window displays additional levels of the structure; selecting an object in the left half of the window displays its contents in the right half of the window.

Under the SIMATIC Manager, you work with the objects in the STEP 7 world. These "logical" objects correspond to "real" objects in your plant. A project contains the entire plant, a station corresponds to a programmable controller. A project may contain several stations connected to one another, for example, via an MPI subnet. A station contains a CPU, and the CPU contains a program, in our case an S7 program. This program, in turn, is a "container" for other objects, such as the object Blocks, which contains, among other things, the compiled blocks.

The STEP 7 objects are connected to one another via a tree structure. Figure 2.2 shows the most important parts of the tree structure (the "main branch", as it were) when you are working with the STEP 7 basic package for S7 applications in offline view. The objects shown in bold type are containers for other objects. All objects in the Figure are available to you in the offline view. These are the objects that are on the programming device's hard disk. If your programming device is online on a CPU (normally a PLC target system), you can switch to the online view by selecting VIEW → ONLINE. This option displays yet another project window containing the objects on the destination device; the objects shown in italics in the Figure are then no longer included.

You can see from the title bar of the active project window whether you are working offline or online. For clearer differentiation, the title bar and the window title can be set to a different color than the offline window. For this purpose, select OPTIONS → CUSTOMIZE and modify the entries in the "View" tab.

Select OPTIONS → CUSTOMIZE to change the SIMATIC Manager's basic settings, such as the session language, the archive program and the storage location for projects and libraries, and configuring the archive program.

Editing sequences

The following applies for the general editing of objects:

To *select an object* means to click on it once with the mouse so that it is highlighted (this is possible in both halves of the project window).

To *name an object* means to click on the name of the selected object (a frame will appear around the name and you can change the name in the window) or select the menu item EDIT → OBJECT PROPERTIES and change the name in the dialog box.

To *open an object*, double-click on that object. If the object is a container for other objects, the SIMATIC Manager displays the contents of the object in the right half of the window. If the object is on the lowest hierarchical level, the SIMATIC Manager starts the appropriate tool for editing the object (for instance, double-clicking on a block starts the editor, allowing the block to be edited).

2.1 STEP 7 Basic Package

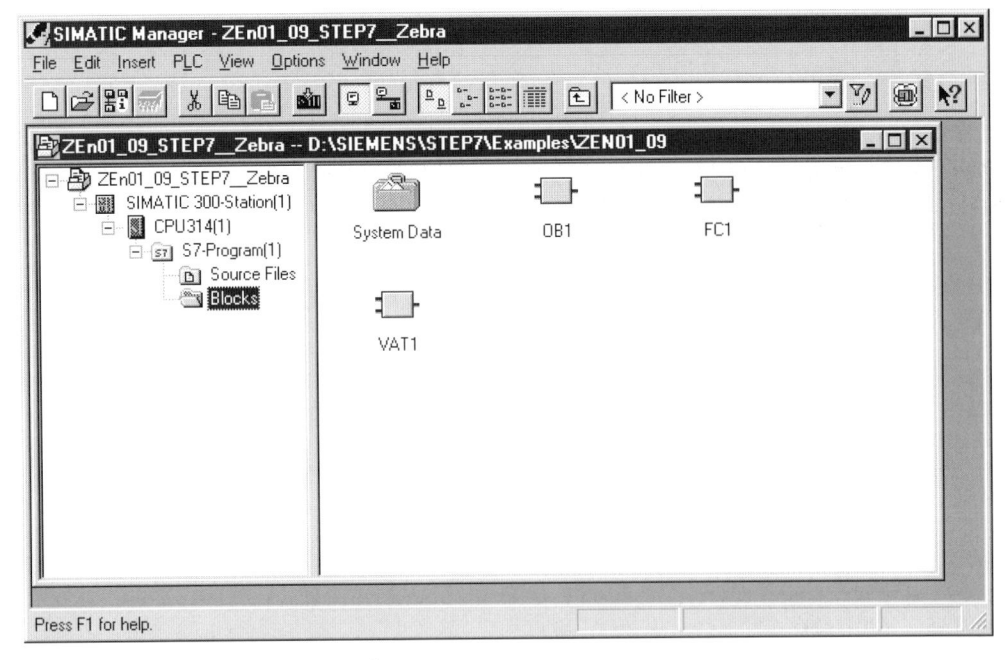

Figure 2.1 SIMATIC Manager Example

In this book, the menu items in the standard menu bar at the top of the window are described as *operator sequences*. Programmers experienced in the use of the operator interface use the icons from the toolbar. The use of the *right mouse button* is very effective. Clicking on an object once with the right mouse button screens a menu showing the current editing options.

2.1.4 Projects and Libraries

In STEP 7, the 'main objects' at the top of the object hierarchy are projects and libraries.

Projects are used for the systematic storing of data and programs needed for solving an automation task. Essentially, these are

▷ the hardware configuration data,

▷ the parameterization data for the modules,

▷ the configuring data for communication via networks,

▷ the programs (code and data, symbols, sources).

The objects in a project are arranged hierarchically. The opening of a project is the first step in editing all (subordinate) objects which that object contains. The following sections discuss how to edit these objects.

Libraries are used for storing reusable program components. Libraries are organized hierarchically. They may contain STEP 7 programs which in turn may contain a user program (a container for compiled blocks), a container for source programs, and a symbol table. With the exception of online connections (no debugging possible), the creation of a program or program section in a library provides the same functionality as in an object.

As supplied, STEP 7 V5 provides the Standard Library containing the following programs:

▷ System Function Blocks
Contains the call interfaces of the system blocks for offline programming integrated in the CPU

▷ S5-S7 Converting Blocks
Contains loadable functions for the S5/S7 converter (replacement of S5 standard function blocks in conjunction with program conversion)

2 STEP 7 Programming Software

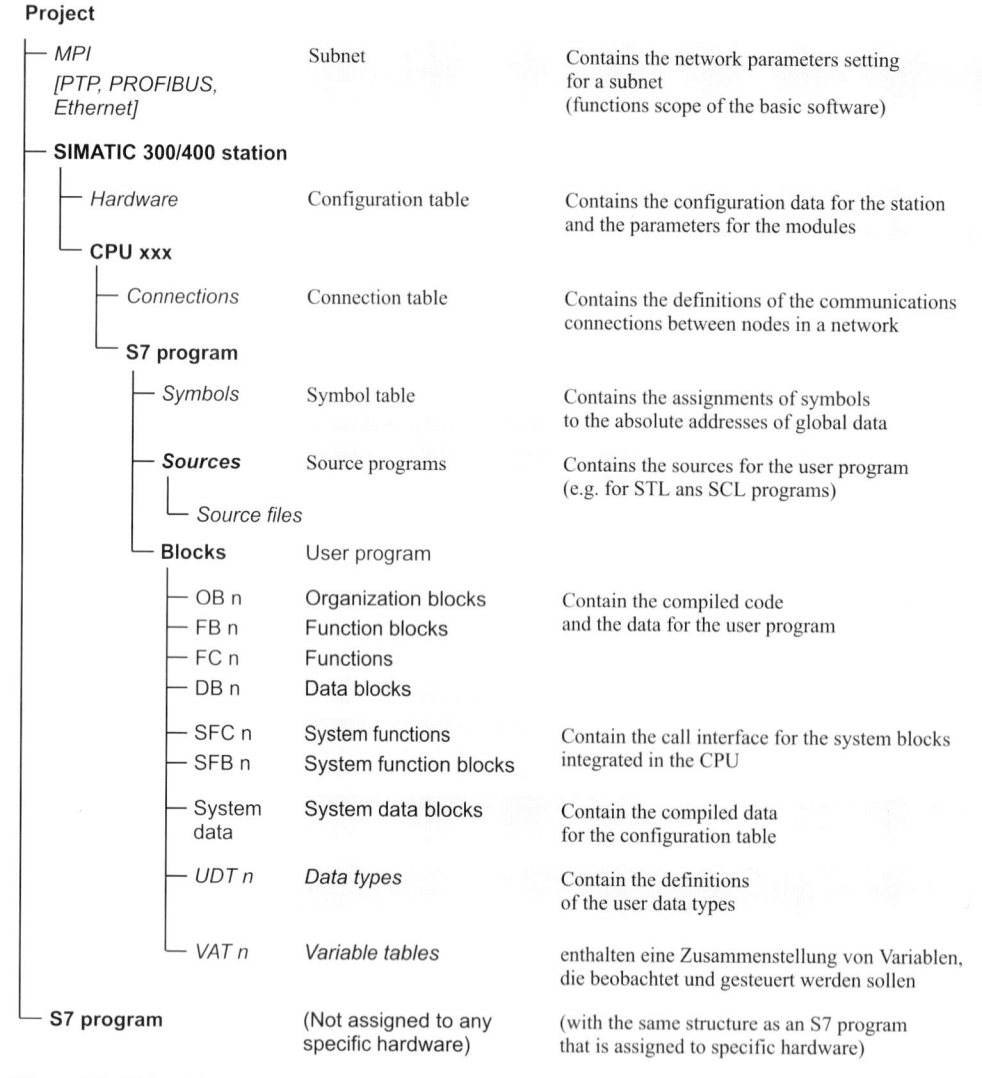

Project
```
├── MPI                         Subnet                     Contains the network parameters setting
│   [PTP, PROFIBUS,                                        for a subnet
│   Ethernet]                                              (functions scope of the basic software)
│
├── SIMATIC 300/400 station
│   │
│   ├── Hardware                Configuration table        Contains the configuration data for the station
│   │                                                      and the parameters for the modules
│   └── CPU xxx
│       │
│       ├── Connections         Connection table           Contains the definitions of the communications
│       │                                                  connections between nodes in a network
│       └── S7 program
│           │
│           ├── Symbols         Symbol table               Contains the assignments of symbols
│           │                                              to the absolute addresses of global data
│           │
│           ├── Sources         Source programs            Contains the sources for the user program
│           │   │                                          (e.g. for STL ans SCL programs)
│           │   └── Source files
│           │
│           └── Blocks          User program
│               │
│               ├── OB n        Organization blocks        Contain the compiled code
│               ├── FB n        Function blocks            and the data for the user program
│               ├── FC n        Functions
│               ├── DB n        Data blocks
│               │
│               ├── SFC n       System functions           Contain the call interface for the system blocks
│               ├── SFB n       System function blocks     integrated in the CPU
│               │
│               ├── System      System data blocks         Contain the compiled data
│               │   data                                   for the configuration table
│               │
│               ├── UDT n       Data types                 Contain the definitions
│               │                                          of the user data types
│               │
│               └── VAT n       Variable tables            enthalten eine Zusammenstellung von Variablen,
│                                                          die beobachtet und gesteuert werden sollen
│
└── S7 program                  (Not assigned to any       (with the same structure as an S7 program
                                specific hardware)         that is assigned to specific hardware)
```

Figure 2.2 Object Hierarchy in a STEP 7 Project

▷ T1-S7 Converting Blocks
Contains additional loadable functions and function blocks for the T1-S7 converter

▷ IEC Function Blocks
Contains loadable functions for editing variables of the complex data types DATE_AND_TIME and STRING

▷ Communication Blocks
Contains loadable functions for controlling CP modules

▷ PID Control Blocks
Contains loadable function blocks for closed-loop control

▷ Organization Blocks
Contains the templates for the organization blocks (essentially the variable declaration for the start information)

You will find an overview of the contents of these libraries in Chapter 33, "Block Libraries". Should you, for example, purchase an S7 module with standard blocks, the associated installation program installs the standard blocks as a library on the hard disk. You can then copy these blocks from the library to your project. A library is opened with FILE → OPEN, and can then be edited in the same way as a project. You can also create your own libraries.

The menu item FILE → NEW... generates a new object at the top of the object hierarchy (project, library). The location in the directory structure where the SIMATIC Manager is to create a project or library must be specified under the menu item OPTIONS → CUSTOMIZE or in the New dialog box.

When you create a project, you can select the project type "Project 2.x". You can also edit projects of this type with STEP 7 V2 (see also Section 2.2.3 "Project Versions").

The INSERT menu is used to add new objects to existing ones (such as adding a new block to a program). Before doing so, however, you must first select the object container in which you want to insert the new object from the left half of the SIMATIC Manager window.

You copy object containers and objects with EDIT → COPY and EDIT → PASTE or, as is usual with Windows, by dragging the selected object with the mouse from one window and dropping it in another. Please note that you cannot undo deletion of an object or an object container in the SIMATIC Manager.

2.1.5 Online Help

The SIMATIC Manager's online help provides information you need during your programming session without the need to refer to hard-copy manuals. You can select the topics you need information on by selecting the HELP menu. The online help option GETTING STARTED, for instance, provides a brief summary on how to use the SIMATIC Manager.

HELP → CONTENTS starts the central STEP 7 Help function from any application. This contains all the basic knowledge.

HELP → CONTEXT-SENSITIVE HELP F1 provides context-sensitive help, i.e. if you press F1, you get information concerning an object selected by the mouse or concerning the current error message.

In the symbol bar, there is a button with an arrow and a question mark. If you click on this button, a question mark is added to the mouse pointer. With this "Help" mouse pointer, you can now click on an object on the screen, e.g. a symbol or a menu command, and you will get the associated online help.

2.2 Editing Projects

When you set up a project, you create "containers" for the resulting data, then you generate the data and fill these containers. Normally, you create a project with the relevant hardware, configure the hardware, or at least the CPU, and receive in return containers for the user program. However, you can also put an S7 program directly into the project container without involving any hardware at all. Note that initializing of the modules (address modifications, CPU settings, configuring connections) is possible only with the Hardware Configuration tool.

We strongly recommend that the entire project editing process be carried out using the SIMATIC Manager. Creating, copying or deleting directories or files as well as changing names (!) with the Windows Explorer within the structure of a project can cause problems with the SIMATIC Manager.

2.2.1 Creating Projects

Project wizard

From STEP 7 V3.2, the *STEP 7 Wizard* helps you in creating a new project. You specify the CPU used and the wizard creates for you a project with an S7 station and the selected CPU as well as an S7 program container, a source container and a block container with the selected organization blocks.

Creating a project with the S7 station

If you want to create a project "manually", this section outlines the necessary actions for you. You will find general information on operator entries for object editing in Section 2.1.3 "SIMATIC Manager".

43

2 STEP 7 Programming Software

Creating a new project

Select FILE → NEW, enter a name in the dialog box, change the project type and storage location if necessary, and confirm with "OK" or RETURN.

Inserting a new station in the project

Select the project and insert a station with INSERT → STATION → SIMATIC 300 STATION (in this case an S7-300).

Configuring a station

Click on the plus box next to the project in the left half of the project window and select the station; the SIMATIC Manager displays the Hardware object in the right half of the window. Double-clicking on *Hardware* starts the Hardware Configuration tool, with which you edit the configuration tables. If the module catalog is not on the screen, call it up with VIEW → CATALOG.

You begin configuring by selecting the rail with the mouse, for instance under "SIMATIC 300" and "RACK 300", "holding" it, dragging it to the free portion in the upper half of the station window, and "letting it go" (drag & drop). You then see a table representing the slots on the rail.

Next, select the required modules from the module catalog and, using the procedure described above, drag and drop them in the appropriate slots. To enable further editing of the project structure, a station requires at least one CPU, for instance the CPU 314 in slot 2. You can add all other modules later. Editing of the hardware configuration is discussed in detail in Section 2.3 "Configuring Stations".

Store and compile the station, then close and return to the SIMATIC Manager. In addition to the hardware configuration, the open station now also shows the CPU.

When it configures the CPU, the SIMATIC Manager also creates an S7 program with all objects. The project structure is now complete.

Viewing the contents of the S7 program

Open the CPU; in the right half of the project window you will see the symbols for the S7 program and for the connection table.

Open the S7 program; the SIMATIC Manager displays the symbols for the compiled user program *(Blocks)*, the container for the source programs, and the symbol table in the right half of the window.

Open the user program *(Blocks)*; the SIMATIC Manager displays the symbols for the compiled configuration data *(System data)* and an empty organization block for the main program (OB 1) in the right half of the window.

Editing user program objects

We have now arrived at the lowest level of the object hierarchy. The first time OB 1 is opened, the window with the object properties is displayed and the editor needed to edit the program in the organization block is opened. You add another empty block for incremental editing by opening INSERT → S7 BLOCK → ... *(Blocks* must be highlighted) and selecting the required block type from the list provided.

When opened, the *System data* object shows a list of available system data blocks. You receive the compiled configuration data. These system data blocks are edited via the *Hardware* object in the container *Station*. You can transfer *System data* to the CPU with PLC → DOWNLOAD and parameterize the CPU in this way.

The object container *Source Files* is empty. With *Source Files* selected, you can select INSERT → S7 SOFTWARE → STL SOURCE FILE to insert an empty source text file or you can select INSERT → EXTERNAL SOURCE FILE to transfer a source text file created, say, with another editor in ASCII format to the *Source Files* container.

Creating a project without an S7 station

If you wish, you can create a program without having first configured a station. To do so, you must generate the container for your program yourself. Select the project and generate an S7 program with INSERT → PROGRAM → S7 PROGRAM. Under this S7 program, the SIMATIC Manager creates the object containers *Sources* and *Blocks*. *Blocks* contains an empty OB 1.

Creating a library

You can also create a program under a library, for instance if you want to use it more than once. In this way, the standard program is always available and you can copy it entire or in part into your current program. Please note that you cannot establish online connections in a library, which means that you can debug a program only within a project.

You cannot copy complete libraries. It is, however, possible to transfer the S7 programs and blocks in a library to other libraries or projects, even from version 2 libraries to version 3 libraries and projects (see also Section 2.2.3 "Project Versions").

2.2.2 Managing, Rearranging and Archiving

The SIMATIC Manager manages projects and libraries in project lists and library lists. When you execute FILE → MANAGE, the SIMATIC Manager shows you a list of all known projects, with name and path. You can then delete projects you no longer want to display from the list ("conceal") or include new projects in the project list ("show"). You manage libraries in the same way.

When it executes FILE → REARRANGE... , the SIMATIC Manager eliminates the gaps created by deletions and optimizes data memory similarly to the way a defragmentation program optimizes the data memory on the hard disk. The reorganization can take some time, depending on the data movements involved.

You can also archive a project or library (FILE → ARCHIVE...). In this case, the SIMATIC Manager stores the selected object in an archive file in compressed form. When archiving, the project or the library must not be under edit; all windows must be closed.

In order to archive a project or library, you need an archive program. STEP 7 contains the archive programs ARJ and PKZIP 2.50, but you may also use other archive programs (*winzip* from version 6.0, *pkzip* from version 2.04g, *JAR* from version 1.02, or *LHARC* from version 2.13).

Projects and libraries cannot be edited in the archived (compressed) state. You can unpack an archived object with FILE → RETRIEVE and then you can edit it further. The retrieved objects are automatically accepted into the project or library management system.

You make the settings for archiving and retrieving on the "Archive" tab under OPTIONS → CUSTOMIZE; e.g. setting the target directory for archiving and retrieving or "Generate archive path automatically" (then no additional specifications are required when archiving because the name of the archive file is generated from the project name).

2.2.3 Project Versions

Since STEP 7 V5 has become available, there are three different versions of SIMATIC projects. STEP 7 V1 creates version 1 projects, STEP 7 V2 version 2 projects, and STEP 7 V3/V4/V5 can be used to create and edit both version 2 and version 3 projects.

If you have a version 1 project, you can convert it into a version 2 project with FILE → OPEN VERSION 1 PROJECT... The project structure with the programs, the compiled version 1 blocks, the STL source programs, the symbol table and the hardware configuration remain unchanged.

Version 2 projects can be created and edited with STEP 7 V3 (Figure 2.3). However, version 2 products still have the function scope of STEP 7 V2; option packages available only in STEP 7 V3 or STEP 7 V4 cannot be used on these projects. The creation of a version 2 project serves a practical purpose when you want to edit it and run it as a V2 project. If you want to convert a version 2 project into a version 3 project, select the menu item FILE → SAVE AS... and then, in the window that appears, select the option "With Consistency Check". You then define the folder (directory), the project name and the project type ("Project" stands for Version 3 project). Version 3 projects can no longer be saved as Version 2 projects.

Program development with STEP 7 is upwards-compatible, so that you can transfer blocks from version 2 projects and libraries to version 3 projects and libraries. However, the reverse, that is, the transfer of blocks from version 3 objects to version 2 objects, is not possible. For this reason, STEP 7 V5 includes the version 2 library *stdlibs* (*V2*) in order to make it possible to copy standard blocks from this library to version 2 projects.

2 STEP 7 Programming Software

STEP 7 V1

V1 project

STEP 7 V2

Use FILE → OPEN VERSION 1 PROJECT to convert a V1 project to V2 with STEP 7 V2 and STEP 7 V3/V4.

Use FILE → NEW → **PROJECT** 2.x to create V2 projects or V2 libraries.

STEP 7 V3/V4/V5

Use FILE → NEW → PROJECT to create new V3 projects or V3 libraries.

V2 project

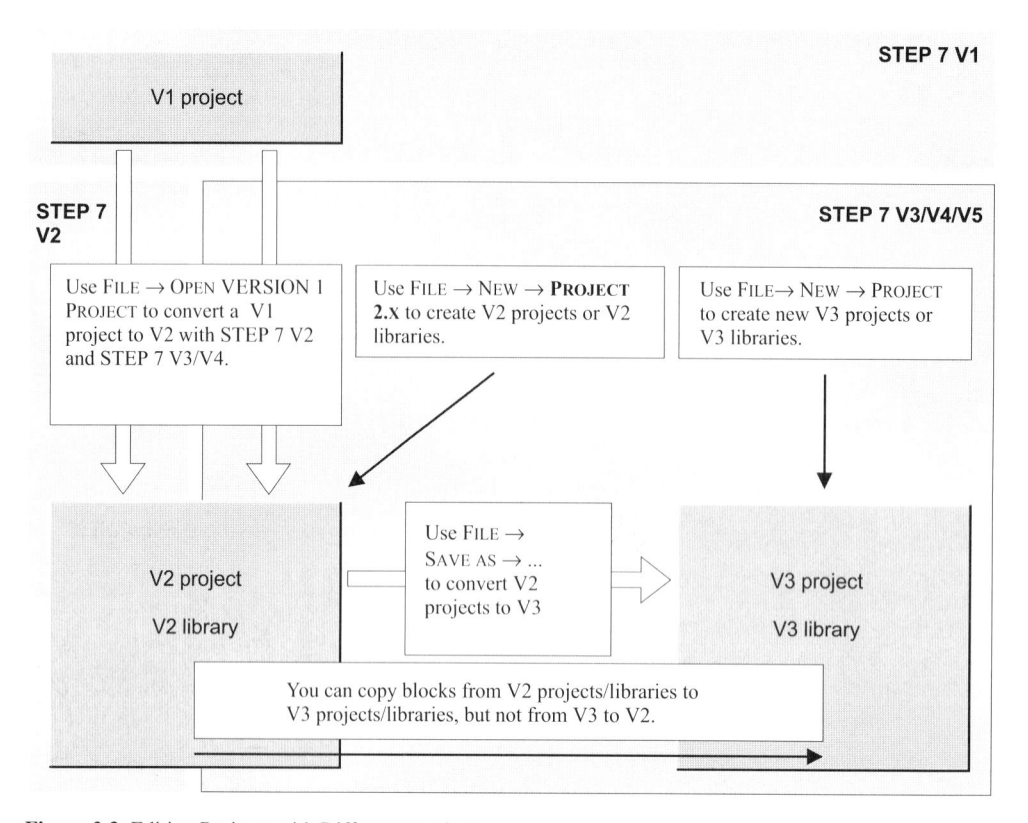

V2 library

Use FILE → SAVE AS → ... to convert V2 projects to V3

V3 project

V3 library

You can copy blocks from V2 projects/libraries to V3 projects/libraries, but not from V3 to V2.

Figure 2.3 Editing Projects with Different Versions

2.3 Configuring Stations

You use the Hardware Configuration tool to plan your programmable controller's configuration. Configuring is carried out offline without connection to the CPU. You can also use this tool to address and parameterize the modules. You can create the hardware configuration at the planning stage or you can wait until the hardware has already been installed.

You start the hardware configuration by selecting the station and then EDIT → OPEN OBJECT or by double-clicking on the Hardware object in the opened container SIMATIC 300/400 Station. You make the basic settings of the hardware configuration with OPTIONS → CUSTOMIZE.

When configuring has been completed, STATION → CONSISTENCY CHECK will show you whether your entries were free of errors. STATION → SAVE stores the configuration tables with all parameter assignment data in your project on the hard disk.

STATION → SAVE AND COMPILE not only saves but also compiles the configuration tables and stores the compiled data in the *System data* object in the offline container *Blocks*. After compiling, you can transfer the configuration data to a CPU with PLC → DOWNLOAD. The object *System data* in the online container *Blocks* represents the current configuration data on the CPU. You can 'return' these data to the hard disk with PLC → UPLOAD.

You export the data of the hardware configuration with STATION → EXPORT. STEP 7 then creates a file in ASCII format that contains the configuration data and parameterization data of the modules. You can choose between a text

2.3 Configuring Stations

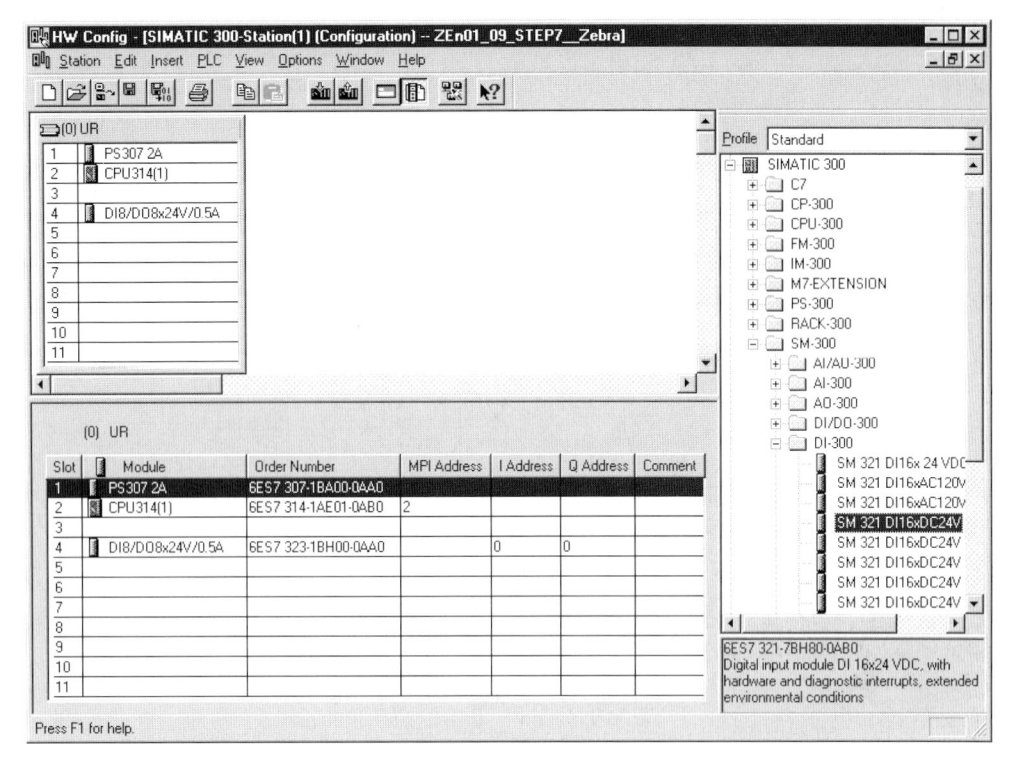

Figure 2.4 Example of a Station Window in the Hardware Configuration

format that contains the data in "readable" English characters, or a compact format with hexadecimal data. You can also import a correspondingly structured ASCII file.

Station window

When opened, the Hardware Configuration displays the station window and the hardware catalog (Figure 2.4). Enlarge or maximize the station window to facilitate editing. In the upper section, it shows the mounting racks in the form of tables and the DP stations in the form of symbols. If there are several racks, you see the connections between the interface modules here and if PROFIBUS is used, you see the configuration of the DP master system. The lower section of the station window shows the configuration table that gives a detailed view of the rack or DP slave selected in the upper section.

Hardware catalog

You can fade the hardware catalog in and out with VIEW → CATALOG. It contains all available mounting racks, modules and interface submodules known to STEP 7. With OPTIONS → EDIT CATALOG PROFILE, you can compile your own hardware catalog that shows only the modules you want to work with – in the structure you select. By double-clicking on the title bar, you can "dock" the hardware catalog onto the right edge of the station window or release it again.

Configuration table

The Hardware Configuration tool works with tables that each represent a mounting rack, a module or a DP station. A configuration table shows the slots with the modules arranged in the slots or the properties of the modules such as their addresses and order numbers. A double-click on a module line opens the properties window of the module and allows parameterization of the module.

2.3.1 Arranging Modules

You begin configuring by selecting and "holding" the rail from the module catalog, for instance under "SIMATIC 300" or "RACK 300", with the mouse, dragging it to the upper half of the station window, and dropping it anywhere in that window (drag&drop). An empty configuration table is screened for the central rack. Next, select the required modules from the module catalog and, in the manner described above, drag and drop them in the appropriate slots. A "No Parking" symbol tells you cannot drop the selected module at the intended slot.

In the case of single-tier S7-300 stations, slot 3 remains empty; it is reserved for the interface module to the expansion rack.

You can generate the configuration table for another rack by dragging the selected rack from the catalog and dropping it in the station window. In S7-400 systems, a non-interconnected rack (or more precisely: the relevant receive interface module) is assigned an interface via the "Link" tab in the Properties window of a Send IM (select module and EDIT → OBJECT PROPERTIES).

The arrangement of distributed I/O stations is described in Section 20.4.2 "Configuring Distributed I/O".

2.3.2 Addressing Modules

When arranging modules, the Hardware Configuration tool automatically assigns a module start address. You can view this address in the lower half of the station window in the object properties for the relevant modules. In the case of the S7-400 CPUs and S7-300 CPUs with integral DP interface, you can change the module addresses. When doing so, please observe the addressing rules for S7-300 and S7-400 systems as well as the addressing capacity of the individual modules.

There are modules that have both inputs and outputs for which you can (theoretically) reserve different start addresses. However, please note carefully the special information provided in the product manuals; the large majority of function and communications modules require the same start address for inputs and outputs.

When assigning the module start address on the S7-400, you can also make the assignment to a subsidiary process image. If there is more than one CPU in the central rack, multiprocessor mode is automatically set and you must assign the module to a CPU.

With VIEW → ADDRESS OVERVIEW, you get a window containing all the module addresses currently in use for the CPU selected.

Modules on the MPI bus or communications bus have an MPI address. You may also change this address. Note, however, that the new MPI address becomes effective as soon as the configuration data are transferred to the CPU.

Symbols for user data addresses

In the Hardware Configuration tool, you can assign to the inputs and outputs symbols (names) that are transferred to the Symbol Table.

After you have arranged and addressed the digital and analog modules, you save the station data. Then you select the module (line) and EDIT → SYMBOLS. In the window that then opens, you can assign a symbol, a data type and a comment to the absolute address for each channel (bit-by-bit for digital modules and word-by-word for analog modules).

The "Symbol = Address" button enters the absolute addresses as symbols in place of the absolute addresses without symbols. The "Apply" button transfers the symbols into the Symbol Table. "OK" also closes the dialog box.

2.3.3 Parameterizing Modules

When you parameterize a module, you define its properties. It is necessary to parameterize a module only when you want to change the default parameters. A requirement for parameterization is that the module is located in a configuration table.

Double-click on the module in the configuration table or select the module and then EDIT → OBJECT PROPERTIES. Several tabs with the specifiable parameters for this module are displayed in the dialog box. When you use this method to parameterize a CPU, you are specifying the run characteristics of your user program.

Some modules allow you to set their parameters at runtime via the user program with the system functions SFC 55 WR_PARM, SFC 56 WR_DPARM and SFC 57 PARM_MOD.

2.3.4 Networking Modules with MPI

You define the nodes for the MPI subsidiary (subnet) with the Module Properties. Select a CPU in the configuration table and open it with EDIT → OBJECT PROPERTIES. The dialog box that then appears contains the "Properties" button in the "Interface" box of the "General" tab. If you click on this button you are taken to another dialog box with a "Parameter" tab where you can find the suitable subnet.

This is also an opportunity to set the MPI address that you have provided for this CPU. Please note that on older S7-300 CPUs, FMs or CPs with MPI connection automatically receive an MPI address derived from the CPU.

The highest MPI address must be greater than or equal to the highest MPI address assigned in the subnet (take account of automatic assignment of FMs and CPs!). It must have the same value for all nodes in the subnet.

Tip: if you have several stations with the same type of CPUs, assign different names (identifiers) to the CPUs in the different stations. They all have the name "CPUxxx(1)" as default so in the subnet they can only be differentiated by their MPI addresses. If you do not want to assign a name yourself, you can, for example, change the default identifier from "CPUxxx(1)" to "CPUxxx(n)" where "n" is equal to the MPI address.

When assigning the MPI address, please also take into account the possibility of connecting a programming device or operator panel (OP) to the MPI network at a later date for service or maintenance purposes. You should connect permanently installed programming devices or OPs direct to the MPI network; for plug-in devices via a spur line, there is an MPI connector with a heavy-gauge threaded-joint socket. Tip: reserve address 0 for a service programming device, address 1 for a service OP and address 2 for a replacement CPU (corresponds to the default addresses).

2.4 Configuring the Network

The basis for communications with SIMATIC is the networking of the S7 stations. The required objects are the subnets and the modules with communications capability in the stations. You can create new subnets and stations with the SIMATIC Manager within the project hierarchy. You then add the modules with communications capability (CPUs and CPs) using the Hardware Configuration tool; at the same time, you assign the communications interfaces of these modules to a subnet. You then define the communications relationships between these modules – the connections – with the Network Configuration tool in the connection table.

The Network Configuration tool allows graphical representation and documentation of the configured networks and their nodes. You can also create all necessary subnets and stations with the Network Configuration tool; then you assign the stations to the subnets and parameterize the node properties of the modules with communications capability.

You can proceed as follows to define the communications relationships via the networking configuration tool:

▷ Open the MPI subnet created as standard in the project container (if it is no longer available, simply create a new subnet with INSERT → SUBNET).

▷ Use the Network Configuration tool to create the necessary stations and – if required – further subnets.

▷ Open the stations and provide them with the modules with communications capability.

▷ Connect the modules with the relevant subnets.

▷ Adapt the network parameters, if necessary

▷ Define the communication connections in the connection table, if required.

2 STEP 7 Programming Software

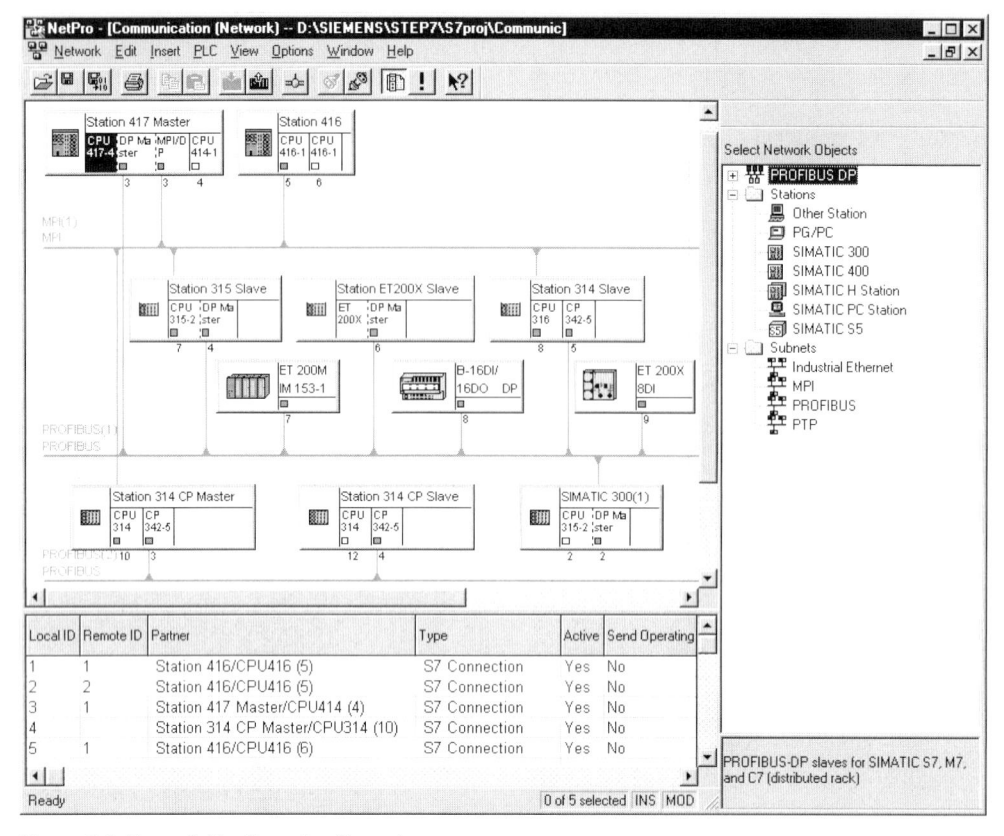

Figure 2.5 Network Configuration Example

You can also configure global data communications within the Network Configuration: select the MPI subnet and then select OPTIONS → DEFINE GLOBAL DATA (see Section 20.5 "Global Data Communications").

NETWORK → SAVE saves an incomplete Network Configuration. You can check the consistency of a Network Configuration with NETWORK → CONSISTENCY CHECK.

You close the Network Configuration with NETWORK → SAVE AND COMPILE.

Network window

To start the Network Configuration, you must have created a project. Together with the project, the SIMATIC Manager automatically creates an MPI subnet.

A double-click on this or any other subnet starts the Network Configuration. You can also reach the Network Configuration if you open the object *Connections* in the *CPU* container.

In the upper section, the Network Configuration window shows all previously created subnets and stations (nodes) in the project with the configured connections.

The connection table is displayed in the lower section of the window if a module with "communications capability", e.g. an S7-400 CPU, is selected in the upper section of the window.

A second window displays the network object catalog with a selection of the available SIMATIC stations, subnets and DP stations. You can fade the catalog in and out with VIEW → CATALOG and you can "dock" it onto the right edge of the network window (double-click on the title bar). With VIEW → ZOOM IN, VIEW → ZOOM OUT and VIEW → ZOOM FACTOR..., you can adjust the clarity of the graphical representation.

50

2.4 Configuring the Network

2.4.1 Configuring the Network View

Selecting and arranging the components

You begin the Network Configuration by selecting a subnet that you select in the catalog with the mouse, hold and drag to the network window. The subnet is represented in the window as a horizontal line. Impermissible positions are indicated with a 'prohibited' sign on the mouse pointer.

You proceed in the same way for the desired stations, at first without connection to the subnet. The stations are still "empty". A double-click on a station opens the Hardware Configuration tool allowing you to configure the station or at least the module(s) with network connection. Save the station and return to the Network Configuration.

The interface of a module with communications capability is represented in the Network Configuration as a small box under the module view. Click on this box, hold and drag it to the relevant subnet. The connection to the subnet is represented as a vertical line.

Proceed in exactly the same way with all other nodes.

You can move created subnets and stations in the network window. In this way, you can also represent your hardware configuration visually.

Setting communications properties

After creating the graphical view, you parameterize the subnets: select the subnets and then EDIT → OBJECT PROPERTIES. The properties window that then appears includes the S7 subnet ID in the "General" tab. The ID consists of two hexadecimal numbers, the project number and the subnet number. You require this S7 subnet ID if you want to go online with the programming device without a suitable project. You set the network properties in the "Network Settings" tab, e.g. the data transfer rate or the highest node address.

When you select the network connection of a node, you can define the network properties of the node with EDIT → OBJECT PROPERTIES, e.g., the node address and the subnet it is connected to, or you can create a new subnet.

On the "Interfaces" tab of the station properties, you can see an overview of all modules with communications capability, with the node addresses and the subnet types used.

You define the module properties of the nodes in a similar way (with the same operator inputs as in the Hardware Configuration tool).

2.4.2 Configuring a DP Master System with the Network Configuration

You can also use the Network Configuration to configure the distributed I/O. Select VIEW → WITH DP SLAVES to display or fade out DP slaves in the network view.

You require the following in order to configure a DP master system

▷ A PROFIBUS subnet (if not already available, drag the PROFIBUS subnet from the network object catalog to the network window).

▷ A DP master in a station (if not already available, drag the station from the network object catalog to the network window, open the station and select a DP master with the Hardware Configuration tool, either integrated in the CPU or as an autonomous module)

▷ The connection from the DP master to the PROFIBUS subnet (either select the subnet in the Hardware Configuration tool or click on the network connection to the DP master in the Network Configuration, "hold" and drag to the PROFIBUS network).

In the network window, select the DP master to which the slave is to be assigned. Find the DP slave in the network object catalog under "PROFIBUS" and the relevant sub-catalog, drag it to the network window and fill out the properties window that appears.

You parameterize the DP slave by selecting it and then selecting EDIT → OPEN OBJECT. The Hardware Configuration is started. Now you can set the user data addresses or, in the case of modular slaves, select the I/O modules (see Section 2.3 "Configuring a Station").

You can only connect an intelligent DP slave to a subnet if you have previously created it (see Section 20.4.2 "Configuring Distributed I/O").

51

In the network object catalog, you can find the type of intelligent DP slave under "Already created stations"; drag it, with the DP master selected, to the network window and fill out the properties window that then appears (as in the Hardware Configuration tool).

With VIEW → HIGHLIGHT → MASTER SYSTEM, you emphasize the assignment of the nodes of a DP master system; first, you select the master or a slave of this master system.

By selecting the PROFIBUS connection and then EDIT → MASTER SYSTEM, you can configure SYNC/FREEZE groups, provided you have a suitable DP master.

2.4.3 Configuring Connections

Connections describe the communications relationships between two devices. Connections must be configured if

▷ you want to establish SFB communications between two SIMATIC S7 devices ("Communication via configured connections") or.

▷ the communications partner is not a SIMATIC S7 device.

Note: You do not require a configured connection for online connection of a programming device to the MPI network for programming or debugging.

Connection table

The communications connections are configured in the connection table. Requirement: you have created a project with all stations that are to exchange data with each other, and you have assigned the modules with communications capability to a subnet.

The object *Connections* in the *CPU* container represents the connection table. A double-click on *Connections* starts the Network Configuration in the same way as a double-click on a subnet in the project container.

To configure the connections, select an S7-400 CPU in the Network Configuration. In the lower section of the network window, you get the connection table (Table 2.1; if it is not visible, place the mouse pointer on the lower edge of the window until it changes shape and then drag the window edge up). You enter a new communication connection with INSERT → CONNECTION or by double-clicking on an empty line.

You create a connection for each "active" CPU. Please note that you cannot create a connection table for an S7-300 CPU; S7-300 CPUs can only be "passive" partners in an S7 connection.

In the "New Connection" window, you select the communications partner in the "Station" and "Module" dialog boxes (Figure 2.6); the station and the module must already exist. You also determine the connection type in this window.

If you want to set more connection properties, activate the check box "Show Properties Dialog".

Connection ID

The number of possible connections is CPU-specific. STEP 7 defines a connection ID for every connection and for every partner. You require this specification when you use communications blocks in your program.

Table 2.1 Connection Table Example

Local ID	Partner ID	Partner	Type	Active connection establishment	Send operating status messages
1	1	Station 416 / CPU416(5)	S7 connection	Yes	No
2	2	Station 416 / CPU416(5)	S7 connection	Yes	No
3		Station 315 / CPU315(7)	S7 connection	Yes	No
4	1	Station 417 / CPU414(4)	S7 connection	Yes	No

2.4 Configuring the Network

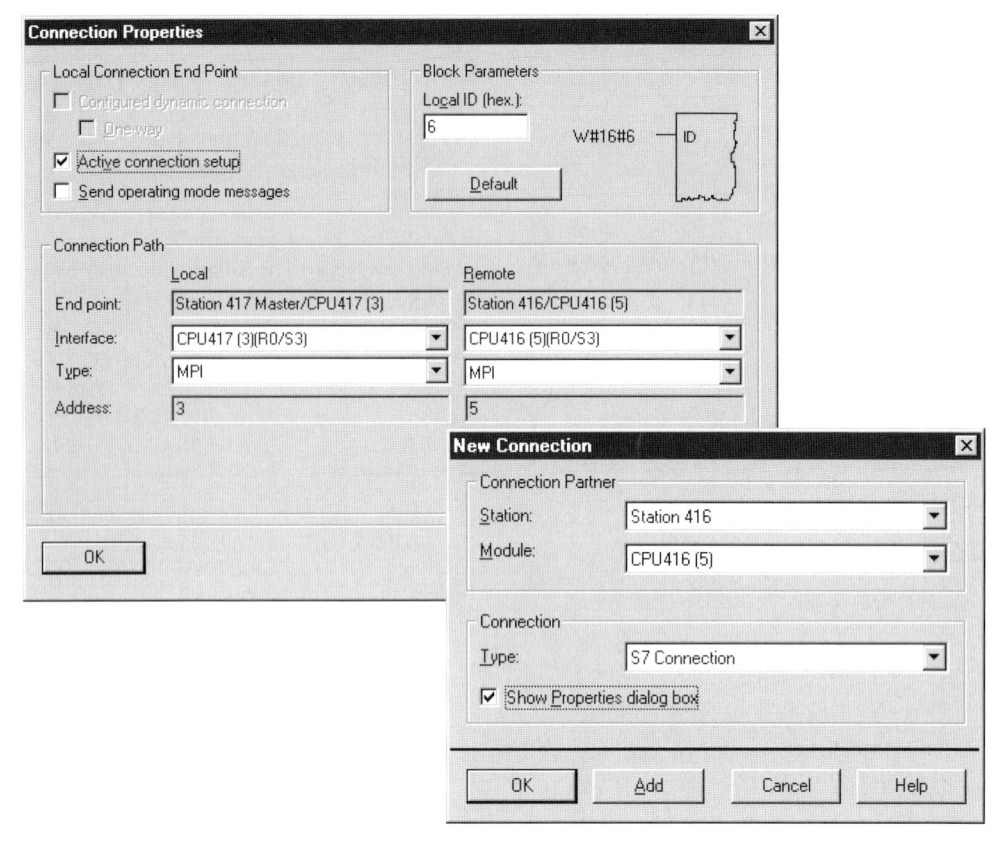

Figure 2.6 Configuring Communications Connections

You can modify the **local ID** (the connection ID of the currently opened module). This is necessary if you have already programmed communications blocks and you want to use the local ID specified there for the connection.

You enter the new local ID as a hexadecimal number. It must be within the following value ranges, depending on the connection type, and must not already be assigned:

▷ Value range for S7 connections:
  0001_{hex} to $0FFF_{hex}$

▷ Value range for PtP connections:
  1000_{hex} to 1400_{hex}

You change the **partner ID** by going to the connection table of the partner CPU and changing (what is then) the local ID: select the connection line and then EDIT → OBJECT PROPERTIES. If STEP 7 does not enter a partner ID, it is a one-way connection (see below).

Partner

This column displays the connection partner. If you want to reserve a connection resource without naming a partner device, enter "unspecified" in the dialog box under Station.

In a **one-way connection**, communication can only be initiated from one partner; example: SFB communications between an S7-400 and S7-300 CPU. Although the SFB communications functions are not available in the S7-300, data can be exchanged by an S7-400 CPU with SFB 14 GET and SFB 15 PUT. In the S7-300, no user program runs for this communication but the data exchange is handled by the operating system.

A one-way connection is configured in the connection table of the "active" CPU. Only then does STEP 7 assign a "Local ID". You also load this connection only in the local station.

53

With a two-way connection, both partners can assume communication actively; e.g. two S7-400 CPUs with the communications functions SFB 8 SEND and SFB 9 BRCV.

You configure a two-way connection only once for one of the two partners. STEP 7 then assigns a "Local ID" and a "Partner ID" and generates the connection data for both stations. You must load each partner with its own connection table.

Connection type

The STEP 7 Basic Package provides you with the following connection types in the Network Configuration:

PtP connection, approved for the subnet PTP (3964(R) and RK 512 procedures) with SFB communications. A PtP (point-to-point) connection is a serial connection between two partners. These can be two SIMATIC S7 devices with the relevant CPs, or a SIMATIC S7 device and a non-Siemens device, e.g. a printer or a barcode reader.

S7 connection, approved for the subnets MPI, PROFIBUS and Industrial Ethernet with SFB communications. An S7 connection is the connection between SIMATIC S7 devices and can include programming devices and human machine interface devices. Data are exchanged via the S7 connection, or programming and control functions are executed.

Fault-tolerant S7 connection, approved for the subnets PROFIBUS and Industrial Ethernet with SFB communications. A fault-tolerant S7 connection is made between fault-tolerant SIMATIC S7 devices and it can also be established to an appropriately equipped PC.

The optional packages "NCM S7 for PROFIBUS" and "NCM S7 for Industrial Ethernet" are available for **parameterizing CPs**. Depending on the NCM software installed, you have additional connection types available for selection: FMS connection, FDL connection, ISO transport connection, ISO-on-TCP connection, UDP connection and email connection.

Active connection buildup

Prior to the actual data transfer, the connection must be built up (initialized). If the connection partners have this capability, you specify here which device is to establish the connection. You do this with the check box "Active Connection Buildup" in the properties window of the connection (select the connection and then EDIT → OBJECT PROPERTIES).

Sending operating state messages

Connection partners with a configured two-way connection can exchange operating state messages. If the local node is to send its operating state messages, activate the relevant check box in the properties window of the connection. In the user program of the partner CPU, these messages can be received with SFB 23 USTATUS.

Connection path

As the connection path, the properties window of the connection displays the end points of the connection and the subnets over which the connection runs. If there are several subnets for selection, STEP 7 selects them in the order Industrial Ethernet before Industrial Ethernet/TCP-IP before MPI before PROFIBUS.

The station and the CPU over which the connection runs are displayed as the end points of the connection. The modules with communications capability are listed under "Interface", specifying the rack number and the slot. If both CPUs are located in the same rack (e.g. S7-400 CPUs in multiprocessor mode), the display box shows "PLC-internal".

Under "Type", you select the subnet over which the connection is to run. If both partners are, for example, connected both to the same MPI subnet and to the same PROFIBUS subnet, "MPI" is specified under "Type". You can now change this specification to "PROFIBUS" and STEP 7 automatically adapts the remaining settings. You will then see the MPI address or PROFIBUS address of the node under "Address".

Connections between projects

For data exchange between two S7 modules belonging to different SIMATIC projects, you enter "unspecified" for connection partner in the connection table (in the local station in both projects).

Please ensure that the connection data agree in both projects (STEP 7 does not check this). After saving and compiling, you load the connection data into the local station in each project.

Connection to non-S7 stations

Within a project, you can also specify stations other than S7 stations as connection partners:

▷ Other stations (non-Siemens devices and also S7 stations in another project)
▷ Programming devices/PCs
▷ SIMATIC S5 stations

A requirement for configuring the connection is that the non-S7 station exists as an object in the project container and you have connected the non-S7 station to the relevant subnet in the station properties (e.g. select the station in the Network Configuration, select EDIT → OBJECT PROPERTIES and connect the station with the desired subnet on the "Interfaces" tab).

2.4.4 Loading the Connection Data

To activate the connections, you must load the connection table into the PLC following saving and compiling (all connection tables into all "active" CPUs).

Requirement: You are in the network window and the connection table is visible. The programming device is a node of the subnet over which the connection data are to be loaded into the modules with communications capability. All subnet nodes have been assigned unique node addresses. The modules to which connection data are to be transferred are in the STOP mode.

With PLC → DOWNLOAD → ..., you transfer the connection and configuration data to the accessible modules. Depending on which object is selected and which menu command is selected, you can choose between the following:

→ SELECTED STATIONS
→ SELECTED AND PARTNER STATIONS
→ SELECTED CONNECTIONS
→ STATIONS ON SUBNET
→ CONNECTIONS AND ROUTERS

The compiled connection data are also a component part of the *System data* in the *Blocks* container. Transfer of the system data and the subsequent startup of the CPUs effectively also transfers the connection data to the modules with communications capability.

For online operation via MPI, a programming device requires no additional hardware. If you connect a PC to a network or if you connect a programming device to an Ethernet or PROFIBUS network, you require the relevant interface module. You parameterize the module with the application "Set PG/PC Interface" in the Windows Control Panel.

2.5 Creating the S7 Program

2.5.1 Introduction

The user program is created under the object *S7 Program*. You can assign this object in the project hierarchy of a CPU or you can create it independently of a CPU. It contains the object *Symbols* and the containers *Source Files* and *Blocks* (Figure 2.7).

With **incremental** program creation, you enter the program direct block-by-block. Entries are checked immediately for syntax. At the same time, the block is compiled as it is saved and then stored in the container *Blocks*. With incremental programming, you can also edit blocks online in the CPU, even during operation. Incremental programming is possible in all the basic languages.

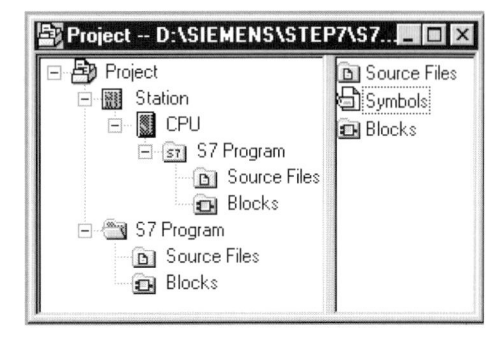

Figure 2.7 Objects for Programming

In the case of **source-oriented** program creation, you write one or more program sources and store these in the container *Source Files*. Program sources are ASCII text files that contain the program statements for one or more blocks, possibly even for the entire program. You compile these sources and you get the compiled blocks in the container *Blocks*. Source-oriented program creation is used in STL; you cannot use source-oriented programming with LAD or FBD, but programs created with LAD or FBD can be stored as source files.

The signal states or the values of addresses are processed in the program. An address is, for example, the input I1.0 (*absolute addressing*). With the help of the Symbol Table under the object *Symbols*, you can assign a symbol (an alphanumeric name, e.g. "Switch motor on") to an address and then access it with this name (*symbolic addressing*). In the properties of the offline object container *Blocks*, you specify whether in the event of a change in the Symbol Table the absolute address or the symbol is to be definitive for the already compiled blocks when next saved (*address priority*).

2.5.2 Symbol Table

In the control program, you work with addresses; these are inputs, outputs, timers, blocks. You can assign absolute addresses (e.g. I1.0) or symbolic addresses (e.g. Start signal). Symbolic addressing uses names instead of the absolute address. You can make your program easier to read by using meaningful names.

In symbolic addressing, a distinction is made between *local* symbols and *global* symbols. A local symbol is known only in the block in which it has been defined. You can use the same local symbols in different blocks for different purposes. A global symbol is known throughout the entire program and has the same meaning in all blocks. You define global symbols in the symbol table (object *Symbols* in the container *S7 Program*).

A global symbol starts with an alpha character and can be up to 24 characters long. A global symbol can also contain spaces, special characters and national characters such as the Umlaut. Exceptions to this are the characters 00_{hex}, FF_{hex} and the inverted commas ("). You must enclose symbols with special characters in inverted commas when programming. In the compiled block, the program editor displays all global symbols in inverted commas. The symbol comment can be up to 80 characters long.

In the symbol table you can assign names to the following addresses and objects:

▷ Inputs I, outputs Q, peripheral inputs PI and peripheral outputs PQ

▷ Memory bits M, timer functions T and counter functions C

▷ Code blocks OBs, FBs, FCs, SFCs, SFBs and data blocks DBs

▷ User-defined data types UDTs

▷ Variable table VAT

Data addresses in the data blocks are included among the local addresses; the associated symbols are defined in the declaration section of the data block in the case of global data blocks and in the declaration section of the function block in the case of instance data blocks.

When creating an S7 program, the SIMATIC Manager also creates an empty symbol table *Symbols*. You open this and can then define the global symbols and assign them to absolute addresses (Figure 2.8). There can be only one single symbol table in an S7 program.

The data type is part of the definition of a symbol. It defines specific properties of the data behind the symbol, essentially the representation of the data contents. For example, the data type BOOL identifies a binary variable and the data type INT designates a digital variable whose contents represent a 16-bit integer. Please refer to Section 3.8 "Variables and Constants" for an overview of the data types used in STEP 7; Chapter 24 "Data Types" contains a detailed description.

With incremental programming, you create the symbol table before entering the program; you can also add or correct individual symbols during program input. In the case of source-oriented programming, the complete symbol table must be available when the program source is compiled.

2.5 Creating the S7 Program

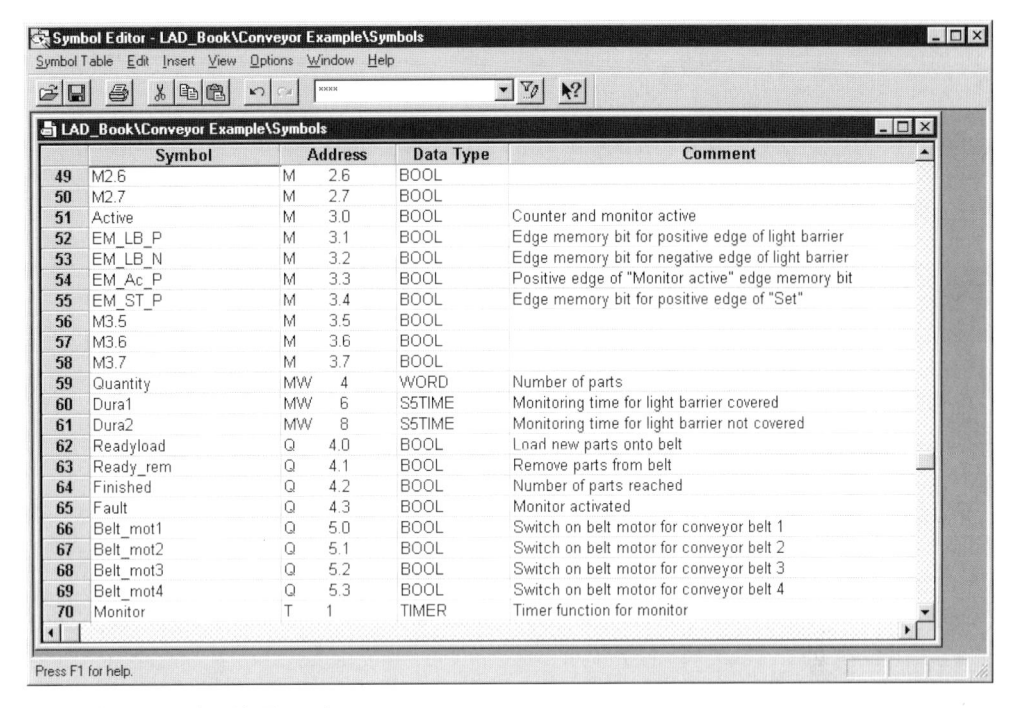

	Symbol	Address	Data Type	Comment
49	M2.6	M 2.6	BOOL	
50	M2.7	M 2.7	BOOL	
51	Active	M 3.0	BOOL	Counter and monitor active
52	EM_LB_P	M 3.1	BOOL	Edge memory bit for positive edge of light barrier
53	EM_LB_N	M 3.2	BOOL	Edge memory bit for negative edge of light barrier
54	EM_Ac_P	M 3.3	BOOL	Positive edge of "Monitor active" edge memory bit
55	EM_ST_P	M 3.4	BOOL	Edge memory bit for positive edge of "Set"
56	M3.5	M 3.5	BOOL	
57	M3.6	M 3.6	BOOL	
58	M3.7	M 3.7	BOOL	
59	Quantity	MW 4	WORD	Number of parts
60	Dura1	MW 6	S5TIME	Monitoring time for light barrier covered
61	Dura2	MW 8	S5TIME	Monitoring time for light barrier not covered
62	Readyload	Q 4.0	BOOL	Load new parts onto belt
63	Ready_rem	Q 4.1	BOOL	Remove parts from belt
64	Finished	Q 4.2	BOOL	Number of parts reached
65	Fault	Q 4.3	BOOL	Monitor activated
66	Belt_mot1	Q 5.0	BOOL	Switch on belt motor for conveyor belt 1
67	Belt_mot2	Q 5.1	BOOL	Switch on belt motor for conveyor belt 2
68	Belt_mot3	Q 5.2	BOOL	Switch on belt motor for conveyor belt 3
69	Belt_mot4	Q 5.3	BOOL	Switch on belt motor for conveyor belt 4
70	Monitor	T 1	TIMER	Timer function for monitor

Figure 2.8 Symbol Table Example

Importing, exporting

Symbol tables can be imported and exported. "Exported" means a file is created with the contents of your symbol table. You can select here either the entire symbol table, a subset limited by filters or only selected lines. For the data format you can choose between pure ASCII text (extension *.asc), sequential assignment list (*.seq), System Data Format (*.sdf for Microsoft Access) and Data Interchange Format (*.dif for Microsoft Excel). You can edit the exported file with a suitable editor. You can also import a symbol table available in one of the formats named above.

Special object properties

With EDIT → SPECIAL OBJECT PROPERTIES, you set attributes for each symbol in the symbol table. These attributes or properties are used in the following:

▷ Human machine interface functions for monitoring with WinCC

▷ Configuring communications

▷ Configuring messages

VIEW → COLUMNS O, M, C makes the settings visible. With OPTIONS → CUSTOMIZE, you specify whether or not the special object properties are to be copied. The special object properties are neither imported nor exported with the symbol table.

2.5.3 Program Editor

For creating the user program, the STEP 7 Basic Package contains a program editor for the LAD, FBD and STL programming languages. You program incrementally with LAD and FBD, that is, you enter an executable block direct; Figure 2.9 shows the possible actions for this.

If you use symbolic addressing for global addresses, the symbols must already be assigned to an absolute address in the case of incremental programming; however, you can enter new symbols or change symbols during program input.

57

LAD/FBD blocks can be "decompiled", i.e. a readable block can be created again from the MC7 code without an offline database (you can read any block from a CPU using a programming device without the associated project). In addition, an STL program source can be created from any compiled block.

Starting the program editor

You reach the program editor when you open a block in the SIMATIC Manager, e.g. by double-clicking on the automatically generated symbol of the organization block OB 1, or via the Windows taskbar with START → SIMATIC → STEP 7 → LAD, STL, FBD – PROGRAM S7 BLOCKS.

You can customize the properties of the program editor with OPTIONS → CUSTOMIZE. On the "Editor" tab, select the properties with which a new block is to be generated and displayed, such as creation language, pre-selection for comments and symbols.

When you open a compiled block in the *Blocks* container (e.g. with a double-click), it is opened for incremental programming.

It is also possible to call blocks that have been written with another programming language such as STL. The user program is generated block by block and every block finally contains executable MC7 code regardless of the programming language used to write it.

With incremental programming, you can execute all programming functions with one exception: if you want to provide block protection (KNOW_HOW_PROTECT), you can only do this via a program source file (see Chapter 24.1 "Block Protection" for more detailed information on this topic).

Incremental programming

With incremental programming, you edit the blocks both in the offline and online *Blocks* container. The editor checks your entries in incremental mode as soon as you have terminated a program line. When the block is closed it is immediately compiled, so that only error-free blocks can be saved.

On the "Create Block" tab under OPTIONS → CUSTOMIZE, you set automatic updating of the reference data when saving a block.

The blocks can be edited both offline in the programming device's database and online in the CPU, generally referred to as the "programmable controller", or "PLC". For this purpose, the SIMATIC Manager provides an offline and an online window; the one is distinguished from the other by the labeling in the title bar.

In the offline window, you edit the blocks right in the PG database. If you are in the editor, you can store a modified block in the offline database with FILE → SAVE and transfer it to the CPU with PLC → DOWNLOAD. If you want to save the opened block under another number or in a different project, or if you want to transfer it to a library or to another CPU, use the menu command FILE → SAVE AS.

To edit a block in the CPU, open that block in the online window. This transfers the block from the CPU to the programming device so that it can be edited. You can write the edited block back to the CPU with PLC → DOWNLOAD. If the CPU is in RUN mode, the CPU will process the edited block in the next program scan cycle. If you want to save a block that you edited online in the offline database as well, you can do so with FILE → SAVE.

Section 2.6.4 "Loading the User Program into the CPU" and Section 2.6.5 "Block Handling" contain further information on online programming. Section 3.3 "Programming Code Blocks" and Section 3.4 "Programming Data Blocks" show you how to enter a LAD/FBD block.

2.5.4 Updating or generating source files

On the "Source Files" tab under OPTIONS → CUSTOMIZE, you can select the option "Generate source file automatically" so that when you save an (incrementally created) block, the program source file is updated or created, if it does not already exist.

You can derive the name of a new source file from the absolute address or the symbolic address. The addresses can be transferred in absolute or symbolic form to the source file.

With the "Execute" button, you select, in the subsequent dialog box, the blocks from which you want to generate a program source file.

You can export source files from the project by selecting EDIT → EXPORT SOURCE FILE in the

2.5 Creating the S7 Prog

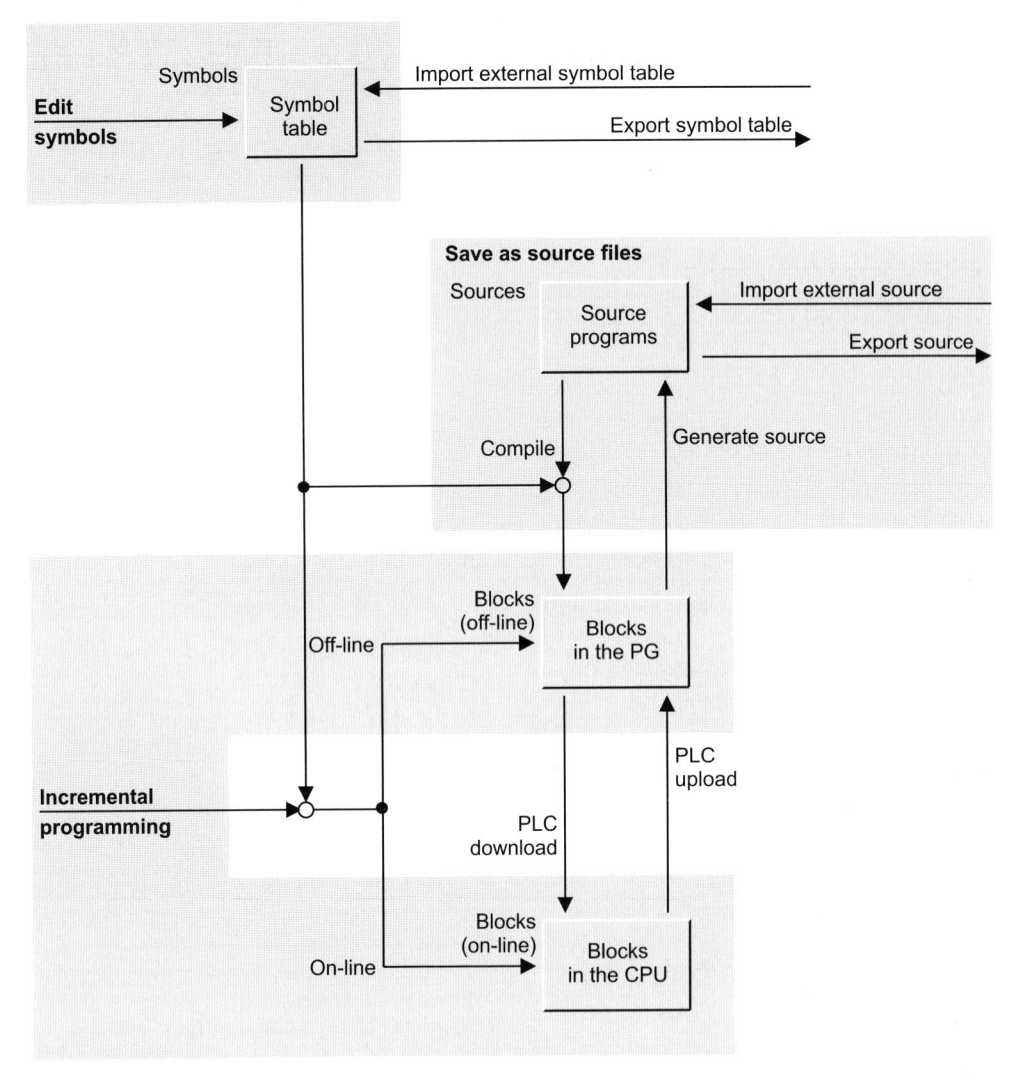

Figure 2.9 Writing Programs with the LAD/FBD Editor

SIMATIC Manager. You can then further process these ASCII files with another text editor, for example. Source files can also be imported back into the *Source Files* container with INSERT → EXTERNAL SOURCE FILE.

If you generate a source file from a block that you have created with LAD or FBD, you can generate a LAD or FBD block again from this source file. You compile the source file by opening it in the SIMATIC MANAGER with a double-click and by then selecting FILE → COMPILE in the program editor. An STL block is created in the *Blocks* container. You open this block and switch to your usual representation with VIEW → LAD or VIEW → FBD. After saving, the block retains this property.

If you selected the setting "Addresses – Symbolic" when creating the source file, you require a complete symbol table for compiling the source file. In this way, you can specify different absolute addresses in the symbol table and, after compilation, you end up with a program with, for example, different inputs and outputs. This allows you to adapt the program to a dif-

59

ferent hardware configuration. For this purpose, it is best to store these source files (which are independent of the hardware addressing) in a library, for example.

Rewiring

The *Rewiring* function allows you to replace addresses in individually compiled blocks or in the entire user program. For example, you can replace input bits I0.0 to I 0.7 with input bits I 16.0 to I 16.7.

Permissible addresses are inputs, outputs, memory bits, timers and counters as well as functions FCs and function blocks FBs.

In the SIMATIC Manager, you select the objects in which you wish to carry out the rewiring; select a single block, a group of blocks by holding Ctrl and clicking with the mouse, or the entire *Blocks* user program. Options → Rewire takes you to a table in which you can enter the old addresses to be replaced and the new addresses. When you confirm with "OK", the SIMATIC Manager then exchanges the addresses.

When "rewiring" blocks, change the numbers of the blocks first and then execute rewiring that changes the calls correspondingly.

A subsequently displayed info file shows you in which block changes were made, and how many.

Further possible methods of rewiring are:

▷ With individually compiled blocks, you can also use the *Address priority* function.

▷ If there is a program source file with symbolic addressing, you change the absolute addresses in the symbol table prior to compiling and you get an "unwired" program after compilation.

2.5.5 Address Priority

In the properties window of the offline object container *Blocks* on the "Blocks" tab, you can set whether the absolute address or the symbol is to have priority for already saved blocks when they are displayed and saved again following a change to the symbol table or to the declaration or assignment of global data blocks.

The default is "Absolute address has priority" (the same behavior as in the previous STEP 7 versions). This default means that when a change is made in the symbol table, the absolute address is retained in the program and the symbol changes accordingly. If "Symbol has priority" is set, the absolute address changes and the symbol is retained.

Example:

The symbol table contains the following:

```
I 1.0 "Limit_switch_up"
I 1.1 "Limit_switch_down"
```

In the program of an already compiled block, input I1.0 is scanned:

I 1.0 "Limit_switch_up"

If the assignments for inputs I1.0 and I1.1 are now changed in the symbol table to:

```
I 1.0 "Limit_switch_down"
I 1.1 "Limit_switch_up"
```

and the already compiled block is read out, then the program contains

I 1.1 **"Limit_switch_up"**

if "Symbol has priority" is set, and if "Absolute address has priority" is set, the program contains

I 1.0 "Limit_switch_down"

If, as a result of a change in the symbol table, there is no longer any assignment between an absolute address and a symbol, the statement will contain the absolute address if "Absolute address has priority" is set (even with symbolic display because the symbol would, of course, be missing); if "Symbol has priority" is set, the statement is rejected as errored (because the mandatory absolute address is missing).

If "Symbol has priority" is set, incrementally programmed blocks with symbolic addressing will retain their symbols in the event of a change to the symbol table. In this way, an already programmed block can be "rewired" by changing the address assignment.

Please note that this "rewiring" does not occur automatically because the already compiled blocks contain the executable MC7 code of the statements with absolute addresses. The change is only made in the relevant blocks – following the relevant message – after they have been opened and saved again.

2.5.6 Reference Data

As a supplement to the program itself, the SIMATIC Manager shows you the reference data, which you can use as the basis for corrections or tests. These reference data include the following:

▷ Cross references
▷ Reserved locations (I, Q, M and T, C)
▷ Program structure
▷ Unused symbols
▷ Addresses without symbols

To generate reference data, select the *Blocks* object and the menu command OPTIONS → REFERENCE DATA → DISPLAY. The representation of the reference data can be changed specifically for each work window with VIEW → FILTER...; you can save the settings for later editing by selecting the "Save as Standard" option. You can display and view several lists at the same time.

With OPTIONS → CUSTOMIZE in the program editor, specify on the "Create Blocks" tab whether or not the reference data are to be updated when compiling a program source file or when saving an incrementally written block.

Please note that the reference data are only available when the data are managed offline; the offline reference data are displayed even if the function is called in a block opened online.

Cross references

The cross-reference list shows the use of the addresses and blocks in the user program. It includes the absolute address, the symbol (if any), the block in which the address was used, how it was used (read or write), and language-related information. For STL, the language-related information column shows the network in which the address was used; for SCL the line and column number. Click on a column header to sort the table by column contents.

If an address is selected, EDIT → GO TO → LINE starts the program editor and displays the address in the programmed environment.

The cross-reference list shows the addresses you selected with VIEW → FILTER... (for instance bit memory). When you double-click on an address, the editor opens the block displayed on that line at the location at which the address appears. STEP 7 then uses the filter saved as "Standard" every time it opens the cross-reference list.

Advantage: the cross references show you whether the referenced addresses were also scanned or reset. They also show you in which blocks addresses are used (possibly more than once).

Assignments

The I/Q/M reference list shows which bits in address areas I, Q and M are assigned in the program. One byte, broken down into bits, appears on each line. Also shown is whether access is by byte, word, or doubleword. The T/C reference list shows the timers and counters used in the program. Ten timers or counters are displayed on a line.

Advantage: the list shows you whether certain address areas were (improperly) assigned or where there are still addresses available.

Program structure

The program structure shows the call hierarchy of the blocks in a user program. The "starting block" for the call hierarchy is specified via filter settings. You have a choice between two different views:

The *tree structure* shows all nesting levels of the block calls. You control the display of nesting levels with the "+" and "–" boxes. The requirements for temporary local data are shown for the entire path following the starting block and/or per call path. Click the right mouse button to fade in a menu field in which you can open the block, switch to the call location, or screen additional block information.

The *parent/child structure* shows two call levels with one block call. Language-related information is also included.

Avantage: which blocks were used? Were all programmed blocks called? What are the blocks' temporary local data requirements? Is the specified local data requirement per priority class (per organization block) sufficient?

2 STEP 7 Programming Software

Unused symbols

This list shows all addresses which have symbol table allocations but were not used in the program. The list shows the symbol, the address, the data type, and the comment from the symbol table.

Advantage: were the addresses in the list inadvertently forgotten when the program was being written? Or are they perhaps superfluous, and not really needed?

Addresses without symbols

This list shows all the addresses used in the program to which no symbols were allocated. The list shows these addresses and how often they were used.

Advantage: were addresses used inadvertently (by accident, or because of a typing error)?

2.6 Online Mode

You create the hardware configuration and the user program on the programming device, generally referred to as the "engineering system" (ES). The S7 program is stored offline on the hard disk here, also in compiled form.

To transfer the program to the CPU, you must connect the programming device to the CPU. You establish an "online" connection. You can use this connection to determine the operating state of the CPU and the assigned modules, i.e., you can carry out diagnostics functions.

2.6.1 Connecting a PLC

The connection between the programming device's MPI interface and the CPU's MPI interface is the mechanical requirement for an online connection. The connection is unique when a CPU is the only programmable module connected. If there are several CPUs in the MPI subnet, each CPU must be assigned a unique node number (MPI address). You set the MPI address when you initialize the CPU. Before linking all the CPUs to one network, connect the programming device to only one CPU a time and transfer the *System Data* object from the offline user program *Blocks* or direct with the Hardware Configuration editor using the menu command PLC → DOWNLOAD. This assigns a CPU its own special MPI address ("naming") along with the other properties.

The MPI address of a CPU in the MPI network can be changed at any time by transferring a new parameter data record containing the new MPI address to the CPU. Note carefully: the new MPI address takes effect immediately. While the programming device adjusts immediately to the new address, you must adapt other applications, such as global data communications, to the new MPI address.

The MPI parameters are retained in the CPU even after a memory reset. The CPU can thus be addressed even after a memory reset. A programming device can always be operated online on a CPU, even with a module-independent program and even though no project has been set up.

If no project has been set up, you establish the connection to the CPU with PLC → DISPLAY ACCESSIBLE NODES. This screens a project window with the structure "*Accessible Nodes*" – "Module (MPI=n)" – "Online User Program *(Blocks)*". When you select the *Module* object, you may utilize the online functions, such as changing the operational status and checking the module status. Selecting the *Blocks* object displays the blocks in the CPU's user memory. You can then edit (modify, delete, insert) individual blocks.

If there is a CPU-independent program in the project window, create the associated online project window. If several CPUs are connected to the MPI and accessible, select EDIT → OBJECT PROPERTIES with the online S7 program selected and set the number of the mounting rack and the CPU's slot on the "Addresses Module" tab.

You can also use this window to change the MPI address of the CPU. Requirement: you must have activated the check box "Allow MPI address change" on the "View" tab under OPTIONS → CUSTOMIZE in the SIMATIC Manager.

If you select the *S7 Program* in the online window all the online functions to the connected CPU are available to you. *Blocks* shows the blocks located in the CPU's user memory. If the blocks in the offline program agree with the blocks in the online program you can edit the blocks in the user memory with the information from the data management system of the programming device (symbolic address, comments).

When you switch a **CPU-assigned program** into online mode, you can carry out program modifications just as you would in a CPU-independent program. In addition, it is now possible for you to configure the SIMATIC station, that is, to set CPU parameters and address and parameterize modules.

2.6.2 Protecting the User Program

With appropriately equipped CPUs, access to the user program can be protected with a password. Everyone in possession of the password has unrestricted access to the user program. For those who do not know the password, you can define 3 protection levels. You set the protection levels with the "Protection" tab of the Hardware Configuration tool when parameterizing the CPU.

Protection level 1: keylock switch position

This protection level is set as default (without password). Here, the user program is protected by the mode selector switch on the front of the CPU. In the RUN-P and STOP positions, you have unrestricted access to the user program; in the RUN position, only read access via the programming device is possible. In this position, you can also remove the keylock switch so that the mode can no longer be changed via the switch.

You can bypass protection via the keylock switch RUN position by selecting the option "Can be revoked with password", e.g. if the CPU, and with it the keylock switch, are not easily accessible or are located at a distance.

Protection level 2: write protection

At this protection level, the user program can only be read, regardless of the position of the keylock switch.

Protection level 3: read/write protection

No access to the user program, regardless of the keylock switch position.

Password protection

If you select protection level 2 or 3 or protection level 2 with "Can be revoked with password", you will be prompted to define a password. The password can be up to 8 characters long.

If you try to access a user program that is protected with a password, you will be prompted to enter the password. Before accessing a protected CPU, you can also enter the password via PLC → ACCESS RIGHTS. First, select the relevant CPU or the S7 program.

In the "Enter Password" dialog box, you can select the option "Use password for other protected modules" to get access to all modules protected with the same password.

Password access authorization remains in force until the last S7 application has been terminated.

Everyone in possession of the password has unrestricted access to the user program in the CPU regardless of the protection level set and regardless of the keylock position.

2.6.3 CPU Information

In online mode, the CPU information listed below is available to you. The menu commands are screened when you have selected a module (in online mode and without a project) or S7 program (in the online project window).

▷ PLC → DIAGNOSE HARDWARE
 (see Section 2.7.1 "Diagnosing the Hardware")

▷ PLC → MODULE INFORMATION
 General information (such as version), diagnostics buffer, memory (current map of work memory and load memory, compression), cycle time (length of the last, longest, and shortest program cycle), timing system (properties of the CPU clock, clock synchronization, run-time meter), performance data (memory configuration, sizes of the address areas, number of available blocks, SFCs, and SFBs), communication (data transfer rate and communication links), stacks in STOP state (B stack, I stack, and L stack).

▷ PLC → OPERATING MODE
 Display of the current operating mode (for instance RUN or STOP), modification of the operating mode

▷ PLC → CLEAR/RESET
Resetting of the CPU in STOP mode

▷ PLC → SET TIME AND DATE
Setting of the internal CPU clock

▷ PLC → CPU Messages
Reporting of asynchronous system errors and of user-defined messages generated in the program with SFC 52 WR_USMSG, SFC 18 ALARM_S and SFC 17 ALARM_SQ.

▷ PLC → DISPLAY FORCE VALUES,
PLC → MONITOR/MODIFY VARIABLES
(see Section 2.7.3 "Monitoring and Modifying Variables" and 2.7.4 "Forcing Variables").

2.6.4 Loading the User Program into the CPU

When you transfer your user program (compiled blocks and configuration data) to the CPU, it is loaded into the CPU's load memory. Physically, load memory can be RAM or flash EPROM and either integrated in the CPU or on a memory card.

If the memory card is a flash EPROM, you can write to it in the programming device and use it as data medium. You plug the memory card into the CPU in the off-circuit state; on power up following memory reset, the relevant data of the memory card are transferred to the work memory of the CPU. With appropriately equipped CPUs, you can also overwrite a flash EPROM card if it is plugged into the CPU, but only with the entire program.

In the case of a RAM load memory, you transfer a complete user program by switching the CPU to the STOP state, performing memory reset and transferring the user program. The configuration data are also transferred. The program in RAM is then lost at memory reset or if the backup battery is switched off.

If you only want to change the configuration data (CPU properties, the configured connections, GD communications, module parameters, and so on), you need only load the *System Data* object into the CPU (select the object and transfer it with menu command PLC → DOWNLOAD. The parameters for the CPU go into effect immediately; the CPU transfers the parameters for the remaining modules to those modules during startup.

Please note that the entire configuration is loaded onto the PLC with the *System data* object. If you use PLC → DOWNLOAD in an application, e.g. in global data communications, only the data edited by the application are transferred.

2.6.5 Block Handling

Transferring blocks

In the case of a RAM load memory, you can also modify, delete or reload individual blocks in addition to transferring the entire program online.

You transfer individual blocks to the CPU by selecting them in the offline window and selecting PLC → DOWNLOAD. With the offline and online windows opened at the same time, you can also drag the blocks with the mouse from one window and drop them in the other.

Special care is needed when transferring individual blocks during operation. If blocks that are not available in the CPU memory are called within a block, you must first load the "lower-level" blocks. This also applies for data blocks whose addresses are used in the loaded block. You load the "highest-level" block last. Then, provided it is called, it will be executed immediately in the next program cycle.

The SIMATIC Manager also allows you to transfer individual blocks or the entire program from the offline container Blocks to the CPU in SCL. Transfer back from the CPU to the hard disk makes little sense since compiled blocks can no longer be edited by the SCL editor. You can only edit the SCL program source file and, form it, generate the compiled blocks.

Modifying blocks online

You can edit blocks incrementally with STL in the online user program (on the CPU), in exactly the same way as in the offline user program. However, if online and offline data management diverge, it may result in the editor being unable to display the additional information of the offline database; these data can then be lost (symbolic identifiers, jump labels, comments, user-defined data types).

Blocks that have been modified online are best stored offline on the hard disk to avoid data inconsistency (e.g. a "time stamp conflict" when the interface of the called block is later than the program in the calling block).

Deleting blocks

If the load memory consists exclusively of RAM, blocks can be modified and deleted. If the user program is located on a flash EPROM, modules can also be deleted and modified provided the additionally available RAM capacity is sufficient. The blocks in flash EPROM are selected as "invalid". However, following memory reset or after power up without battery backup, the blocks are transferred again from the flash EPROM load memory to the work memory.

You can delete a flash EPROM memory card only in the programming device.

Compressing

When you load a new or modified block into the CPU, the CPU places the block in load memory and transfers the relevant data to work memory. If there is already a block with the same number, this "old block" is declared invalid (following a prompt for confirmation) and the new block "added on at the end" in memory. Even a deleted block is "only" declared invalid, not actually removed from memory.

This results in gaps in user memory which increasingly reduce the amount of memory still available. These gaps can be filled only by the *Compress* function. When you compress in RUN mode, the blocks currently being executed are not relocated; only in STOP mode can you truly achieve compression without gaps.

The current memory allocation can be displayed in percent with the menu command PLC → MODULE INFORMATION, on the *Memory* tab. The dialog box which then appears also has a button for preventive compression.

You can initiate event-driven compressing per program with the call SFC 25 COMPRESS.

Data blocks offline/online

The data addresses in a data block can be assigned an *initial value* and an *actual value* (see also Section 3.6 "Programming Data Blocks". If a data block is loaded into the CPU, the initial values are transferred to load memory and the actual values are transferred to work memory. Every value change made to a data address per program corresponds to a change to the actual value.

When you load a data block from the CPU, its values are taken from work memory, for work memory is where all actual values are to be found. You can view the actual values at the time they are read out with VIEW → DATA VIEW. If you modify an actual value in the data block and write the block back to the CPU, the modified value is placed in work memory.

When a Flash EPROM memory card is used as load memory, the blocks on the memory card are transferred to work memory following a CPU memory reset. The data block retains the initial values originally programmed for them. The same applies on power-up without battery backup. On the S7-300, you can prevent this from happening to a data area by declaring that area retentive.

A data block generated with the property UN-LINKED is not transferred to work memory; it remains in load memory. A data block with this property can only be read with SFC 20 BLK-MOV.

2.7 Testing the Program

After establishing a connection to a CPU and loading the user program, you can test (debug) the program as a whole or in part, such as individual blocks. You initialize the variables with signals and values, e.g. with the help of simulator modules and evaluate the information returned by your program. If the CPU goes to the STOP state as a result of an error, you can get support in finding the cause of the error from the CPU information among other things.

Extensive programs are debugged in sections. If, for example, you only want to debug one block, load this block into the CPU and call it in OB 1. If OB 1 is organized is in such a way that the program can be debugged section by section "from beginning to end", you can select the blocks or program sections for debugging by using jump functions to skip those calls or program sections that are not to be debugged.

With the PLCSIM optional software, you can simulate a CPU on the programming device and so debug your program without additional hardware.

2.7.1 Diagnosing the Hardware

In the event of a fault, you can fetch the diagnostics information of the faulty modules with the help of the function "Diagnose Hardware". You connect the programming device to the MPI bus and start the SIMATIC Manager.

If the project associated with the plant configuration is available in the programming device database, open the online project window with VIEW → ONLINE. Otherwise, select PLC → DISPLAY ACCESSIBLE NODES and select the CPU. Now you can get a quick overview of the faulty modules with PLC → DIAGNOSE HARDWARE (default). The Hardware Configuration supplies the detailed diagnostics information from the modules in the online view; this can be set in the SIMATIC Manager on the "View" tab under OPTIONS → CUSTOMIZE.

You can get information on the status and operating state of the modules accessible online in the form of a project view (display of the stations reporting errors), a station view (modules reporting errors) and a module view (display of the available diagnostics information).

2.7.2 Determining the Cause of a STOP

If the CPU goes to STOP because of an error, the first measure to take in order to determine the reason for the STOP is to output the diagnostics buffer. The CPU enters all messages in the diagnostic buffer, including the reason for a STOP and the errors which led to it. To output the diagnostic buffer, switch the PG to online, select an S7 program, and choose the *Diagnostics Buffer* tab with the menu command PLC → MODULE INFORMATION. The last event (the one with the number 1) is the cause of the STOP, for instance "STOP because programming error OB not loaded". The error which led to the STOP is described in the preceding message, for example "FC not loaded". By clicking on the message number, you can screen an additional comment in the next lower display field. If the message relates to a programming error in a block, you can open and edit that block with the "Open Block" button.

If the cause of the STOP is, for example, a programming error, you can ascertain the surrounding circumstances with the *Stacks* tab. When you open *Stacks*, you will see the B stack (block stack), which shows you the call path of all non-terminated blocks up to the block containing the interrupt point. Use the "I stack" button to screen the interrupt stack, which shows you the contents of the CPU registers (accumulators, address register, data block register, status word) at the interrupt point at the instant the error occurred. The L stack (local data stack) shows the block's temporary local data, which you select in the B stack by clicking with the mouse.

2.7.3 Monitoring and Modifying Variables

One excellent resource for debugging user programs is the monitoring and modifying of variables with VAT variable tables. Signal states or values of variables of elementary data types can be displayed. If you have access to the user program, you can also modify variables, i.e. change the signal state or assign new values.

Caution: you must ensure that no dangerous states can result from modifying variables!

Creating a variable table

For monitoring and modifying variables, you must create a VAT variable table containing the variables and the associated data formats. You can generate up to 255 variable tables (VAT 1 to VAT 255) and assign them names in the symbol table. The maximum size of a variable table is 1024 lines with up to 255 characters (Figure 2.10).

You can generate a VAT offline by selecting the user program *Blocks* and then INSERT → S7 BLOCK → VARIABLE TABLE, and you can generate an unnamed VAT online by selecting *S7 Program* and selecting PLC → MONITOR/MODIFY VARIABLES.

You can specify the variables with either absolute or symbolic addresses and choose the data

2.7 Testing the Program

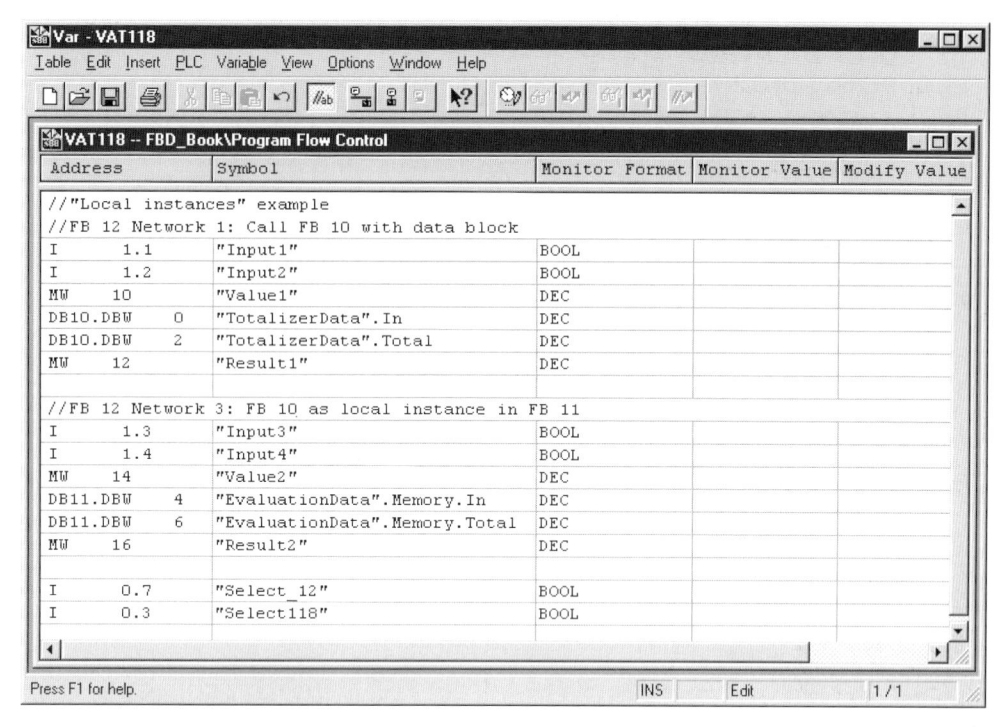

```
Var - VAT118                                                      _ □ ×
Table  Edit  Insert  PLC  Variable  View  Options  Window  Help

VAT118 -- FBD_Book\Program Flow Control                           _ □ ×
Address           Symbol                      Monitor Format  Monitor Value  Modify Value
//"Local instances" example
//FB 12 Network 1: Call FB 10 with data block
I       1.1       "Input1"                    BOOL
I       1.2       "Input2"                    BOOL
MW      10        "Value1"                    DEC
DB10.DBW   0      "TotalizerData".In          DEC
DB10.DBW   2      "TotalizerData".Total       DEC
MW      12        "Result1"                   DEC

//FB 12 Network 3: FB 10 as local instance in FB 11
I       1.3       "Input3"                    BOOL
I       1.4       "Input4"                    BOOL
MW      14        "Value2"                    DEC
DB11.DBW   4      "EvaluationData".Memory.In     DEC
DB11.DBW   6      "EvaluationData".Memory.Total  DEC
MW      16        "Result2"                   DEC

I       0.7       "Select_12"                 BOOL
I       0.3       "Select118"                 BOOL

Press F1 for help.                             INS    Edit        1 / 1
```

Figure 2.10 Variable Table Example

type (display format) with which a variable is to be displayed and modified (with VIEW → MONITOR FORMAT or by clicking the right mouse button directly on the monitor format). Use comment lines to give specific sections of the table a header. You may also stipulate which columns are to be displayed. You can change variable or display format or add or delete lines at any time. You save the variable table in the *Blocks* object container with TABLE → SAVE.

Establishing an online connection

To operate a variable table that has been created offline, switch it online with PLC → CONNECT TO → .... You must switch each individual VAT online and you can clear down the connection again with PLC → DISCONNECT.

Trigger conditions

In the variable table, select VARIABLE → SET TRIGGER to set the trigger point and the trigger conditions separately for monitoring and modifying. The trigger point is the point at which the CPU reads values from the system memory or writes values to the system memory. You specify whether reading and writing is to take place once or periodically.

If monitoring and modifying have the same trigger conditions, monitoring is carried out before modifying. If you select the trigger point "Start of cycle" for modifying, the variables are modified after updating of the process input image and before calling OB 1. If you select the trigger point "End of cycle" for monitoring, the status values are displayed after termination of OB 1 and before output of the process output image.

Monitoring variables

Select the Monitor function with the menu command VARIABLE → MONITOR. The variables in the VAT are updated in accordance with the specified trigger conditions. Permanent monitoring allows you to follow changes in the

67

values on the screen. The values are displayed in the data format which you set in the Monitor Format column. VARIABLE → UPDATE MONITOR VALUES updates the monitor values once only and immediately without regard to the specified trigger conditions.

Modifying variables

Use VARIABLE → MODIFY to transfer the specified values to the CPU dependent on the trigger conditions. Enter values only in the lines containing the variables you want to modify. You can expand the commentary for a value with "//" or with VARIABLE → MODIFY VALUE VALID; these values are not taken into account for modification. You must define the values in the data format which you set in the Monitor Format column.

Only the values visible on starting the modify function are modified. The ESC key terminates a permanent modify function.

VARIABLE → ACTIVATE MODIFY VALUES transfers the modify values only once and immediately, without regard to the specified trigger conditions.

2.7.4 Forcing Variables

With appropriately equipped CPUs, you can specify fixed values for certain variables. The user program can no longer change these values ("forcing"). Forcing is permissible in any CPU operating state and is executed immediately.

Caution: you must ensure that no dangerous states can result from forcing variables!

The starting point for forcing is a variable table (VAT). Create a VAT, enter the addresses to be forced and establish a connection to the CPU. You can open a window containing the force values by selecting VARIABLE → DISPLAY FORCE VALUES.

If there are already force values active in the CPU, these are indicated in the force window in bold type. You can now transfer some or all addresses from the variable table to the force window or enter new addresses. You save the contents of a force window in a VAT with TABLE → SAVE AS.

The following address areas can be provided with a force value:

▷ Inputs I (process image)
  (S7-300 and S7-400(

▷ Outputs Q (process image)
  (S7-300 and S7-400(

▷ Peripheral inputs PI
  (S7-400 only(

▷ Peripheral outputs PQ
  (S7-300 and S7-400(

▷ Memory bits M
  (S7-400 only(

You start the force job with VARIABLE → FORCE. The CPU accepts the force values and permits no more changes to the forced addresses.

While the force function is active, the following applies:

▷ All read accesses to a forced address via the user program (e.g. load) and via the system program (e.g. updating of the process image) always yield the force value.

▷ On the S7-400, all write accesses to a forced address via the user program (e.g. transfer) and via the system program (e.g. via SFCs) remain without effect. On the S7-300, the user program can overwrite the force values.

Forcing on the S7-300 corresponds to cyclic modifying: after the process input image has been updated, the CPU overwrites the inputs with the force value; before the process output image is output, the CPU overwrites the outputs with the force value.

Note: forcing is not terminated by closing the force window or the variable table, or by breaking the connection to the CPU! You can only delete a force job with VARIABLE → DELETE FORCE.

Forcing is also deleted by memory reset or by a power failure if the CPU is not battery-backed. When forcing is terminated, the addresses retain the force values until overwritten by either the user program or the system program.

Forcing is effective only on I/O assigned to a CPU. If, following restart, forced peripheral inputs and outputs are no longer assigned (e.g. as a result of reparameterizing), the relevant peripheral inputs and outputs are no longer forced.

Error handling

If the access width when reading is greater than the force width (e.g. forced byte in a word), the unforced component of the address value is read as usual. If a synchronization error occurs here (access or area length error) the "error substitute value" specified by the user program or by the CPU is read or the CPU goes to STOP.

If, when writing, the access width is greater than the force width (e.g. forced byte in a word), the unforced component of the address value is written to as usual. An errored write access leaves the forced component of the address unchanged, i.e. the write protection is not revoked by the synchronization error.

Loading forced peripheral inputs yields the force value. If the access width agrees with the force width, input modules that have failed or have not (yet) been plugged in can be "replaced" by a force value.

The input I in the process image belonging to a forced peripheral input PI is not forced; it is not preassigned and can still be overwritten. When updating the process image, the input receives the force value of the peripheral input.

When forcing peripheral outputs PQ, the associated output Q in the process image is not updated and not forced (forcing is only effective "externally" to the module outputs). The outputs Q are retained and can be overwritten; reading the outputs yields the written value (not the force value). If an output module is forced and if this module fails or is removed, it will receive the force value again immediately on reconnection.

The output modules output signal state "0" or the substitute value with the OD signal (disable output modules at STOP, HOLD or RESTART) – even if the peripheral outputs are forced (exception: analog modules without OD evaluation continue to output the force value). If the OD signal is deactivated, the force value becomes effective again.

If, in STOP mode, the function *Enable PQ* is activated, the force values also become effective in STOP mode (due to deactivation of the OD signal). When *Enable PQ* is terminated, the modules are set back to the "safe" state (signal state "0" or substitute value); the force value becomes effective again at the transition to RUN.

2.7.5 Enabling Peripheral Outputs

In STOP mode, the output modules are normally disabled by the OD signal; with the Enable peripheral outputs function, you can deactivate the OD signal so that you can modify the output modules even at CPU STOP. Modifying is carried out via a variable table. Only the peripheral outputs assigned to a CPU are modified. Possible application: wiring test of the output at STOP and without user program.

Caution: you must ensure that no dangerous states can result from enabling the peripheral outputs!

Create a variable table and enter the peripheral outputs (PQ) and the modify values. Switch the variable table online with PLC → CONNECT TO →..., and stop the CPU if necessary, e.g. with PLC → OPERATING MODE and "STOP".

You deactivate the OD signal VARIABLE → ENABLE PERIPHERAL OUTPUTS; the module outputs now have signal state "0" or the substitute value or force value. You modify the peripheral outputs with VARIABLE → ACTIVATE MODIFY VALUES. You can change the modify value and modify again.

You can switch the function off again by selecting VARIABLE → ENABLE PERIPHERAL OUTPUTS again, or by pressing the ESC key. The OD signal is then active again, the module outputs are set to "0" and the substitute value or the force value is reset.

If STOP is exited while "enable peripheral outputs" is still active, all peripheral inputs are deleted, the OD signal is activated at the transition to RESTART and deactivated again at the transition to RUN.

2.7.6 LAD/FBD Program Status

With the *Program status* function, the program editor provides an additional test method for the user program. The editor shows you the binary signal flow and digital values within a network.

The block whose program you want to debug is in the CPU's user memory and is called and edited there. Open this block, for example by double-clicking on it in the SIMATIC Manager's online window. The editor is started and shows the program in the block.

Select the network you want to debug. Activate the Program Status function with DEBUG → MONITOR. Now you can see the binary signal flow in the block window and you can follow the changes in it. You define the representation (e.g. color) in the program editor with OPTIONS → CUSTOMIZE on the "LAD/FBD" tab. You can deactivate the Program Status function again by selecting DEBUG → MONITOR again.

You set the trigger conditions with DEBUG → CALL ENVIRONMENT. You require this setting if the block to be debugged is called more than once in your program. You can initiate status recording either by specifying the order of calls or by making it dependent on the opened data block. If the block is called only once, select "No condition".

You can modify variables in the Program Status function. Select the address to be modified and select DEBUG → MODIFY ADDRESS.

The recording of the program status information requires additional execution time in the program cycle. For this reason, you can choose two operating modes for debugging purposes: debug mode and process mode .In *debug mode*, all debugging functions can be used without restriction. You would select this, for example, to debug blocks without connection to the system, because this can significantly increase the cyclic execution time. In process mode, care is taken to keep the increase in cycle time to a minimum and this results in debugging restrictions such as in program loops (not all loop passes are displayed).

On appropriately equipped CPUs, you set the operating mode on the "Protection" tab during CPU parameterization. If the debug mode has been defined during CPU parameterization, you can only change it by reparameterizing. Otherwise, it can be changed in the dialog box displayed. The set operating mode is displayed with DEBUG → OPERATION.

3 SIMATIC S7 Program

This chapter shows you the structure of the user program for the SIMATIC S7-300/400 CPUs starting from the different priority classes (program execution types) via the component parts of a user program (blocks) right up to the variables and data types. The focus of this chapter is the description of block programming with LAD and FBD. Following this is a description of the data types.

You define the structure of the user program right back at the design phase when you adapt the technological and functional conditions; it is decisive for program creation, program test and startup. To achieve effective programming, it is therefore necessary to devote special attention to the program structure.

3.1 Program Processing

The overall program for a CPU consists of the operating system and the user program.

The *operating system* is the totality of all instructions and declarations which control the system resources and the processes using these resources, and includes such things as data backup in the event of a power failure, the activation of priority classes, and so on. The operating system is a component of the CPU to which you, as user, have no write access. However, you can reload the operating system from a memory card, for instance in the event of a program update.

The *user program* is the totality of all instructions and declarations for signal processing, through which a plant (process) is affected in accordance with the defined control task.

3.1.1 Program Processing Methods

The user program may be composed of program sections which the CPU processes in dependence on certain events. Such an event might be the start of the automation system, an interrupt, or detection of a program error (Figure 3.1). The programs allocated to the events are divided into *priority classes*, which determine the program processing order (mutual interruptibility) when several events occur.

The lowest-priority program is the *main program*, which is processed cyclically by the CPU. All other events can interrupt the main program at any location, the CPU then executes the associated interrupt service routine or error handling routine and returns to the main program.

A specific *organization block (OB)* is allocated to each event. The organization blocks represent the priority classes in the user program. When an event occurs, the CPU invokes the assigned organization block. An organization block is a part of a user program which you yourself may write.

Before the CPU begins processing the main program, it executes a startup routine. This routine can be triggered by switching on the mains power, by actuating the mode switch on the CPU's front panel, or via the programming device. Program processing following execution of the startup routine always starts at the beginning of the main program in S7-300 systems (complete restart); in S7-400 systems, it is also possible to resume the program scan at the point at which it was interrupted (warm restart).

The main program is in organization block OB 1, which the CPU always processes. The start of the user program is identical to the first network in OB 1. After OB 1 has been processed (end of program), the CPU returns to the operating system and, after calling for the execution of various operating system functions, such as the updating of the process images, it once again calls OB 1.

71

3 SIMATIC S7 Program

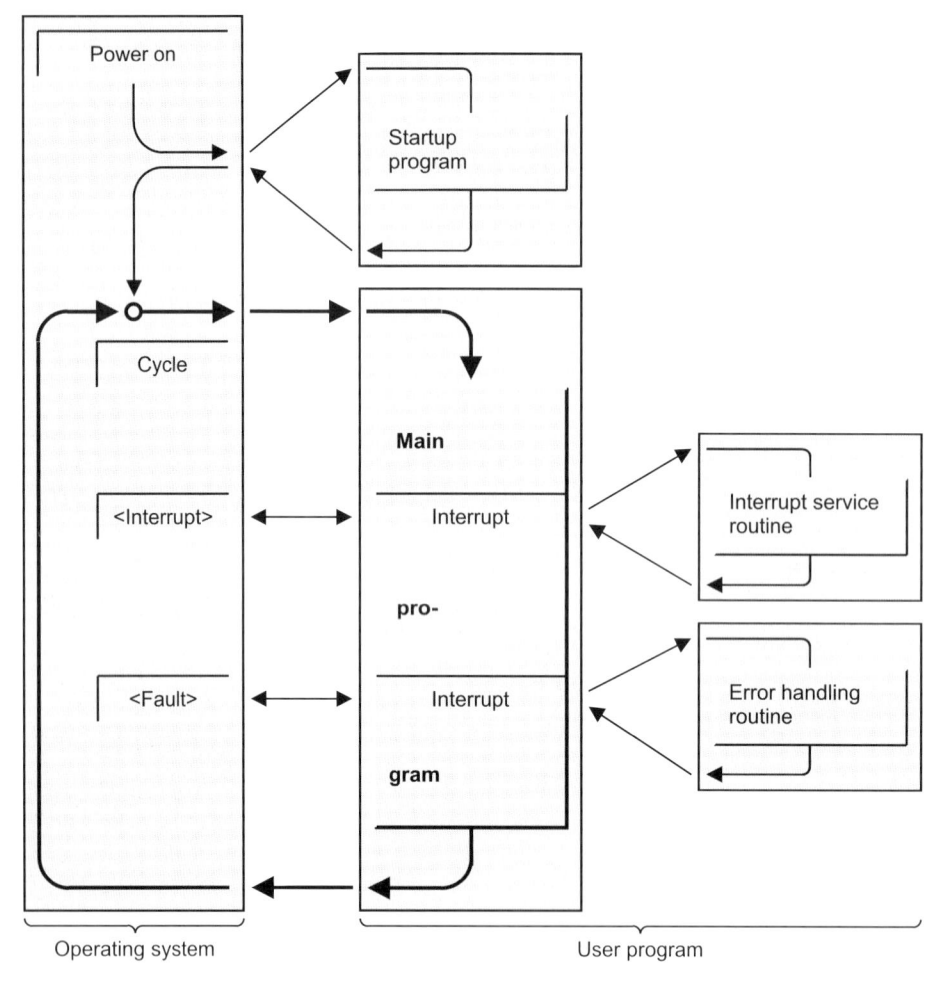

Figure 3.1 Methods of Processing the User Program

Events which can intervene in the program are interrupts and errors. Interrupts can come from the process (hardware interrupts) or from the CPU (watchdog interrupts, time-of-day interrupts, etc.). As far as errors are concerned, a distinction is made between synchronous and asynchronous errors. An asynchronous error is an error which is independent of the program scan, for example failure of the power to an expansion unit or an interrupt that was generated because a module was being replaced. A synchronous error is an error caused by program processing, such as accessing a non-existent address or a data type conversion error. The type and number of recorded events and the associated organization blocks are CPU-specific; not every CPU can handle all possible STEP 7 events.

3.1.2 Priority Classes

Table 3.1 lists the available SIMATIC S7 organization blocks, each with its priority. In some priority classes, you can change the assigned priority when you parameterize the CPU. The Table shows the lowest and highest possible priority classes; each CPU has a different low/high range; a specific CPU occupies a section of this overview.

Table 3.1 SIMATIC S7 Organization Blocks

Organization block	Called under the following circumstances	Priority Default	Priority Modifiable
Free cycle OB 1	Cyclic call via the operating system	1	No
TOD interrupts OB 10 to OB 17	At a specific time of day or at regular intervals (e.g. monthly)	2	2 to 24
Time-delay interrupts OB 20 to OB 23	After a programmable time, controlled by the user program	3 to 6	2 to 24
Watchdog interrupts OB 30 to OB 38	Regularly at programmable intervals (e.g. every 100 ms)	7 to 15	2 to 24
Process interrupts OB 40 to OB 47	On interrupt signals from I/O modules	16 to 23	2 to 24
Multiprocessor interrupt OB 60	Event-driven via the user program in multiprocessor mode	25	No
Redundancy errors OB 70, OB 72	Redundancy loss via I/O errors, CPU redundancy error	25 / 28	2 to 26 / 2 to 28
Asynchronous errors OB 80, OB 81 to OB 84, 86, 87, OB 85	Errors not involved in program execution (e.g. time errors, SE errors, diagnostics interrupt, insert/remove-module interrupt, rack/station failure)	$26^{2)}$ / $26^{2)}$ / $26^{2)}$	26 / 2 to 26 / 24 to 26
Background processing OB 90	Minimum cycle time duration not yet reached	$29^{1)}$	No
Startup routine OB 100, OB 101, OB 102	Programmable controller startup	27	No
Synchronous errors OB 121, OB 122	Errors connected with program execution (e.g. I/O access errors)	Priority of the OBs causing the errors	

[1] see text [2] at startup: 28

Organization block OB 90 (background processing) executes alternately with organization block OB 1, and can, like OB 1, be interrupted by all other interrupts and errors.

The startup routine may be in organization block OB 100 (complete restart) and OB 101 (warm restart), and has priority 27. Asynchronous errors occurring in the startup routine have priority class 28. Diagnostic interrupts are regarded as asynchronous errors.

You determine which of the available priority classes you want to use when you parameterize the CPU. Unused priority classes (organization blocks) must be assigned priority 0. The relevant organization blocks must be programmed for all priority classes used; otherwise the CPU will invoke OB 85 ("Program Processing Error") or go to STOP.

For each priority class selected, temporary local data (L stack) must be available in sufficient volumes (see Section 18.1.5 "Temporary Local Data" for more details).

3.1.3 Specifications for Program Processing

The CPU's operating system normally uses default parameters. You can change these defaults when you parameterize the CPU (in the *Hardware* object) to customize the system to suit your particular requirements. You can change the parameters at any time.

Every CPU has its own specific number of parameter settings. The following list provides an overview of all STEP 7 parameters and their most important settings.

▷ Startup
Specifies the type of startup (cold restart, complete restart, warm restart); monitoring of Ready signals or module parameterization; maximum amount of time which may elapse before a warm restart

▷ Cycle/Clock Memory
Enable/disable cyclic updating of the process image; specification of the cycle monitoring time and minimum cycle time; amount of cycle time, in percent, for communication; number of the clock memory byte; size of the process images

▷ Retentive Memory
Number of retentive memory bytes, timers and counters; specification of retentive areas for data blocks

▷ Memory
max. number of temporary local data in the priority classes (organization blocks); max size of the L stack and number of communications jobs

▷ Interrupts
Specification of the priority for hardware interrupts, time-delay interrupts, asynchronous errors and (available soon) communication interrupts

▷ Time-of-Day Interrupts
Specification of the priority, specification of the start time and periodicity

▷ Cyclic Interrupts
Specification of the priority, the time cycle and the phase offset

▷ Diagnostics/Clock
Indicate the cause of a STOP; type and interval for clock synchronization, correction factor

▷ Protection
Specification of the protection level; defining a password

▷ Multicomputing
Specification of the CPU number

▷ Integrated I/O
Activation and parameterization of the integrated I/O

On startup, the CPU puts the user parameters into effect in place of the defaults, and they remain in force until changed.

3.2 Blocks

You can subdivide your program into as many sections as you want to in order to make it easier to read and understand. The STEP 7 programming languages support this by providing the necessary functions. Each program section should be self-contained, and should have a technological or functional basis. These program sections are referred to as "Blocks". A block is a section of a user program which is defined by its function, structure or intended purpose.

3.2.1 Block Types

STL provides different types of blocks for different tasks:

▷ User blocks
Blocks containing user program and user data

▷ System blocks
Blocks containing system program and system data

▷ Standard blocks
Turnkey, off-the-shelf blocks, such as drivers for FMs and CPs

User blocks

In extensive and complex programs, "structuring" (subdividing) of the program into blocks is recommended, and in part necessary. You may choose among different types of blocks, depending on your application:

Organization blocks (OBs)

These blocks serve as the interface between operating system and user program. The CPU's operating system calls the organization blocks when specific events occur, for example in the event of a hardware or time-of-day interrupt. The main program is in organization block OB 1. The other organization blocks have permanently assigned numbers based on the events they are called to handle.

Function blocks (FBs)

These blocks are parts of the program whose calls can be programmed via block parameters. They have a variable memory which is located

Example:

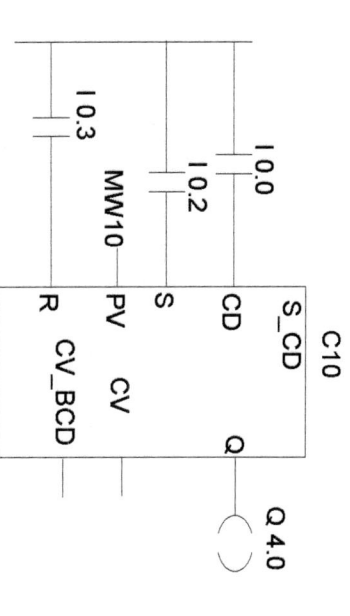

```
           C10
        ┌───────┐
      ──┤ S_CD  │
   I0.0 │       │
      ──┤ CD  Q ├──○── Q4.0
   I0.2 │       │
      ──┤ S     │
  MW10──┤ PV CV │
   I0.3 │       │
      ──┤ R CV_BCD
        └───────┘
```

If I0.2 changes from "0" to "1", the counter is preset with the value of MW10. If the signal state of I0.0 changes from "0" to "1", the value of counter C10 will be decremented by one - unless the value of C10 is equal to "0". Q4.0 is "1" if C10 is not equal to zero.

in a data block. This data block is permanently allocated to the function block, or, to be more precise to the function block *call*. It is even possible to assign a different data block (with the same data structure but containing different values) to each function block call. Such a permanently assigned data block is called an instance data block, and the combination of function block call and instance data block is referred to as a call instance, or "instance" for short. Function blocks can also save their variables in the instance data block of the calling function block; this is referred to as a "local instance".

Functions (FCs)

Functions are used to program frequently recurring or complex automation functions. They can be parameterized, and return a value (called the function value) to the calling block. The function value is optional, in addition to the function value, functions may also have other output parameters. Functions do not store information, and have no assigned data block.

Data blocks (DBs)

These blocks contain your program's data. By programming the data blocks, you determine in which form the data will be saved (in which block, in what order, and in what data type). There are two ways of using data blocks: as global data blocks and as instance data blocks. A global data block is, so to speak, a "free" data block in the user program, and is not allocated to a code block. An instance data block, however, is assigned to a function block, and stores part of that function block's local data.

The number of blocks per block type and the length of the blocks is CPU-dependent. The number of organization blocks, and their block numbers, are fixed; they are assigned by the CPU's operating system. Within the specified range, you can assign the block numbers of the other block types yourself. You also have the option of assigning every block a name (a symbol) via the symbol table, then referencing each block by the name assigned to it.

System blocks

System blocks are components of the operating system. They can contain programs (system functions (SFCs) or system function blocks (SFBs)) or data (system data blocks (SDBs)). System blocks make a number of important system functions accessible to you, such as manipulating the internal CPU clock, or various communications functions

You can call SFCs and SFBs, but you cannot modify them, nor can you program them yourself. The blocks themselves do not reserve space in user memory; only the block calls and the instance data blocks of the SFBs are in user memory.

SDBs contain information on such things as the configuration of the automation system or the parameterization of the modules. STEP 7 itself generates and manages these blocks. You, however, determine their contents, for instance when you configure the stations. As a rule, SDBs are located in load memory. You cannot access them from your user program.

Standard blocks

In addition to the functions and function blocks you create yourself, off-the-shelf blocks (called "standard blocks") are also available. They can either be obtained on a storage medium or are contained in libraries delivered as part of the STEP 7 package (for example IEC functions, or functions for the S5/S7 converter).

Chapter 33 "Block Libraries" contains an overview of the standard blocks supplied in the *Standard Library*.

3.2.2 Block Structure

Essentially, code blocks consist of three parts (Figure 3.2):

▷ The block header,
which contains the block properties, such as the block name

▷ The declaration section,
in which the block-local variables are declared, that is, defined

▷ The program section,
which contains the program and program commentary

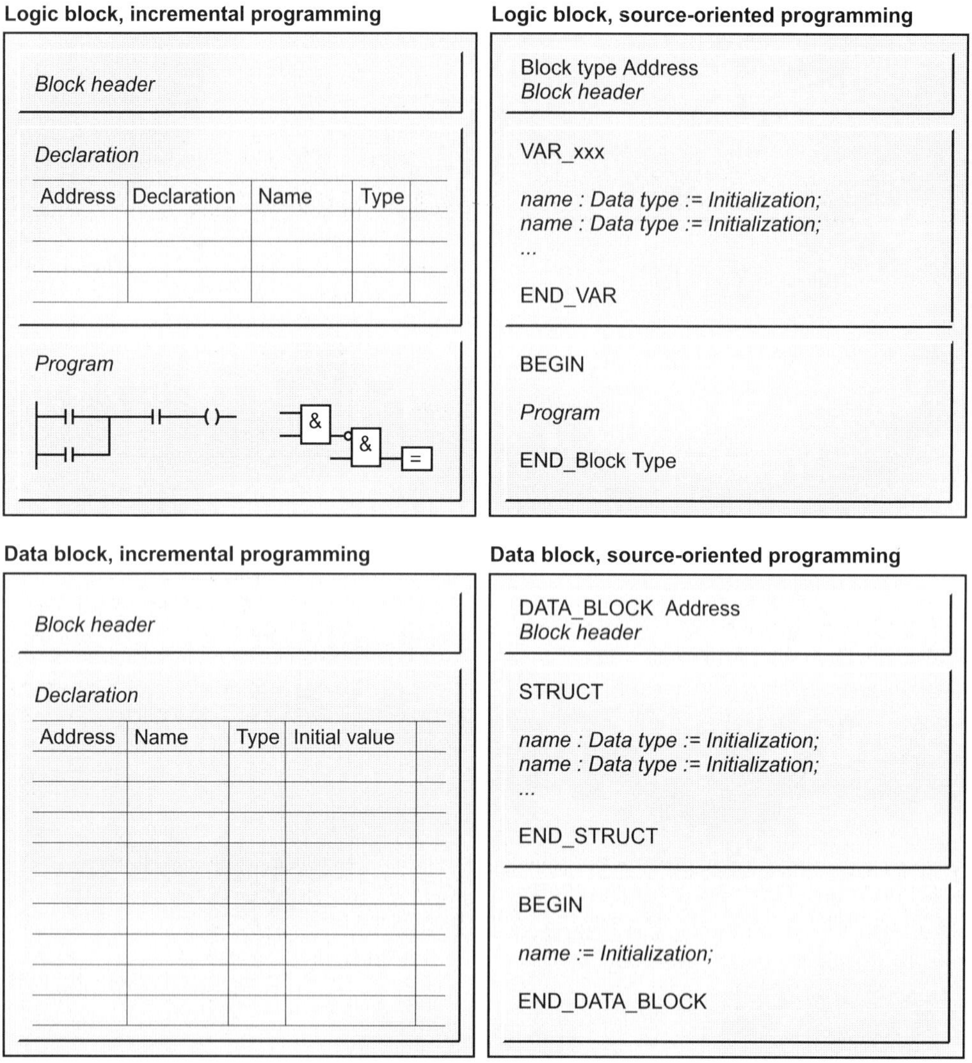

3 SIMATIC S7 Program

Logic block, incremental programming

Block header

Declaration

Address	Declaration	Name	Type

Program

Logic block, source-oriented programming

Block type Address
Block header

VAR_xxx

name : Data type := Initialization;
name : Data type := Initialization;
...

END_VAR

BEGIN

Program

END_Block Type

Data block, incremental programming

Block header

Declaration

Address	Name	Type	Initial value

Data block, source-oriented programming

DATA_BLOCK Address
Block header

STRUCT

name : Data type := Initialization;
name : Data type := Initialization;
...

END_STRUCT

BEGIN

name := Initialization;

END_DATA_BLOCK

Figure 3.2 Structure of a Block

A data block is similarly structured:

▷ The block header contains the block properties

▷ The declaration section contains the definitions of the block-local variables, in this case the data addresses with data type specification

▷ The initialization section, in which initial values can be specified for individual data addresses

In incremental programming, the declaration section and the initialization section are combined. You define the data addresses and their data types in the "declaration view", and you can initialize each data address individually in the "data view" (see below).

3.2.3 Block Properties

The block properties, or attributes, are contained in the block header. You can view and modify the block attributes from the Editor with the menu command FILE → PROPERTIES (Figure 3.3).

The tab "General - Part 2" shows the memory allocation for the block in bytes:

▷ Local data: Allocation in the local data stack (temporary local data)
▷ MC 7: size of the block (code only)
▷ Load memory requirement
▷ Work memory requirement

The *KNOW HOW Protection* attribute is used for block protection. If a block is KNOW HOW-protected, the program in that block can not be viewed, printed out or modified. The Editor shows only the block header and the declaration table with the block parameters. When you input to a source file, you can protect every block yourself with the keyword KNOW_HOW_PROTECT. When you do this to a block, no one can view the compiled version of that block, not even you (make sure you keep the source file in a safe place!).

The block header of any standard block which comes from Siemens contains the "*Standard Block*" attribute.

The attribute "*DB is write-protected in the PLC*" is an attribute for data blocks only. It means that you can only read that data block in your program. Output of an error message prevents the overwriting of the data in that data block. This write protection feature must not be confused with block protection. A data block with block protection can be read out and written to in the user program, but its data can no longer be viewed with a programming or operator monitoring device.

A data block that has the *Unlinked* attribute is only in load memory; it is not "execution-relevant". You cannot write to data blocks in load memory, and you can read them only with system function SFC 20 BLKMOV.

Further specifications on the tab "General – Part 2": *Name* identifies the block; this is not the same as the symbolic address. Different blocks can have the same name. With *Family*, you assign a common feature to a group of blocks. The block name and family are displayed when you insert blocks and when you select a block in the dialog window of the program elements catalog. *Author* is the creator of

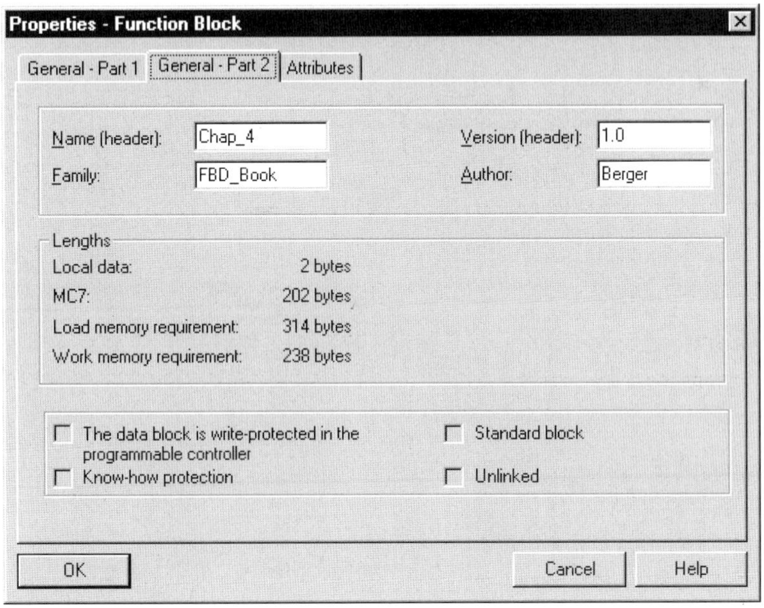

Figure 3.3 Block Properties

the block. Name, Family and Author can consist of up to 8 characters. Alphanumeric characters and the underscore are permissible. *Version* is entered twice, with two digits from 0 to 15.

"General – Part 1" tab: the editor records the modification date of the block in two time stamps: for the program code and for the interface, i.e. the block parameters and the static local data. Please note that the modification date of the interface must be the same or less (older) than the modification date of the program code in the calling block. If this is not the case, the editor signals a "time stamp conflict" when outputting the calling block.

Blocks can be created or compiled as version 1 or version 2. This is only of practical significance for the function blocks. If the property "multi-instance capability" is activated, which is normally the case, you are dealing with a version 2 block. If "multi-instance capability" is de-activated, you cannot call the block yourself as a local instance and you cannot call another function block within this block as a local instance. The advantage of a version 1 block is the restricted use of instance data in the case of indirect addressing (only significant in STL programming).

"Attributes" tab: blocks may have system attributes. System attributes control and coordinate functions between applications, for example in the SIMATIC PCS7 control system.

Program length, memory requirement

The memory requirement of a compiled block is shown in the block properties (select the block in the SIMATIC Manager and then select the "General – Part 2" tab under EDIT → OBJECT PROPERTIES). You will see the load and working memory requirement for this block.

The length of the user program is held in the Properties of the offline container *Blocks* (select *Blocks* and then EDIT → OBJECT PROPERTIES). You will find the information "Size in work memory" and "Size in load memory" on the "Blocks" tab.

Please note that the configuration data (system data blocks) are missing from the size specification for load memory. With the *Blocks* container open, you can see the load memory requirement of the system data in the detailed view (displayed as table). In the status line, the SIMATIC Manager gives the sum of all blocks that you select if you press the Ctrl key.

With the programming device switched online, the SIMATIC Manager shows you the current assignment of the CPU memory on the "Memory" tab under PLC → MODULE INFORMATION.

3.3 Programming Code Blocks

Section 2.5 "Creating the S7 Program" contains an introduction to program creation and to operating the program editor.

3.3.1 Generating blocks

You begin block programming by opening a block either by double-clicking on the block in the SIMATIC Manager's project window or by selecting FILE → OPEN in the Editor. If the block does not yet exist, you can generate it in the following ways:

▷ In the SIMATIC Manager by selecting the *Blocks* object in the left half of the project window and generating a new block with INSERT → S7 BLOCK → ... . You will see a dialog box with the block header (number of the block, language, block attributes). Choose the "STL" language. You can enter the remaining attributes later.

▷ In the Editor with menu command FILE → NEW, which displays a dialog box with the block header (number of the block, language, block attributes). After closing the dialog box you can program the contents of the block.

You can enter the information for the block header when you generate the block or you can enter the block attributes later in the Editor by opening the block and selecting the menu command FILE → PROPERTIES.

Block window

When a code block is opened, three windows are displayed (Figure 3.4):

3.3 Programming Code Blocks

▷ The variable declaration table at the top
It is here that you define the block-local variables

▷ The program window
This is where you enter the program

▷ The program elements catalog
In STL, this catalog contains the available blocks

Variable declaration table

The variable declaration table is in the window above the program window. If it is not visible, position the mouse pointer to the upper line of demarcation for the program window, click on the left mouse button when the mouse pointer changes its form, and pull down. You will see the variable declaration table, which is where you define the block-local variables (see Table 3.2). Not every type of variable can be programmed in every kind of code block. If you do not use a variable type, the corresponding line remains empty.

The declaration for a variable consists of the name, the data type, a default value, if any, and a variable comment (optional). Not all variables can be assigned a default value (for instance, it is not possible for temporary local data). The default values for functions and function blocks are described in detail in Chapter 19, "Block Parameters".

The order of the declarations in code blocks is fixed (as shown in the table above), while the order within a variable type is arbitrary. You can save room in memory by bundling binary variables into blocks of 8 or 16 and BYTE variables into pairs. The Editor stores a (new) BOOL or BYTE variable at a byte boundary and a variable of another data type at a word boundary (beginning at a byte with an even address).

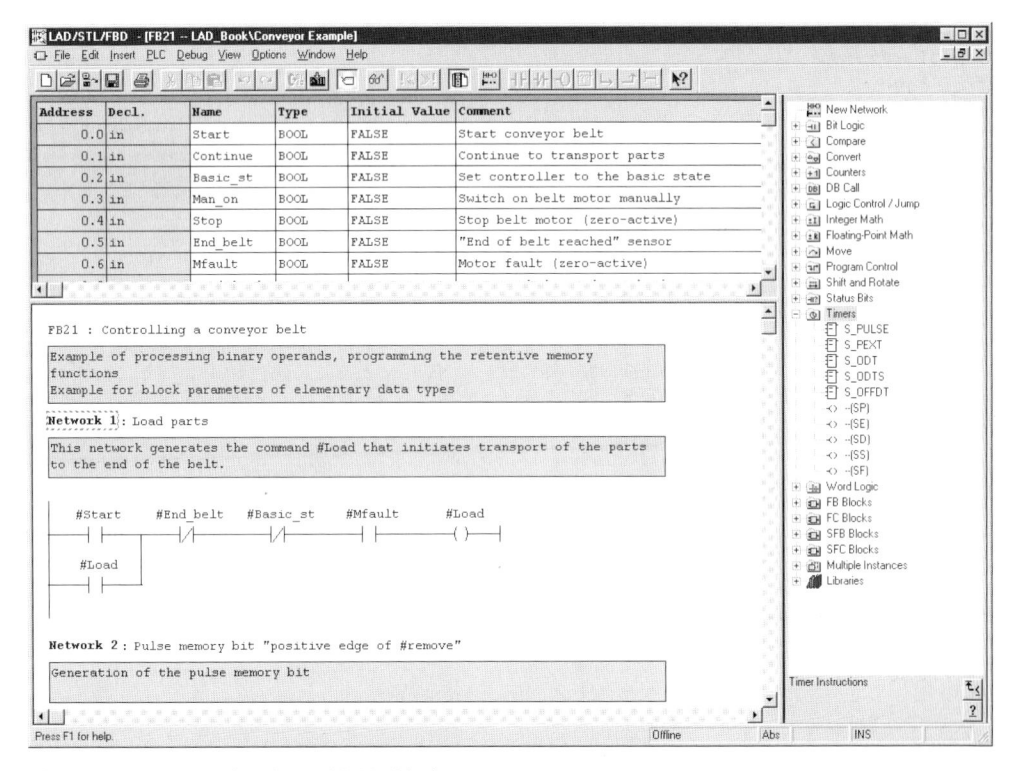

Figure 3.4 Example of an Opened LAD Block

3 SIMATIC S7 Program

Table 3.2 Variable Types in the Declaration Section

Variable Type	Declaration	Possible in Block Type		
Input parameters	in	-	FC	FB
Output parameters	out	-	FC	FB
In-out parameters	in_out	-	FC	FB
Static local data	stat	-	-	FB
Temporary local data	temp	OB	FC	FB

Program window

In the program window, you will see - depending on the Editor's default settings - the fields for the block title and the block comment and, if it is the first network, the fields for the network title, the network comment, and the field for the program entry. In the program section of a code block, you control the display of comments and symbols with the menu commands VIEW → COMMENT, VIEW → SYMBOLIC REPRESENTATION and VIEW → SYMBOL INFORMATION. You can change the size of the display with VIEW → ZOOM IN, VIEW → ZOOM OUT and VIEW → ZOOM FACTOR.

Programming networks

You can subdivide a LAD/FBD program into networks which each represent a current path or a logic operation. The Editor numbers the networks automatically, beginning with 1. You may give each network a *network title* and a *network comment*. During editing, you can select each network directly with the menu command EDIT → GO TO → ... . Subdivision into networks is optional.

To enter the program code, click once below the window for the network comment, or, if you have set "Display with Comments", click once below the shaded area for network comments. You will see a framed empty window. You can begin entering your program anywhere within this window. The chapters below show you what a LAD current path or an FBD logic operation looks like.

You program a new network with INSERT → NETWORK. The Editor then inserts an empty network behind the currently selected network.

You need not terminate a block with a special statement, simply stop making entries. However, you can program a last (empty) network with the title "Block End", providing an easily seen visual end of the block (an advantage, particularly in the case of exceptionally long blocks).

Network templates

Just as you store blocks in a library to reuse them in other programs, you also save network templates in order to copy them again and again in, for example, other blocks.

To save network templates, create a library containing at least one S7 program and the *Source Files* container.

You program the networks that you want to use as templates quite "normally" in (any) block. Then you replace the addresses that are to change with the dummy characters %00 to %99. You can also vary the network title and the network comment in this way.

The dummy characters replacing the addresses are presented in red because a block cannot be stored in this form. This is not significant because following saving of the network template(s), this block can be rejected (close the block without saving).

After entering the dummy characters, mark the network by clicking on the network number at top left before the network title. You can also combine several networks to form one template; hold down the Ctrl key while you click on further network numbers.

Now select CREATE → NETWORK TEMPLATE... In the dialog box that then appears, you can assign meaningful comments to all the dummy characters. In the next dialog box, you assign a name for the network template and you define the storage location (Source Files container in a library).

If you want to use the network templates, open the relevant library in the Program Elements Catalog and then select the desired network template (double-click or drag to the Editor window). A dialog box appears automatically and here you replace the dummy characters with valid entries. The network template is inserted after the selected network.

Absolute addressing

Absolute addressing references addresses and block parameters with the address ID and the bit/byte address. If there are three red question marks in the network in the place of addresses and parameters, you must replace this character string with valid addresses. If there are three black points, replacement is optional.

The Program Editor checks that the data types of the addresses and parameters are correct. You can deactivate some of these type checks (in the Program Editor under Options Æ Customize, "LAD/FBD" tab, "Type check for addresses" option).

Symbolic addressing

If you want to use symbolic names for global operands in incremental programming, these names must already be assigned to absolute addresses in the Symbol Table. While entering the program with the Program Editor, you can call up the Symbol Table for editing with OPTIONS → SYMBOL TABLE and then you can change symbols or enter new symbols.

You activate display of the symbol addresses with VIEW → DISPLAY → SYMBOLIC REPRESENTATION. The menu point VIEW → DISPLAY → SYMBOL INFORMATION provides, for each network, a list of the symbol-to-absolute-address assignments for each symbol used in the network.

While entering the symbols, you can view a list of all the symbols in the symbol table with INSERT → SYMBOL (or right mouseclick and INSERT → SYMBOL) and you can then transfer one of the symbols with a click of the mouse. The list is displayed automatically if you have set VIEW → DISPLAY → SYMBOL SELECTION.

Decompiling

When the Program Editor opens a compiled block, it executes "decompilation" into the LAD/FBD representation. It uses the non-execution-relevant program sections in the programming device database to represent the likes of symbols, comments and jump labels. If the information from the programming device database is missing at the decompiling stage, the Editor uses substitute symbols.

Networks that cannot be decompiled into LAD or FBD are represented in STL.

Adapting block calls

If the block call no longer matches the block interface at program output, for example, because a block parameter has been changed in the meantime or because the block interface is of a later date than the call ("Time stamp conflict"), you can adapt the block call with the menu point EDIT → CALL → UPDATE. In the same way, you can adapt invalid uses of user-defined data types UDTs.

With EDIT → CALL → CHANGE TO MULTI-INSTANCE CALL and EDIT → CALL → CHANGE TO FB/DB CALL, you can change function block calls into local instance calls or into calls with data block. After changing the block calls, you must regenerate the affected instance data blocks.

Program Elements Catalog

If the Program Elements Catalog is not visible, you can screen it with VIEW → CATALOG or with INSERT → PROGRAM ELEMENTS.

The program elements catalog is located in its own window that you can dock onto the right-hand edge of the editor window and remove again (double-click on the title bar of the catalog window in each case).

The program elements catalog supports programming in LAD and FBD by providing the available graphic elements and also the blocks already located in the offline Blocks container, as well as the already-programmed multi-instances and the available libraries (Figure 3.5).

3.3.2 Editing LAD Elements

Programming in general

The program consists of individual LAD elements arranged in series or parallel to one another. Programming of a current path, or rung, begins on the left power rail. You select the location in the rung at which you want to insert an element, then you select the program elements you want

▷ with the corresponding function key (for example F2 for a normally open (NO) contact),

▷ with the corresponding button on the function bar,

▷ from the Program Elements Catalog (with INSERT → PROGRAM ELEMENTS or VIEW → CATALOG).

You terminate a rung with a coil or a box (see Section 3.3 "Programming Code Blocks" for further details).

Most program elements must be assigned memory locations (variables). The easiest way to do this is to first arrange all program elements, and then label them.

Contacts

Binary addresses such as inputs are scanned using contacts. The scanned signal states are combined according to the arrangement of the contacts in a serial or parallel layout.

"Current flows" through a normally open contact if the scanned binary address has signal state "1" (the contact is activated) "current flows" through a normally closed contact if the scanned binary address has signal state "0" (the contact is not activated). You can also scan status bits or negate the result of the logic operation (NOT contact).

Coils

Coils are used to control binary addresses, such as outputs. A simple coil sets the binary addresses when current flows in the coil, and resets it when power no longer flows.

There are coils with additional labels, such as Set and Reset coils, which serve a special function. You can also use coils to control timers

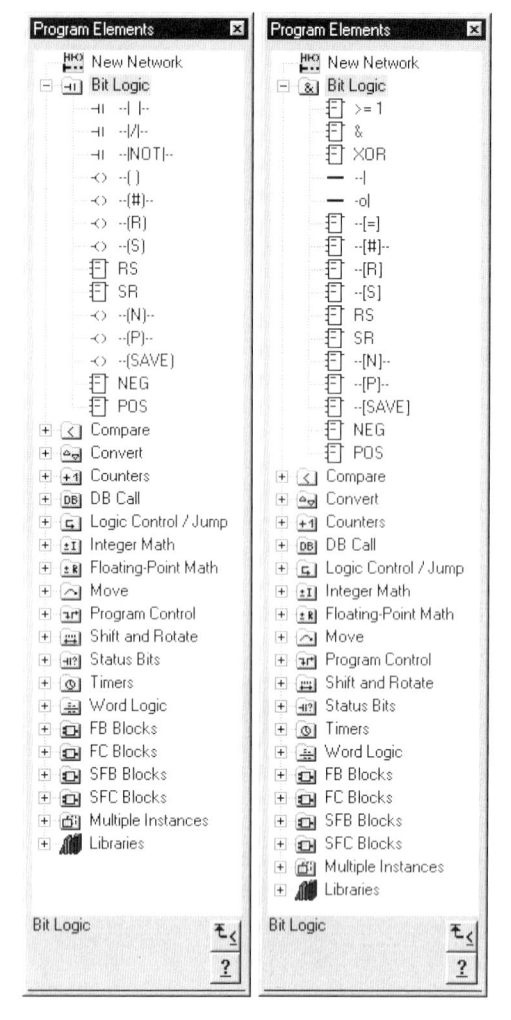

Figure 3.5
Program Elements Catalog for LAD and FBD

and counters, call blocks without parameters, execute jumps in the program, and so on.

Boxes

Boxes represent LAD elements with complex functions. STEP 7 provides "standard boxes" of two different types: without EN/ENO mechanism (such as memory functions, timer and counter functions, comparison boxes), and with EN/ENO (such as MOVE, arithmetic and math functions, data type conversions). When you

Contacts

NO contact   Binary address
—| |—

NC contact   Binary address
—|/|—

Contact with special function
(e.g. negation)   —|NOT|—

Coils

Simple coil   Binary address
—( )—

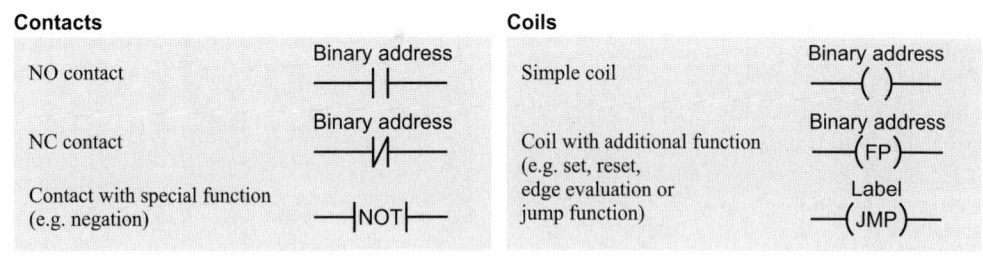

Coil with additional function
(e.g. set, reset,
edge evaluation or   Binary address
jump function)   —(FP)—

Label
—(JMP)—

Boxes

Standard boxes without EN/ENO
(e.g. timer and counter functions)

```
      Timer address
     ┌──S_PULSE──┐
  ─ S           Q ─
  ─ TV         BI ─
               BCD ─
  ─ R
     └───────────┘
```

Standard boxes without EN/ENO
(e.g. arithmetic functions)

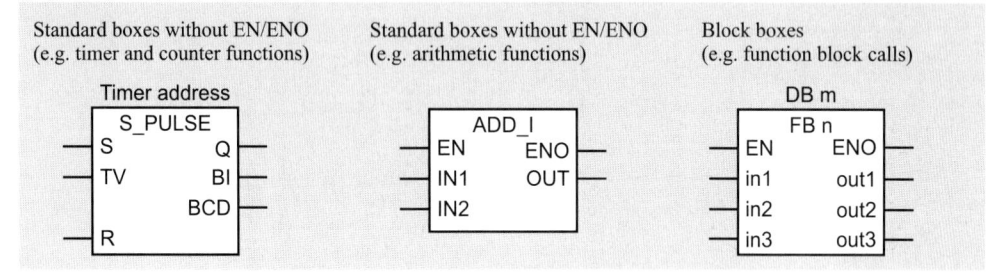

```
     ┌──ADD_I──┐
  ─ EN     ENO ─
  ─ IN1    OUT ─
  ─ IN2
     └─────────┘
```

Block boxes
(e.g. function block calls)

```
       DB m
     ┌──FB n──┐
  ─ EN    ENO ─
  ─ in1   out1 ─
  ─ in2   out2 ─
  ─ in3   out3 ─
     └────────┘
```

Figure 3.6 Examples of LAD Program Elements

call code blocks (FCs, FBs, SFCs and SFBs), LAD also represents the calls as boxes with EN/ENO. LAD also provides an "empty box" in which you can enter the desired function when programming.

Layout restrictions

The LAD editor sets up a network according to the "main rung" principle. This is the upper-most branch, which begins directly on the left power rail and must terminate with a coil or a box. All LAD elements can be located in this rung. In parallel branches which do not begin on the left power rail, there are sometimes restrictions in the choice of program elements.

Additional restrictions dictate that no LAD element may be "short-circuited" with an "empty" parallel branch, and that no "power" may flow

```
Network 2: Parts ready to remove
```
```
When the parts have reached the end of the belt, they are ready for removal.
```

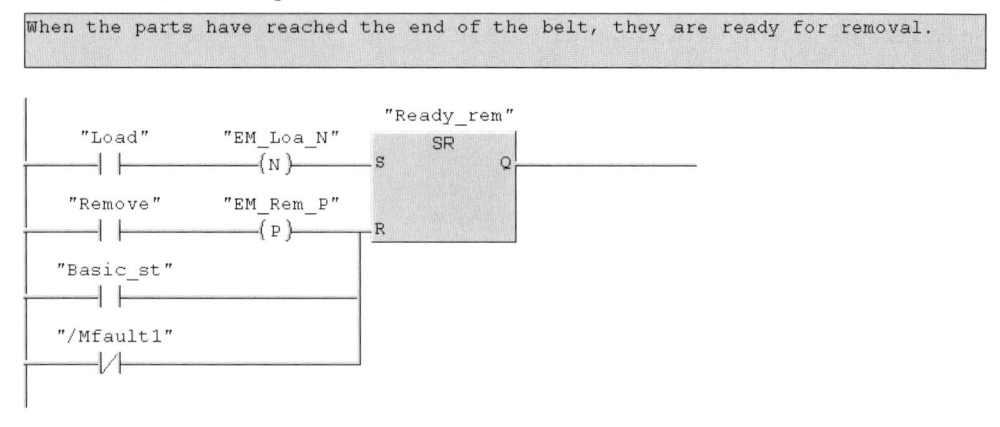

```
                                          "Ready_rem"
      "Load"         "EM_Loa_N"               SR
   ───| |──────────────( N )──────────S              Q────────────
      "Remove"       "EM_Rem_P"
   ───| |──────────────( P )──────────R
      "Basic_st"
   ───| |──────────────
      "/Mfault1"
   ───|/|──────────────
```

Figure 3.7 Example of a LAD Network in "3-Dimensional" Representation

3 SIMATIC S7 Program

through an element from right to left (a parallel branch must be closed to the branch in which it was opened). Any further rules applying to the layout of special LAD elements are discussed in the relevant chapters.

When using boxes as program elements, you can

▷ program a single box per network
▷ arrange boxes in T branches in branches that start at the left power rail
▷ arrange boxes in series by switching the ENO output of one box to the EN input of the following box
▷ switch boxes in parallel in branches on the left power rail via its ENO output

With the arrangement of the boxes, you evaluate the signal states of the ENO outputs: if you terminate the ENO outputs with a coil, "power" flows into the coil if all the boxes have all been processed without errors in the case of series connection, or if one of the boxes has been processed without errors in the case of parallel connection (see also Section 15.4 "Using the Binary Result").

3.3.3 Editing FBD Elements

Programming in general

The program consists of individual program elements that are connected together via the binary signal flow to form logic operations or networks. You begin programming a logic operation by selecting the programming elements on the left of the logic operation

▷ with the function key, (e.g. F2 for the AND function)
▷ via the menu (INSERT → FBD ELEMENT → AND BOX) or
▷ from the Program Elements Catalog (with INSERT → PROGRAM ELEMENTS or VIEW → CATALOG)

You terminate a binary logic operation in the simplest case with an assign box (see Section 3.3 "Programming Code Blocks").

Most program elements must be assigned memory locations (variables). The easiest way to do this is to first arrange all program elements, then label them.

Binary functions

And function,
OR function,
Exclusive OR function

Negation of the scan
and of the result of
logic operation

Simple boxes

Assign, set, reset,
midline output,
edge evaluation

Control timers
and counters

Jumps, Master Control
Relay, etc.

Binary address

Counter address

Destination

Complex boxes

Standard boxes with EN/ENO
(e.g. timer and counter functions)

Standard boxes without EN/ENO
(e.g. arithmetic functions)

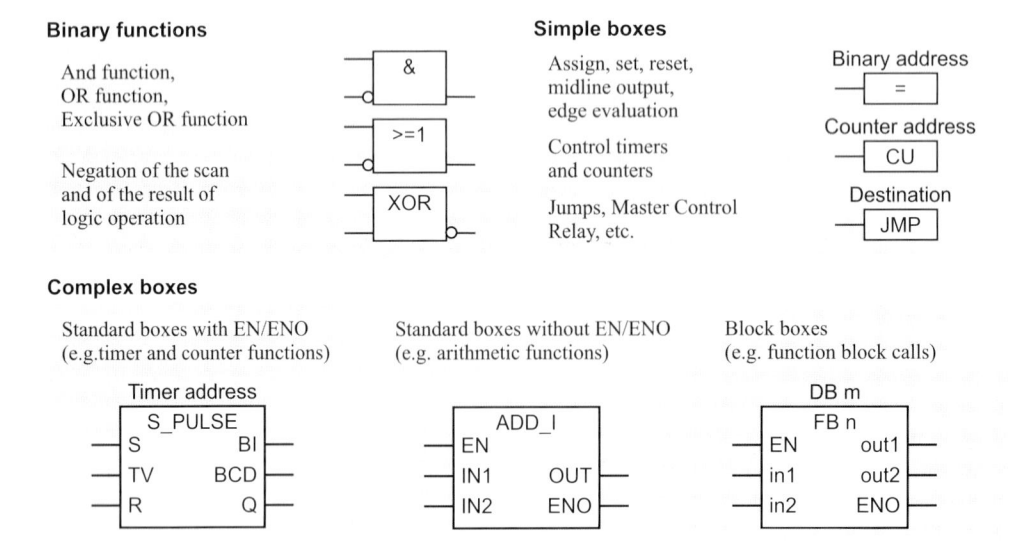

Block boxes
(e.g. function block calls)

Figure 3.8 Examples of FBD Program Elements

3.3 Programming Code Blocks

Binary functions

You scan the binary addresses such as inputs and combine the scanned signal states using the binary functions AND, OR and exclusive OR. Each binary input of a box also scans the binary address at the input.

The scanning of an address can be negated so that scan result "1" can be obtained for status "0" of the address. You can also scan status bits or the result of a logic operation within a logic operation.

Simple boxes

You control binary addresses such as outputs with simple boxes. Simple boxes generally have only one input and may have an additional label.

There are simple boxes for controlling a binary address, evaluating an edge, controlling timer and counter addresses, calling blocks without parameters, executing jumps in the program, and so on.

Complex boxes

Complex boxes represent program elements with complex functions. STEP 7 provides "standard boxes" in two versions:

▷ without EN/ENO mechanism (such as memory functions, timers and counters, comparison boxes) and

▷ with EN/ENO (such as MOVE, arithmetic and math functions, data type conversion).

If you call code blocks (FCs, FBs, SFCs and SFBs), FBD also represents the calls as boxes with EN/ENO.

FBD also provides an "empty box" in which you can enter the desired function when programming.

Layout restrictions

The FBD editor sets up a network from left to right and from the top down. From the left, the inputs lead to the functions and the outputs exit to the right.

A logic operation always has a "terminating function". In its simplest form, this is an assignment of the result of the logic operation to a binary address.

With the help of a T branch of a logic operation, you can program further "terminating functions" for a logic operation ("multiple output"). Following a T branch however, the selection of programmable elements is restricted. For example, you cannot arrange edge evaluations and call boxes following a T branch. Any further rules applying to the layout of special FBD elements are discussed in the relevant chapters.

When using boxes as program elements, you can

▷ program a single box per network

▷ arrange boxes in T branches in branches that start at the left power rail

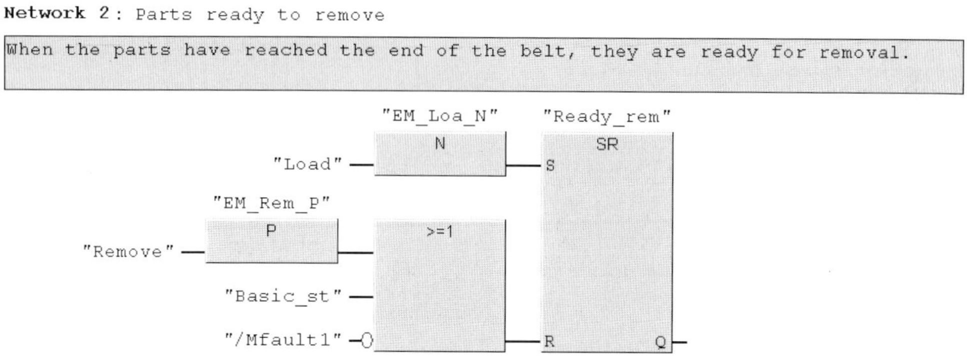

```
Network 2: Parts ready to remove
```
When the parts have reached the end of the belt, they are ready for removal.

```
                          "EM_Loa_N"      "Ready_rem"
                              N              SR
                "Load" ─┤          ├──   ─┤S         
                "EM_Rem_P"
                    P           >=1
   "Remove" ─┤         ├────┤       ├─
             "Basic_st" ─┤
             "/Mfault1" ─○┤         ├─R       Q├─
```

Figure 3.9 Example of an FBD Network in "3-Dimensional" Representation

3 SIMATIC S7 Program

▷ arrange boxes in series by switching the ENO output of one box to the EN input of the following box

▷ AND or OR boxes via the ENO output.

In the case of boxes switched in series, you can control their processing as a group (see also Section 15.4 "Using the Binary Result"). You evaluate the error messages of the boxes by combining the ENO outputs: ANDing of the ENO outputs is fulfilled if all boxes have been processed without error, and Oring of the ENO outputs is fulfilled if one of the boxes has been processed without error.

3.4 Programming Data Blocks

Section 2.5 "Creating the S7 Program" gives an introduction to program creation and the use of the program editor. Data blocks are programmed in the same way in LAD and FBD.

3.4.1 Creating Blocks

You begin block programming by opening a block, either with a double-click on the block in the project window of the SIMATIC Manager or by selecting FILE → Open in the editor. If the block does not yet exist, create it as follows:

▷ In the SIMATIC Manager: select the object *Blocks* in the left-hand portion of the project window and create a new data block with INSERT → S7 BLOCK → DATA BLOCK. You will see a dialog box with the block properties (number of the block, creation language, name, version, etc.). The creation language is fixed at "DB". You can also enter the remaining block properties later.

▷ In the editor: with FILE → New, you get a dialog box in which you can enter the desired block under "Object name". After closing the dialog box, you can program the block contents.

You can fill out the header of a block as you create it or you can add the block properties at a later point. You program later additions to the block header in the editor by selecting FILE → PROPERTIES while the block is open.

3.4.2 Types of Data Blocks

When you first open a new data block, you will see the window "New Data Block"; you must now decide which type to assign to the data block.

You can choose between three options by clicking on one of the following:

▷ "Data block"
Creation as a global data block; you declare the data addresses when programming the data block in this case

▷ "Data block with assigned user-defined data type"
Creation as a data block of user-defined data type; you declare the data structure as a user-defined data type UDT in this case

▷ "Data block with assigned function block"
Creation as an instance data block; here, the data structure that you have declared when programming the relevant function block is transferred.

3.4.3 Block Window

Figure 3.10 shows an open data block window. You can choose between two views:

▷ Declaration view (in this view, you enter the data addresses, provide them with a data type and specify an initial value) and

▷ Data view (in this view, you specify an actual value).

When programming a global data block, you can provide every data address with an initial value. The variables have a standard default value of zero, the smallest value or blanks, depending on data type.

An instance data block generated from a function block accepts the default from the declaration section of the function block as its initial value.

If you generate a data block from a user-defined data type UDT, the data block contains the initialization values (default values) from the UDT as initial values.

The editor displays a data block in two views: declaration view and data view.

In *Declaration view* (VIEW → DECLARATION VIEW), you define all the data addresses and

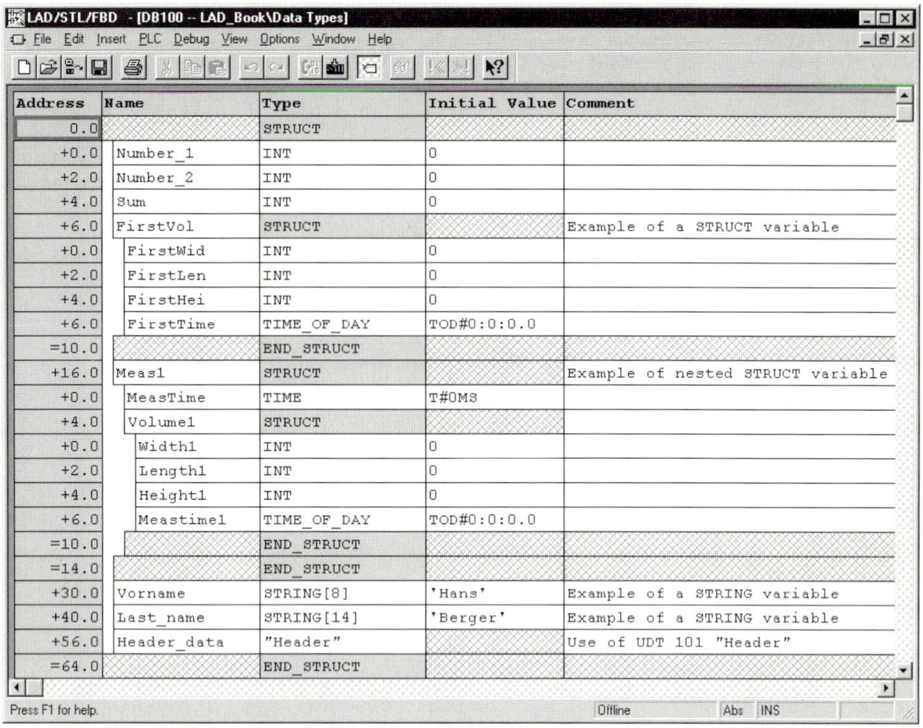
LAD/STL/FBD - [DB100 -- LAD_Book\Data Types]

File Edit Insert PLC Debug View Options Window Help

Address	Name	Type	Initial Value	Comment
0.0		STRUCT		
+0.0	Number_1	INT	0	
+2.0	Number_2	INT	0	
+4.0	Sum	INT	0	
+6.0	FirstVol	STRUCT		Example of a STRUCT variable
+0.0	FirstWid	INT	0	
+2.0	FirstLen	INT	0	
+4.0	FirstHei	INT	0	
+6.0	FirstTime	TIME_OF_DAY	TOD#0:0:0.0	
=10.0		END_STRUCT		
+16.0	Meas1	STRUCT		Example of nested STRUCT variable
+0.0	MeasTime	TIME	T#0MS	
+4.0	Volume1	STRUCT		
+0.0	Width1	INT	0	
+2.0	Length1	INT	0	
+4.0	Height1	INT	0	
+6.0	Meastime1	TIME_OF_DAY	TOD#0:0:0.0	
=10.0		END_STRUCT		
=14.0		END_STRUCT		
+30.0	Vorname	STRING[8]	'Hans'	Example of a STRING variable
+40.0	Last_name	STRING[14]	'Berger'	Example of a STRING variable
+56.0	Header_data	"Header"		Use of UDT 101 "Header"
=64.0		END_STRUCT		

Press F1 for help. Offline Abs INS

Figure 3.10 Example of an Opened Data Block (Declaration View)

you also see the variables as you have defined them, for example, an array (field) or a user data type as a single variable.

In the *Data view* (VIEW → DATA VIEW), the editor displays each variable and each component or element of a field or a structure individually. Now you see an additional column with the actual value. The actual value is the value a data address in CPU work memory has or will have. As standard, the editor assumes the initial value as the actual value.

You can modify the actual value individually for every data address: you generate several instance data blocks from one function block. However, for each function block call (for each FB/DB pair) you want only a slight difference in the pre-assignment of individual instance data. You can now edit each data block with VIEW → DATA VIEW, and you can enter the values valid for this data block in the Actual value column. With EDIT → INITIALIZE DATA BLOCK, you cause the editor to replace the actual values again with the initial values.

3.5 Variables, Constants and Data Types

3.5.1 General Remarks Concerning Variables

A variable is a value with a specific format (Figure 3.11). Simple variables consist of an address (such as input 5.2) and a data type (such as BOOL for a binary value). The address, in turn, comprises an address identifier (such as I for input) and an absolute storage location (such as 5.2 for byte 5, bit 2). You can also reference an address or a variable symbolically by assigning the address a name (a symbol) in the symbol table.

A bit of data type BOOL is referred to as a *binary address* (or *binary operand).* Addresses comprising one, two or four bytes or variables with the relevant data types are called *digital operands.*

3 SIMATIC S7 Program

A variable consists of the address and the data type. It is addressed symbolically.

Variable
   |
Address + Data type
   |
Address identifier + Memory location

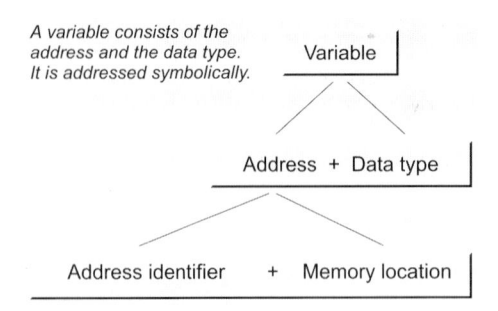

Figure 3.11 Structure of a Variable

Variables, which you declare within a block, are referred to as (block-) local variables. These include the block parameters, the static and temporary local data, even the data addresses in global data blocks. When these variables are of an elementary data type, they can also be accessed as operands (for instance static local data as DI operands, temporary local data as L operands, and data in global data blocks as DB operands).

Local variables, however, can also be of complex data type (such as structures or arrays). Variables with these data types require more than 32 bits, so that they can no longer, for example, be loaded into the accumulator. And for the same reason, they cannot be addressed with "normal" STL statements. There are special functions for handling these variables, such as the IEC functions, which are provided as a standard library with STEP 7 (you can generate variables of complex data type in block parameters of the same data type).

If variables of complex data type contain components of elementary data type, these components can be treated as though they were separate variables (for example, you can load a component of an array consisting of 30 INT values into the accumulator and further process it).

Constants are used to preset variables to a fixed value. The constant is given a specific prefix depending on the data type.

3.5.2 Addressing Variables

When addressing variables, you may choose between absolute addressing and symbolic addressing. Absolute addressing uses numerical addresses beginning with zero for each address area. Symbolic addressing uses alphanumeric names, which you yourself define in the symbol table for global addresses or in the declaration section for block-local addresses. An extension of absolute addressing is indirect addressing, in which the addresses of the memory locations are not computed until runtime.

Absolute addressing of variables

Variables of elementary data type can be referenced by absolute addresses.

The absolute address of an input or output is computed from the module start address, which you set or had set in the configuration table and the type of signal connection on the module. A distinction is made between binary signals and analog signals.

Binary signals

A binary signal contains one bit of information. Examples of binary signals are the input signals from limit switches, momentary-contact switches and the like which lead to digital input modules, and output signals which control lamps, contactors, and the like via digital output modules.

Analog signals

An analog signal contains 16 bits of information. An analog signal corresponds to a "channel", which is mapped in the controller as a word (2 bytes) (see below). Analog input signals (such as voltages from resistance thermometers) are carried to analog input modules, digitized, and made available to the controller as 16 information bits. Conversely, 16 bits of information can control an indicator via an analog output module, where the information is converted into an analog value (such as a current).

The information width of a signal also corresponds to the information width of the variable in which the signal is stored and processed. The information width and the interpretation of the information (for instance the positional weight), taken together, produce the *data type* of the variable. Binary signals are stored in variables of data type BOOL, analog signals in variables of data type INT.

3.5 Variables, Constants and Data Types

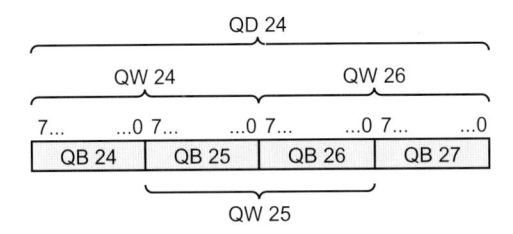

```
          QD 24
   ┌────────────────────┐
      QW 24        QW 26
   ┌─────────┐  ┌─────────┐
  7... ...0 7... ...0 7... ...0 7... ...0
  │ QB 24 │ QB 25 │ QB 26 │ QB 27 │
              └─────────┘
                QW 25
```

Figure 3.12
Byte Contents in Words and Doublewords

The only determining factor for the addressing of variables is the information width. In STEP 7, there are four widths, which can be accessed with absolute addressing:

▷ 1 bit   Data type BOOL
▷ 8 bits Data type BYTE or another data type with 8 bits
▷ 16 bits Data type WORD or another data type with 16 bits
▷ 32 bits Data type DWORD or another data type with 32 bits

Variables of data type BOOL are referenced via an address identifier, a byte number, and – separated by a decimal point – a bit number. Numbering of the bytes begins at zero for each address area. The upper limit is CPU-specific. The bits are numbered from 0 to 7. Examples:

I 1.0   Input bit no. 0 in byte no. 1
Q 16.4   Output bit no. 4 in byte no. 16

Variables of data type BYTE have as absolute address the address identifier and the number of the byte containing the variable. The address identifier is supplemented by a B. Examples:

IB 2   Input byte no. 2
QB 18   Output byte no. 18

Variables of data type WORD consist of two bytes (a word). They have as absolute address the address identifier and the number of the low-order byte of the word containing the variable. The address identifier is supplemented by a W. Examples:

IW 4   Input word no. 4; contains bytes 4 and 5
QW 20   Output word no. 20; contains bytes 20 and 21

Variables of data type DWORD consist of four bytes (a doubleword). They have as absolute address the address identifier and the number of the low-order byte of the word containing the variable. The address identifier is supplemented by a D. Examples:

ID 8   Input doubleword no. 8; contains bytes 8, 9, 10 and 11
QD 24   Output doubleword no. 24; contains bytes no. 24, 25, 26 and 27

Addresses for the data area include the data block. Examples:

DB 10.DBX 2.0
    Data bit 2.0 in data block DB 10

DB 11.DBB 14
    Data byte 14 in data block DB 11

DB 20.DBW 20
    Data word 20 in data block DB 20

DB 22.DBD 10
    Data doubleword 10 in data block DB 22

Additional information on addressing the data area can be found in section 18.2.2, "Accessing the Data Area".

Symbolic Addressing of Variables

Symbolic addressing uses a name (called a symbol) in place of an absolute address. You yourself choose this name. Such a name must begin with a letter and may comprise up to 24 characters. A distinction is made between upper and lower case. A keyword is not permissible as a symbol in STL; to use a keyword as a symbol in SCL, insert the hash character (#) before the name.

The name, or symbol, must be allocated to an absolute address. A distinction is made between global symbols and symbols that are local to a block.

Global symbols

You may assign names in the symbol table to the following objects:

▷ Data blocks and code blocks
▷ Inputs, outputs, peripheral inputs and peripheral outputs
▷ Memory bits, timers and counters
▷ User data types
▷ Variable tables

A global symbol may also include spaces, special characters and country-specific characters such as the umlaut. Exceptions to this rule are the characters 00_{hex} and FF_{hex}. When using symbols containing special characters, you must put the symbols in quotation marks in the program. In compiled blocks, the STL Editor always shows global symbols in quotation marks.

You can use global symbols throughout the program; each such symbol must be unique within a program.

Editing, importing and exporting of global symbols are described in Section 2.5.2 "Symbol Table".

Block-local symbols

The names for the local data are specified in the declaration section of the relevant block. These names may contain only letters, digits and the underline character.

Local symbols are valid only within a block. The same symbol (the same variable name) may be used in a different context in another block. The Editor shows local symbols with a leading "#". When the Editor cannot distinguish a local symbol from an address, you must precede the symbol with a "#" character during input.

Local symbols are available only in the programming device database (in the offline container *Blocks*). If this information is missing on decompilation, the Editor inserts a substitute symbol.

Using symbol names

If you use symbolic names while programming with the incremental Editor, they must have already been allocated to absolute addresses. You also have the option of entering new symbolic names in the symbol table while programming with the incremental Editor. Once the new symbolic names have been entered, you can use them immediately while writing the rest of your program. If you are using a text file to input your program, you must make the absolute addresses available at compilation time.

In the case of arrays, the individual components are accessed via the array name and a subscript,

for example MSERIES[1] for the first component. In LAD and FBD, the index is a constant INT value.

In structures, each subidentifier is separated from the preceding subidentifier by a decimal point, for instance FRAME.HEADER.CNUM. Components of user data types are addressed exactly like structures.

Data addresses

Symbolic addressing of data uses complete addressing including the data block. Example: the data block with the symbolic address MVALUES contains the variables MVALUE1, MVALUE2 and MTIME. These variables can be addressed as follows:

```
"MVALUES".MVALUE1
"MVALUES".MVALUE2
"MVALUES".MTIME
```

Please refer to Section 18.2.2 "Accessing the Data Area" for further information.

3.5.3 Overview of Data Types

Data types stipulate the characteristics of data, essentially the representation of the contents of a variable, and the permissible ranges. STEP 7 provides predefined data types, which you can combine into user-defined data types.

The data types are available on a global basis, and can be used in every block. LAD and FBD use the same data types.

Depending on structure and application, the data types are classified as follows:

▷ Elementary data types
▷ Complex data types
▷ User data types
▷ Parameter types

Table 3.3 shows the properties of these data type classes.

On the diskette that accompanies this book, you will find examples of the declaration and use of variables of all data types in the libraries "LAD_Book" and "FBD_Book" under the "Data Types" program.

3.5 Variables, Constants and Data Types

Table 3.3 Subdivision of the Data Types

Elementary Data Types	Complex Data Types	User Data Types	Parameter Data Types
BOOL, BYTE, CHAR, WORD, INT, DATE, DWORD, DINT, REAL, S5TIME, TIME, TOD	DT, STRING ARRAY, STRUCT	UDT Global data blocks Instances	TIMER, COUNTER, BLOCK_DB, BLOCK_SDB, BLOCK_FC, BLOCK_FB, POINTER, ANY
Data types comprising no more than one doubleword (32 bits)	Data types that can comprise more than one doubleword (DT, STRING) or which consist of several components	Structures or data areas which can be assigned a name	Block parameters
Can be mapped to operands referenced with absolute and symbolic addressing	Can be mapped only to variables that are addressed symbolically		Can be mapped only to block parameters (symbolic addressing only)
Permitted in all address areas	Permitted in data blocks (as global data and instance data), as temporary local data and as block parameters		Permitted in conjunction with block parameters

3.5.4 Elementary Data Types

Elementary data types can reserve a bit, a byte, a word or a doubleword.

Table 3.5 shows the elementary data types. For many data types, there are two constant representations that you can use equally (e.g. TIME# or T#). The table contains the minimum value for a data type in the upper line and the maximum value in the lower line.

Declaration of elementary data types

Table 3.4 shows some examples of the declaration of variables of elementary data types. *Name* is the identifier for a block-local variable (up to 24 characters, alphanumeric and underscore only). You enter the associated data type in the *Type* column.

With the exception of the temporary local data and block parameters of functions, you can assign an *initial value* to the variables. Use the syntax suitable for the data type for this purpose. *Comments* are optional.

BOOL, BYTE, WORD, DWORD, CHAR

A variable of data type BOOL represents a bit value, for example I 1.0. Variables with data types BYTE, WORD and DWORD are bit strings comprising 8, 16 and 32 bits, respectively. The individual bits are not evaluated.

Special forms of these data types are the BCD numbers and the count as used in conjunction with counter functions, as well as data type CHAR, which represents an ASCII character.

Table 3.4 Examples of Declaration and Initial Value for Elementary Data Types

Name	Type	Initial Value	Comments
Automatic	BOOL	FALSE	Initial value is signal state "0"
Manual_off	BOOL	TRUE	Initial value is signal state "1"
Measured_value	DINT	L#0	Initial value of a DINT variable
Memory	WORD	W#16#FFFF	Initial value of a WORD variable
Waiting_time	S5TIME	S5T#20s	Initial value of an S5 time variable

3 SIMATIC S7 Program

Table 3.5 Overview of Elementary Data Types

Data Type	(Width)	Description	Example for Constant Notation
BOOL	(1 bit)	Bit	FALSE TRUE
BYTE	(8 bits)	8-bit hexadecimal number	B#16#00, 16#00 B#16#FF, 16#FF
CHAR	(8 bits)	One character (ASCII)	Printable character, e.g. 'A'
WORD	(16 bits)	16-bit hexadecimal number	W#16#0000, 16#0000 W#16#FFFF, 16#FFFF
		16-bit binary number	2#0000_0000_0000_0000 2#1111_1111_1111_1111
		Count value, 3 decades BCD	C#000 C#999
		Two 8-bit unsigned decimal numbers	B(0,0) B(255,255)
DWORD	(32 bits)	32-bit hexadecimal number	DW#16#0000_0000, 16#0000_0000 DW#16#FFFF_FFFF, 16#FFFF_FFFF
		32-bit binary number	2#0000_0000...0000_0000 2#1111_1111...1111_1111
		Four 8-bit unsigned decimal numbers	B(0,0,0,0) B(255,255,255,255)
INT	(16 bits)	Fixed-point number	−32 768 +32 767
DINT	(32 bits)	Fixed-point number	L#−2 147 483 648 L#+2 147 483 647 ("L#" may be omitted if the number is outside the INT number range)
REAL	(32 bits)	Floating-point number (for value range see text)	+1.234567E+02 in exponential representation
			123.4567 as decimal number
S5TIME	(16 bits)	Time value in SIMATIC format	S5T#0ms S5TIME#2h46m30s
TIME	(32 bits)	Time value in IEC format	T#−24d20h31m23s647ms TIME#24d20h31m23s647ms
			T#−24.855134d TIME#24.855134d
DATE	(16 bits)	Date	D#1990-01-01 DATE#2168-12-31
TIME_OF_DAY	(32 bits)	Time of day	TOD#00:00:00 TIME_OF_DAY#23:59:59.999

BCD numbers

BCD numbers have no special identifier. Simply enter a BCD number with the data type 16# (hexadecimal) and use only digits 0 to 9.

BCD numbers occur in coded processing of time values and counts and in conjunction with conversion functions. Data type S5TIME# is used to specify a time value for starting a timer (see below), data type 16# or C# for specifying a count value. A C# count value is a BCD number between 000 and 999, whereby the sign is always 0.

As a rule, BCD numbers have no sign. In conjunction with the conversion functions, the sign of a BCD number is stored in the leftmost (highest) decade, so that there is one less decade for the number.

When a BCD number is in a 16-bit word, the sign is in the uppermost decade, whereby only bit position 15 is relevant. Signal state "0" means that the number is positive. Signal state "1" stands for a negative number. The sign has no affect on the contents of the individual decades. An equivalent assignment applies for a 32-bit word.

The available value range is 0 to ± 999 for a 16-bit BCD number and 0 to ± 9 999 999 for a 32-bit number.

Char

A variable with data type CHAR (character) reserves one byte. Data type CHAR represents a single character in ASCII format. Example: 'A'.

You can use any printable character in apostrophes. Some special characters require use of the notation shown in Table 3.6. Example: '$$' represents a dollar sign in ASCII code

The MOVE function allows you to use two or four ASCII characters enclosed in apostrophes as a special form of data type CHAR for writing ASCII characters in a variable

INT

A variable with data type INT is stored as an integer (16-bit fixed-point number). Data type INT has no special identifier.

A variable with data type INT reserves one word. The signal states of bits 0 to 14 represent the digit positions of the number; the signal state of bit 15 represents the sign (S). Signal state "0" means that the number is positive, signal state "1" that it is negative. A negative number is represented as two's complement. The permissible number range is from +32 767 (7FFF$_{hex}$) to –32 768 (8000$_{hex}$).

DINT

A variable with data type DINT is stored as an integer (32-bit fixed-point number). An integer is stored as a DINT variable when it exceeds 32 767 or falls below –32 768 or when the number is preceded by type identifier L#.

A variable with data type DINT reserves one doubleword. The signal states of bits 0 to 30 represent the digit positions of the number; the sign is stored in bit 31. Bit 31 is "0" for a positive and "1" for a negative number. Negative numbers are stored as two's complement. The number range is
from +2 147 483 647 (7FFF FFFF$_{hex}$)
to –2 147 483 648 (8000 0000$_{hex}$).

REAL

A variable of data type REAL represents a fraction, and is stored as a 32-bit floating-point number. An integer is stored as a REAL variable when you add a decimal point and a zero.

In exponent representation, you can precede the "e" or "E" with an integer number or fraction with seven relevant digits and a sign. The digits that follow the "e" or "E" represent the exponent to base 10. STEP 7 handles the conversion of the REAL variable into the internal representation of a floating-point number.

REAL variables are divided into numbers, which can be represented with complete accuracy ("normalized" floating-point numbers) and those with limited accuracy ("denormalized" floating-point numbers). The value range of a normalized floating-point number lies between:

$-3.402\,823 \times 10^{+38}$ to $-1.175\,494 \times 10^{-38}$
±0
$+1.175\,494 \times 10^{-38}$ to $+3.402\,823 \times 10^{+38}$

A denormalized floating-point number may be in the following range:

$-1.175\,494 \times 10^{-38}$ to $-1.401\,298 \times 10^{-45}$
and
$+1.401\,298 \times 10^{-45}$ to $+1.175\,494 \times 10^{-38}$

Table 3.6 Special Characters for CHAR

CHAR	Hex	Description
$$	24$_{hex}$	Dollar sign
$'	27$_{hex}$	Apostrophe
$L or $l	0A$_{hex}$	Line feed (LF)
$P or $p	0C$_{hex}$	New page (FF)
$R or $r	0D$_{hex}$	Carriage return (CR)
$T or $t	09$_{hex}$	Tabulator

3 SIMATIC S7 Program

Data type CHAR

Byte m

| 7 | 6 | 5 | 4 | 3 | 2 | 1 | 0 |

ASCII Code

BCD number, 3 decades

| Byte m | Byte m+1 |

| 15 | 12 | 11 | 8 | 7 | 4 | 3 | 0 |

Sign — 10^2 — 10^1 — 10^0

BCD number, 7 decades

| Byte m | Byte m+1 | Byte m+2 | Byte m+3 |

| 31 | 28 | 27 | 24 | 23 | 20 | 19 | 16 | 15 | 12 | 11 | 8 | 7 | 4 | 3 | 0 |

Sign — 10^6 — 10^5 — 10^4 — 10^3 — 10^2 — 10^1 — 10^0

Data type INT

15 14 0

| S | $2^{14} 2^{13}$ ... | ... $2^2\ 2^1\ 2^0$ |

Data type DINT

31 30 ... ...16 15 0

| S | $2^{30} 2^{29}$ ... | ... 2^{16} | 2^{15} ... | ... $2^2\ 2^1\ 2^0$ |

Data type REAL

31 30 ... ...23 22 0

| S | 2^7 ... | ... 2^0 | $2^{-1}\ 2^{-2}$ ... | ... 2^{-23} |

Exponent — Mantissa

Data type S5TIME

15 14 0

| 10^0 | 10^2 | 10^1 | 10^0 |

Time base — Timer value

Data type DATE

15 14 0

| $2^{15}\ 2^{14}\ 2^{13}$ ... | ... $2^2\ 2^1\ 2^0$ |

Data type TIME

31 30 ... ...16 15 0

| S | $2^{30}\ 2^{29}$ ... | ... 2^{16} | 2^{15} ... | ... $2^2\ 2^1\ 2^0$ |

Data type TIME_OF_DAY

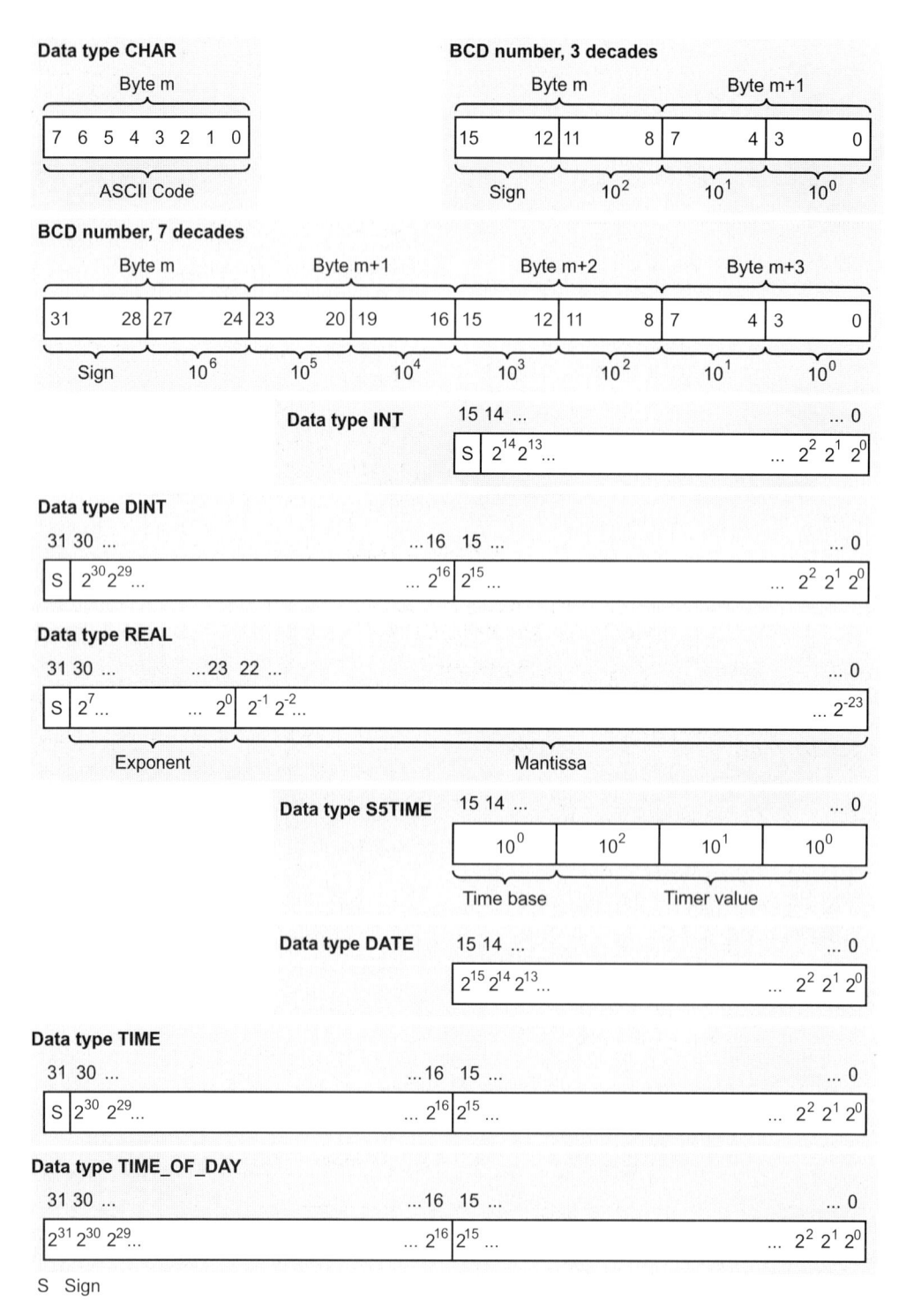

31 30 ... ...16 15 0

| $2^{31}\ 2^{30}\ 2^{29}$ ... | ... 2^{16} | 2^{15} ... | ... $2^2\ 2^1\ 2^0$ |

S Sign

Figure 3.13 Structure of the Variables of Elementary Data Type

3.5 Variables, Constants and Data Types

The S7-300 CPUs (with the exception of the CPU 318) cannot calculate with denormalized floating-point numbers. The bit pattern of a denormalized number is interpreted as a zero. If a result falls within this range, it is represented as zero, and status bits OV and OS are set (overflow).

A variable of data type REAL consists internally of three components, namely the sign (bit 31), the 8-bit exponent to base 2 (bits 23 to 30), and the 23-bit mantissa (bits 0 to 22). The sign may assume the value "0" (positive) or "1" (negative). Before the exponent is stored, a constant value (bias, +127) is added to it so that it shows a value range of from 0 to 255. The mantissa represents the fractional portion of the number. The integer portion of the mantissa is not saved, as it is either always 1 (in the case of normalized floating-point numbers) or always 0 (in the case of denormalized floating-point numbers). Table 3.7 shows the internal range of a floating-point number.

S5TIME

A variable with data type S5TIME is used in the basic languages STL, LAD and FBD to set the SIMATIC timers. It reserves one 16-bit word with 1 + 3 decades.

The time is specified in hours, minutes, seconds and milliseconds. STEP 7 handles conversion into internal representation. Internal representation is as BCD number in the range 000 to 999. The time interval can assume the following values: 10 ms (0000), 100 ms (0001), 1 s (0010), and 10 s (0011). The duration is the product of time interval and time value.

Examples:

S5TIME#500ms (= 0050_{hex})

S5T#2h46m30s (= 3999_{hex})

DATE

A variable with data type DATE is stored in a word as an unsigned fixed-point number. The contents of the variable correspond to the number of days since 01.01.1990. Its representation shows the year, month and day, separated from one another by a hyphen. Examples:

DATE#1990-01-01 (= 0000_{hex})

D#2168-12-31 (= $FF62_{hex}$)

TIME

A variable with data type TIME reserves one double word. Its representation contains the information for days (d), hours (h), minutes (m), seconds (s) and milliseconds (ms), whereby individual items of this information may be omitted. The contents of the variable are interpreted in milliseconds (ms) and stored as a signed 32-bit fixed-point number.

Table 3.7 Range Limits of a Floating-Point Number

Sign	Exponent	Mantissa	Description
0	255	Not equal to 0	Not a valid floating-point number
0	255	0	+ infinite
0	1 to 254	Arbitrary	Positive normalized floating-point number
0	0	Not equal to 0	Positive denormalized floating-point number
0	0	0	+ zero
1	0	0	– zero
1	0	Not equal to 0	Negative denormalized floating-point number
1	1 ... 254	Arbitrary	Negative normalized floating-point number
1	255	0	– infinite
1	255	Not equal to 0	Not a valid floating-point number

Examples:

TIME#24d20h31m23s647ms
(= 7FFF_FFFF$_{hex}$)

TIME#0ms (= 0000_0000$_{hex}$)

T#–24d20h31m23s648ms (= 8000_0000$_{hex}$)

A "decimal representation" is also possible for TIME, e.g. TIME#2.25h or T#2.25h.

Examples:

TIME#0.0h (=0000_0000 $_{hex}$)

TIME#24.855134d (=7FFF_FFFF $_{hex}$)

TIME_OF_DAY

A variable of data type TIME_OF_DAY reserves one double word. It contains the number of milliseconds since the day began (0:00 o'clock) in the form of an unsigned fixed-point number. Its representation contains the information for hours, minutes and seconds, separated by a colon. The milliseconds, which follow the seconds and are separated from them by a decimal point, may be omitted.

Examples:

TIME_OF_DAY#00:00:00 (= 0000_0000$_{hex}$)

TOD#23:59:59.999 (= 0526_5BFF$_{hex}$)

3.5.5 Complex Data Types

STEP 7 defines the following four complex data types:

▷ DATE_AND_TIME
  Date and time (BCD-coded)

▷ STRING
  Character string with up to 254 characters

▷ ARRAY
  Array variable (combination of variables of the same type)

▷ STRUCT
  Structure variable (combination of variables of different types)

The data types are pre-defined, with the length of the data type STRING (character string) and the combination and size of the data types ARRAY and STRUCT (structure) being defined by the user.

You can declare variables of complex data types only in global data blocks, in instance data blocks, as temporary local data or as block parameters.

Variables of complex data types can only be applied at block parameters as complete variables.

There are IEC functions for processing variables of data types DT and STRING, e.g. extraction of the date and conversion to the DATE representation or combining two character strings to one variable. These IEC functions are loadable standard FC blocks that you can find in the *Standard Library* under the *IEC Function Blocks* program.

DATE_AND_TIME

The data type DATE_AND_TIME represents a time consisting of the date and the time of day. You can also use the abbreviation DT in place of DATE_AND_TIME

The individual components of a DT variable are ASCII coded (Figure 3.14).

Table 3.8 Examples of the Declaration of DT Variables and STRING Variables

Name	Type	Initial Value	Comments
Date1	DT	DT#1990-01-01-00:00:00	DT variable minimum value
Date2	DATE_AND_TIME	DATE_AND_TIME#2089-12-31-23:59:59.999	DT variable maximum value
First_name	STRING[10]	'Jack'	STRING variable, 4 out of 10 char. specified
Last_name	STRING[7]	'Daniels'	STRING variable, all 7 char. specified
NewLine	STRING[2]	'RL'	STRING variable, special char. specified
BlankString	STRING[16]	''	STRING variable, no specification

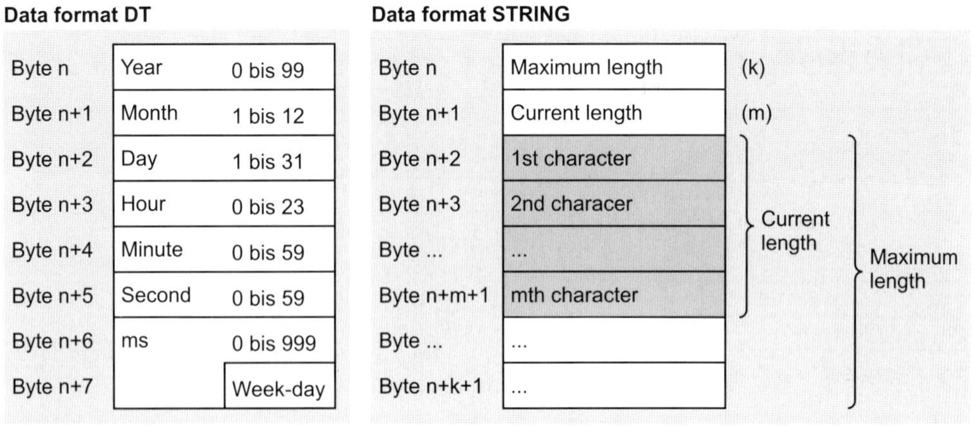

3.5 Variables, Constants and Data Types

Data format DT

Byte n	Year	0 bis 99
Byte n+1	Month	1 bis 12
Byte n+2	Day	1 bis 31
Byte n+3	Hour	0 bis 23
Byte n+4	Minute	0 bis 59
Byte n+5	Second	0 bis 59
Byte n+6	ms	0 bis 999
Byte n+7		Week-day

Data format STRING

Byte n	Maximum length	(k)
Byte n+1	Current length	(m)
Byte n+2	1st character	
Byte n+3	2nd characer	
Byte ...	...	
Byte n+m+1	mth character	
Byte ...	...	
Byte n+k+1	...	

Current length / Maximum length

Week-day
 from 1 = Sunday
 to   7 = Saturday

Figure 3.14 Structure of a DT and a STRING Variable

STRING

The data type STRING represents a character string consisting of up to 254 characters. You specify the maximum permissible number of characters in square brackets following the keyword STRING.

This specification can also be omitted; the Editor then uses a length of 254 bytes. In the case of functions FCs, the Editor does not permit specification of the length or it demands the standard length of 254.

A variable of data type STRING occupies two bytes more of memory than the declared maximum length.

Pre-assignment is carried out with ASCII-coded characters between single inverted commas or with a prefixed dollar sign in the case of certain characters (see data type CHAR).

If the initial or pre-assigned value is shorter than the declared maximum length, the remaining character locations are not reserved. When a variable of data type STRING is post-processed, only the currently reserved character locations are taken into consideration. It is also possible to define an "empty string" as the initial value. Figure 3.14 shows the structure of a STRING variable.

ARRAY

Data type ARRAY represents an array or field comprising a fixed number of elements of the same data type.

You specify the range of field indices in square brackets following the data type ARRAY. The initial value on the left must be less than or equal to the final value on the right. Both indices are INT numbers in the range −32,768 to +32,767. A field can have up to 6 dimensions each of whose limits are separated by a comma.

The data type of the individual field components is located in the line under the data type ARRAY. All data types except ARRAY are permissible; it can also be a user-defined data type.

Pre-assignment

At the declaration stage, you can pre-assign values to individual field components (not as a block parameter in a function, as an in/out parameter in a function block or as a temporary variable). The data type of the pre-assignment value must match the data type of the field.

You do not require to pre-assign all field components; if the number of pre-assignment values is less than the number of field components, only the first components are pre-assigned. The number of pre-assignment values must not be greater than the number of field components.

Table 3.9 Examples of an Array Declaration

Name	Type	Initial Value	Comments
Meas. val.	ARRAY[1..24]	0.4, 1.5, 11 (2.6, 3.0)	Array variable with 24 REAL elements
	REAL		
TOD	ARRAY[-10..10]	21 (TOD#08:30:00)	TOD array with 21 elements
	TIME_OF_DAY		
Result	ARRAY[1..24,1..4]	96 (L#0)	Two-dimensional array with 96 elements
	DINT		
Char.	ARRAY[1..2,3..4]	2 ('a'), 2 ('b')	Two-dimensional array with 4 elements
	CHAR		

The pre-assignment values are each separated by a comma. Multiple pre-assignment with the same values is specified within round brackets with a preceding repetition factor.

Application

You can apply a field as a complete variable at block parameters of data type ARRAY with the same structure or at a block parameter of data type ANY. For example, you can copy the contents of a field variable using the system function SFC 20 BLKMOV. You can also specify individual field components at a block parameter if the block parameter is of the same data type as the components.

If the individual field components are of elementary data types, you can process them with "normal" LAD or FBD functions.

A field component is accessed with the field name and an index in square brackets. The index is a fixed value in LAD and FBD and cannot be modified at runtime (no variable indexing possible).

Multi-dimensional fields or arrays

Fields can have up to 6 dimensions. Multi-dimensional fields are analogous to one-dimensional fields. At the declaration stage, the ranges of the dimensions are written in square brackets, each separated by a comma.

Structure of the variables

An ARRAY variable always begins at a word boundary, that is, at a byte with an even address. ARRAY variables occupy the memory up to the next word boundary.

Components of data type BOOL begin in the least significant bit; components of data type BYTE and CHAR begin in the right-hand byte. The individual components are listed in order.

In multi-dimensional fields, the components are stored line-wise (dimension-wise) starting with the first dimension). With bit and byte components, a new dimension always starts in the next byte, and with components of other data types a new dimension always starts in the next word (in the next even byte).

STRUCT

The data type STRUCT represents a data structure consisting of a fixed number of components that can each be of a different data type.

You specify the individual structure components and their data types under the line with the variable name and the keyword STRUCT. All data types can be used including other structures.

Pre-assignment

At the declaration stage, you can pre-assign values to the individual structure components (not as a block parameter in a function, as an in/out parameter in a function block or as a temporary variable). The data types of the pre-assignment values must match the data types of the components.

Application

You can apply a complete variable at block parameters of data type STRUCT with the same structure or at a block parameter of data type ANY. For example, you can copy the contents of a STRUCT variable with the system function SFC 20 BLKMOV. You can also specify an individual structure component at a block param-

3.5 Variables, Constants and Data Types

Table 3.10 Example of Declaring a Structure

Name	Type	Initial Value	Comment
MotCont	STRUCT		Simple structure variable with 4 components
On	BOOL	FALSE	Variable MotCont.On of type BOOL
Off	BOOL	TRUE	Variable MotCont.Off of type BOOL
Delay	S5TIME	S5TIME#5s	Variable MotCont.Delay of type S5TIME
maxSpeed	INT	5000	Variable MotCont.maxSpeed of type INT
	END_STRUCT		

eter if the block parameter is of the same data type as the component.

If the individual structure components are of elementary data types, you can process them with "normal" LAD or FBD functions.

A structure component is accessed with the structure name and the component name separated by a dot.

Structure of the variables

A STRUCT variable always begins at a word boundary, that is, at a byte with an even address; following this, the individual components are located in the memory in the order of their declaration. STRUCT variables occupy the memory up to the next word boundary.

Components of data type BOOL begin in the least significant bit; components of data type BYTE and CHAR begin in the right-hand byte. Components of other data types begin at a word boundary.

A nested structure is a structure as a component of another structure. A nesting depth of up to 6 structures is possible. All components can be accessed individually with "normal" LAD or FBD functions provided they are of elementary data type. The individual names are each separated by a dot.

3.5.6 Parameter Types

Parameter types are data types for block parameters (Table 3.11). The length specifications in the Table refer to the memory requirements for block parameters for function blocks. You can also use TIMER and COUNTER in the symbol table as data types for timers and counters.

3.5.7 User Data Types

A user data type (UDT) corresponds to a structure (combination of components of any data type) with global validity. You can use a user data type if a data structure occurs frequently in your program or you want to assign a name to a data structure.

UDTs have global validity; i.e., once declared, they can be used in all blocks. UDTs can be addressed symbolically; you assign the absolute address in the symbol table. The data type of a UDT (in the symbol table) is identical with the absolute address.

If you want to give a variable the data structure defined in the UDT, assign the UDT to it at declaration like a "normal" data type. The UDT can be absolutely addressed (UDT 0 to UDT 65,535) or symbolically addressed.

You can also define a UDT for an entire data type. When programming the data block, you assign this UDT to the block as a data structure.

The example "Message Frame Data" in Section 26.4 shows you how to work with user data types.

Programming UDTs

You create a user-defined data type either in the SIMATIC Manager by selecting the *Blocks* object and then INSERT → S7 BLOCK → DATA TYPE, or in the editor by selecting FILE → NEW and entering "UDTn" in the "Object name" line.

A double-click on the *UDT* object in the program window opens a declaration table that looks exactly like the declaration table of a data block. A UDT is programmed in exactly the same way as a data block, with individual lines for Name, Type, Initial value and Comments. The only difference is that switching to the data view is not possible. (With a UDT, you do not create any variables but only a collection of data types; for this reason, there can be no actual values here.)

The initial values you program in the UDT are transferred to the variables at declaration.

Table 3.11 Overview of Parameter Types

Parameter Type	Description		Examples of Actual Addresses
TIMER	Timer	16 bits	T 15 or symbol
COUNTER	Counter	16 bits	C 16 or symbol
BLOCK_FC	Function	16 bits	FC 17 or symbol
BLOCK_FB	Function block	16 bits	FB 18 or symbol
BLOCK_DB	Data block	16 bits	DB 19 or symbol
BLOCK_SDB	System data block	16 bits	SDB 100 or symbol
POINTER	DB pointer	48 bits	P#M10.0 (pointer) P#DB20.DBX22.2 (pointer) MW 20 (address) I 1.0 (address)
ANY	ANY pointer	80 bits	P#DB10.DBX0.0 WORD 20 or any variable

Basic Functions

This section of the book describes those functions of the LAD and FBD programming languages which represent a certain "basic functionality". These functions allow you to program a PLC on the basis of contactor or relay controls.

In a ladder diagram (LAD), the arrangement of the contacts in series and parallel circuits determines the combining of binary signal states. In a function block diagram (FBD), boxes analogous to electronic switching systems represent the boolean functions AND and OR.

The memory functions hold onto an RLO so that it can, for example, be scanned and processed further in another part of the program.

The move functions are used to exchange the values of individual operands and variables or to copy entire data areas.

The timing relays in contactor control systems are timers in programmable controllers. The timers integrated in the CPU allow you, for example, to program waiting and monitoring times.

Finally, the counters can count up and down in the range 0 to 999.

This section of the book describes the functions for the operand areas for inputs, outputs, and memory bits. Inputs and outputs are the link to the process or plant. The memory bits correspond to auxiliary contactors which store binary states. The subsequent sections of the book describe the remaining operand areas, on which you can also use binary logic. Essentially, these are the data bits in global data blocks as well as the temporary and static local data bits.

In Chapter 5, "Memory Functions", you will find a programming example for the binary logic operations and memory functions, and in Chapter 8, "Counters", an example for timers and counters. In both cases, the example is in an FC function without block parameters. You will find the same examples as function blocks (FBs) with block parameters in Chapter 19, "Block Parameters".

4   Binary Logic Operations

Series and parallel circuits (LAD), AND, OR and exclusive OR functions (FBD): negation; taking account of the sensor type

5   Memory functions

LAD coils; FBD boxes; midline outputs; edge evaluation; conveyor belt example

6   Move functions

MOVE box, system functions for moving data

7   Timers

Starting 5 different kinds of timer, resetting and scanning a timer; IEC timers

8   Counters

Setting a counter; up and down counting; resetting and scanning a counter; IEC counters; feed example

4 Binary Logic Operations

4.1 Series and Parallel Circuits (LAD)

Binary signal states are combined in LAD through series and parallel connection of contacts. Series connection corresponds to an AND function and parallel connection to an OR function. You use the contacts to check the signal states of the following binary operands:

▷ Input and output bits, memory bits
▷ Timers and counters
▷ Global data bits
▷ Temporary local data bits
▷ Static local data bits
▷ Status bits (evaluation of calculation results)

You can reference an operand via a contact using either an absolute or a symbolic address. LAD uses only NO contacts (scan for signal state "1") and NC contacts (scan for signal state "0").

A rung may consist of a single contact, but it may also consist of a large number of contacts connected together. A rung must always be terminated, for example with a coil. The coil controls a binary operand with the RLO (the "power flow") of the rung.

The examples shown in this chapter are also on the diskette accompanying this book in function block FB 104 of the "Basic Functions" program, library "LAD_Book". For incremental programming, you will find the elements for binary logic operations in the Program Element Catalog (with VIEW → CATALOG [Ctrl – K] or with INSERT → PROGRAM ELEMENTS) under "Bit Logic".

4.1.1 NO Contact and NC Contact

In order to explain the bit logic combinations in a ladder diagram, we will refer below as graphically as possible to "contact closed", "power flowing", and "coil energized". If "power" is flowing at a point in the ladder diagram, this means that the bit logic combination applies up to this point; the result of the logic operation (RLO) is "1". If "power" is flowing in a single coil, the coil is energized; the associated binary operand then carries signal state "1".

LAD has two kinds of contacts for scanning bit operands: the NO contact and the NC contact.

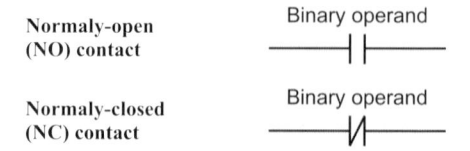

Normaly-open (NO) contact	Binary operand ─┤ ├─
Normaly-closed (NC) contact	Binary operand ─┤/├─

Normally open (NO) contact

A normally open contact corresponds to a scan for signal state "1". If the scanned binary operand has signal state "1", the NO contact is activated, so it closes and "power flows".

The example in Figure 4.1 (left side) shows sensor S1 connected to input I 1.0 and scanned by an NO contact. If sensor S1 is open, input I 1.0 is "0" and no power flows through the NO contact. Contactor K1, controlled by output Q 4.0, does not switch on.

If sensor S1 is now activated, input I 1.0 has signal state "1". Power flows from the left power rail through the NO contact into the coil, and contactor K1, which is connected to output Q 4.0, is activated. The NO contact scans the input for signal state "1" and then closes, regardless of whether the sensor at the input is an NO or NC contact.

Normally closed (NC) contact

Power flows through an NC contact if the binary operand has the signal state "0". If the signal state is "1", an NC contact "opens" and the flow of power is interrupted.

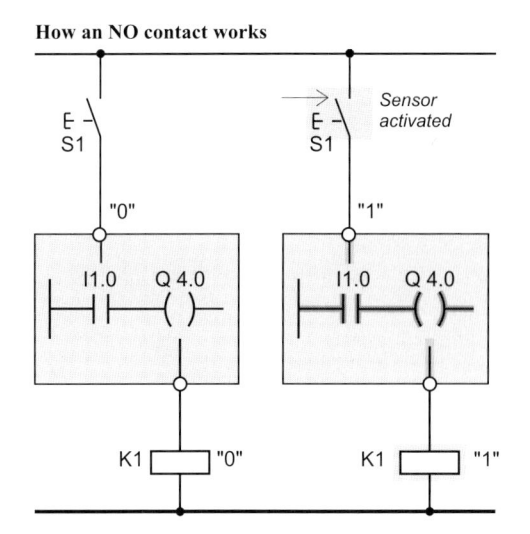

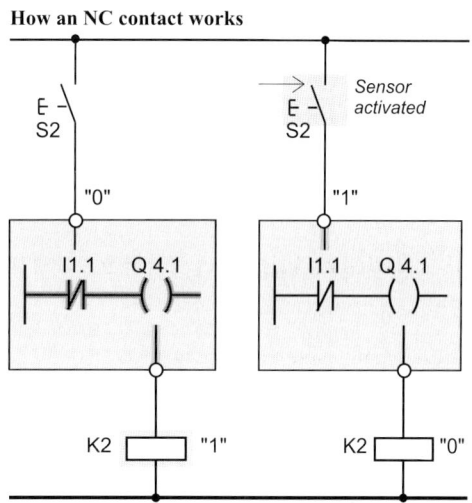

4.1 Series and Parallel Circuits (LAD)

Figure 4.1 NO Contacts and NC Contacts

In the example in Figure 4.1 (right side), power flows through the NC contact if sensor S2 is not closed (input I 1.1 has signal state "0"). Power also flows in the coil and energizes contactor K2 at output Q 4.1.

If sensor S2 is now activated, input I 1.1 has signal state "1" and the NC contact opens. The power flow is interrupted and contactor K2 releases.

The NC contact checks the input for signal state "0" and then remains closed, regardless of whether the sensor at the input is an NO or NC contact (also see Section 4.4, "Taking Account of the Sensor Type").

4.1.2 Series Circuits

In series circuits, two or more contacts are connected in series. Power flows through a series circuit when all contacts are closed.

Figure 4.2 shows a typical series circuit. In network 1, the series circuit has three contacts; any binary operands can be scanned. All contacts are NO contacts. If the associated operands all have signal state "1" (that is, if the NO contacts are activated), power flows through the rung to the coil. The operand controlled by the coil is set to "1". In all other cases, no power flows and the operand *Coil1* is reset to "0".

Network 2 shows a series circuit with one NC contact. Power flows through an NC contact if the associated operand has signal state "0" (that is, the NC contact is not activated). So power only flows through the series circuit in the example if the operand *Contact4* has signal state "1" and the operand *Contact5* has signal state "0".

4.1.3 Parallel Circuits

When two or more contacts are arranged one under the other, we refer to a parallel circuit. Power flows through a parallel circuit if one of the contacts is closed.

Figure 4.2 shows a typical parallel circuit. In network 3, the parallel circuit consists of three contacts; any binary operands can be scanned. All contacts are NO contacts. If one of the operands has signal state "1", power flows through the rung to the coil. The operand controlled by the coil is set to "1". If all operands scanned have signal state "0", no power flows to the coil and the operand *Coil3* is reset to "0".

Network 4 shows a parallel circuit with one NC contact. Power flows through an NC contact if the associated operand is "0", that is, power flows through the series circuit in the example if the operand *Contact4* has signal state "1" or the operand *Contact5* has signal state "0".

In LAD, you can also program a branch in the middle of the rung (for an example, see Figure 4.3 network 8). You then get a parallel branch

103

4 Binary Logic Operations

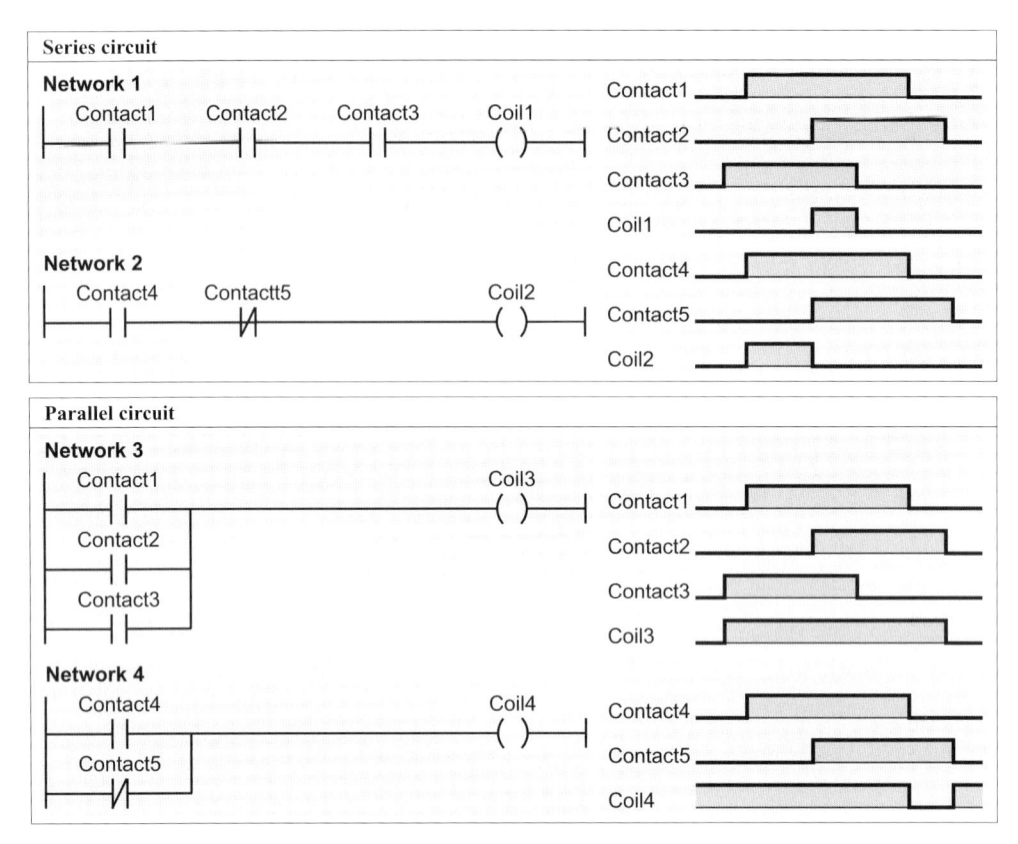

Figure 4.2 Series and Parallel Circuits

that does not begin at the left power rail. Use of LAD program elements is restricted to this parallel branch; your attention is drawn to this in the relevant chapters. An "open" parallel circuit is called a "T-branch".

4.1.4 Combinations of Binary Logic Operations

You can combine series and parallel circuits, for example, by arranging several series circuits in parallel or several parallel circuits in series. You can combine series and parallel circuits even when both types are complex in nature (Figure 4.3).

Connecting series circuits in parallel

Instead of contacts, you can also arrange series circuits one under the other. Figure 4.3 shows two examples. In network 5, power flows into the coil if *Contact1* and *Contact2* are closed or if *Contact3* and *Contact4* are closed. In the lower rung (network 6), power flows if *Contact5* or *Contact6* and *Contact7* or *Contact0* are closed.

Connecting parallel circuits in series

Instead of contacts, you can also arrange parallel circuits in series. Figure 4.3 shows two examples. In network 7, power flows into the coil if either *Contact1* or *Contact3* and either *Contact2* or *Contact4* are closed. To allow power to flow in the lower example (network 8), *Contact5*, *Contact0* and either *Contact6* or *Contact7* must be closed.

4.1.5 Negating the Result of the Logic Operation

The NOT contact negates the RLO. You can use this contact, for example, to run a series circuit

4.1 Series and Parallel Circuits (LAD)

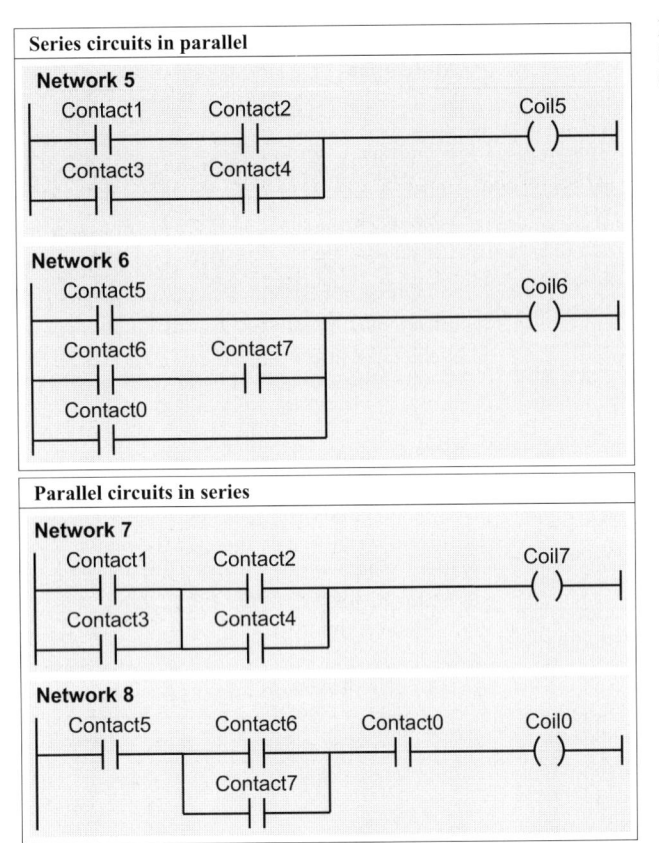

Figure 4.3
Series and Parallel Circuits in Combination

Series circuits in parallel

Network 5: Contact1 — Contact2 — Coil5; parallel branch Contact3 — Contact4

Network 6: Contact5 — Coil6; parallel branches: Contact6 — Contact7; Contact0

Parallel circuits in series

Network 7: (Contact1 ∥ Contact3) — (Contact2 ∥ Contact4) — Coil7

Network 8: Contact5 — (Contact6 ∥ Contact7) — Contact0 — Coil0

---

negated to a coil (Figure 4.4, Network 9). Power will then only flow into the coil if there is no power in the NOT contact, that is, if either *Contact1* or *Contact2* is open (see the Figure in the adjacent pulse diagram).

The same applies by analogy for network 10, in which a NOT contact is inserted after a parallel circuit. Here, *Coil10* is set if neither of the contacts is closed.

You can insert NOT instead of another contact into a branch that begins at the left power rail. Inserting a NOT contact in a parallel branch that begins in the middle of a rung is not permissible.

NOT contact (Negating the result of the logic operation RLO)

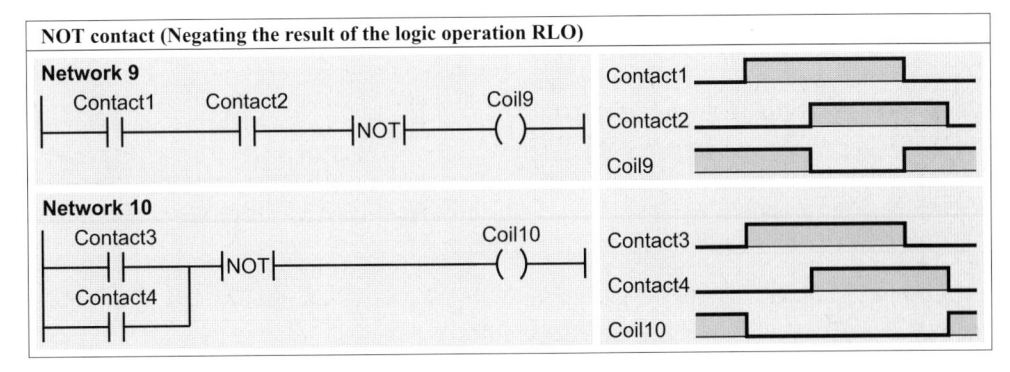

Network 9: Contact1 — Contact2 — NOT — Coil9
Pulse diagram: Contact1, Contact2, Coil9

Network 10: (Contact3 ∥ Contact4) — NOT — Coil10
Pulse diagram: Contact3, Contact4, Coil10

Figure 4.4 Examples of a NOT Contact

4.2 Binary Logic Operations (FBD)

In FBD, the logic operations performed on binary signal states take the form of AND, OR and Exclusive OR functions. The operands whose signal states you want to scan and combine are written at the inputs of these functions. You can scan the following operands:

- Input and output bits, memory bits (discussed in this section)
- Timers and counters
- Global data bits
- Temporary local data bits
- Static local data bits
- Status bits (evaluation of calculation results)

Every binary operand can be addressed absolutely or symbolically. When scanning a binary operand, or within a binary logic circuit, you can negate the result of the logic operation with the negation symbol (which is a circle).

In FBD, you program one binary logic circuit per network. The logic circuit may consist of only one or of a very large number of interconnected functions. A logic circuit, or logic operation, must always be terminated, for example with an assign statement. The assign controls a binary operand with the result of the logic operation.

The examples shown in this chapter are also in function block FB 104 of the "Basic Functions" program, library "FBD_Book", on the diskette accompanying the book. For incremental programming, you will find the elements for binary logic operations in the Program Element Catalog (VIEW → CATALOG [Ctrl – K] or with INSERT → PROGRAM ELEMENTS) under "Bit Logic".

4.2.1 Elementary Binary Logic Operations

FBD uses the binary functions AND, OR, and Exclusive OR. All functions may (theoretically) have any number of function inputs. If an input leads directly to the function element, the signal state of the operand scanned is used directly in the logic operation; if the input has a negation character (a circle), the signal state of the scanned operand is negated prior to execution of the logic operation (see below).

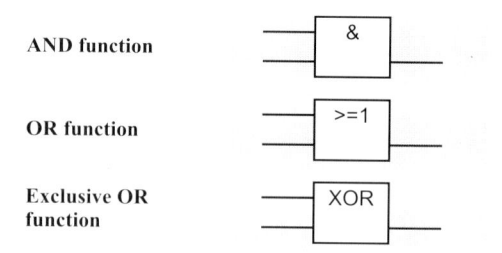

AND function

OR function

Exclusive OR function

The number of binary functions and the scope of a binary function are theoretically unlimited; in practice, however, limits are set by the length of a block or the size of the CPU's main memory.

Scanning and assigning signal states

Before the binary functions perform logic operations on signal states, they scan the binary operands at the function inputs. An operand can be scanned for "1" or "0". If scanned for "1", the function input leads directly to the box. A scan for "0" is recognizable by the negation character at the function input.

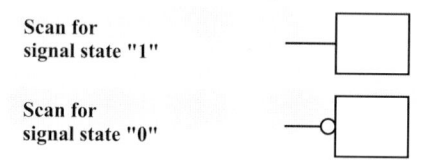

Scan for signal state "1"

Scan for signal state "0"

A scan for "1" produces a scan result of "1" when the signal state of the binary operand scanned is "1"; it produces a scan result of "0" when the signal state of the binary operand is "0". A scan for signal state "0" negates the scan result, that is, the scan result is "1" when the status of the binary operand scanned is "0". The binary functions combine the *scan result*, which is, at it were, the result applied "directly" to the box. As far as functionality is concerned, these two methods of scanning binary operands allow you to treat NO contacts and NC contacts identically.

Here an example: "0" is applied to the input module for a non-activated NO contact (Figure 4.5). A scan for signal state "1" forwards this status to a function box. To effect the same for an NC contact, you have to scan an input with an NC contact for signal state "0" (must include a circle for negation). The signal state "1" applied to the input module for a non-activated

4.2 Binary Logic Operations (FBD)

Scan for signal state "1"

[Diagram: E-S1 (NO contact, not activated) → I1.0 "0" → Function box "0"]
[Diagram: E-S1 (NO contact, Sensor activated) → I1.0 "1" → Function box "1"]

Scan for signal state "0"

[Diagram: E-S2 (NC contact, not activated) → I1.1 "0" → Function box "1"]
[Diagram: E-S2 (NC contact, Sensor activated) → I1.1 "1" → Function box "0"]

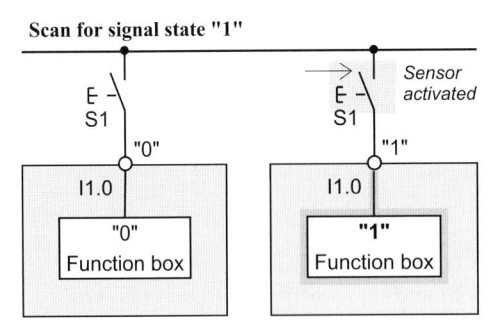

Figure 4.5 Scanning for Signal State "1" and "0"

NC contact is then converted into signal state "0" at the function box.

If you now activate both the NO and NC contacts, the function box will show signal state "1" in both cases. Additional information can be found in Section 4.2, "Taking Account of the Sensor Type".

You must always connect the output of a binary function; in the simplest case, simply connect the output to an Assign box (also see Chapter 5, "Memory Functions"). With this result of the logic operation, you can also start a timer, execute a digital operation, call a block, and so on. The next chapter provides all the information you need.

To assign the signal state of a binary operand directly to another binary operand without performing any additional logic operations, for example to connect an input directly to an output, the AND function is normally used, although it would also be possible to use an OR or Exclusive OR.

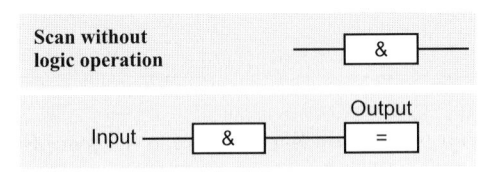

[Diagram: Scan without logic operation — & box]
[Diagram: Input — & box — Output =]

Simply select the AND function, connecting only one function input and removing the other.

AND function

The AND function combines two binary states with one another and produces an RLO of "1" when both states (both scan results) are "1". If the AND function has several inputs, the scan results of all inputs must be "1" for the collective RLO to be "1". In all other cases, the AND function produces an RLO of "0" at its function output.

Figure 4.6 shows an example of an AND function. In Network 1, the AND function has three inputs, each of which can be connected to any binary operand. All operands are scanned for signal state "1", so that the signal state of the operands is directly ANDed. If all the operands that were scanned have a signal state of "1", the AND function sets the operand *Output1* to "1" via the Assign box (see next section). In all other cases, the AND condition is not fulfilled and operand *Output1* is reset to "0".

Network 2 shows an AND function with a negated input. Negation of the input is indicated by a circle. The scan result for a negated operand is "1" when this operand is "0", that is, the AND condition in the example is fulfilled when the operand *Input4* is "1" and the operand *Input5* is "0".

OR function

The OR function combines two binary signal states and returns an RLO of "1" when one of these states (one of the scan results) is "1". If the OR function has several inputs, the scan result of only one input need be "1" in order for the result of the logic operation (RLO) to be "1". The OR function returns an RLO of "0" when the scan results of all inputs are "0".

Figure 4.6 shows an example of an OR function. In Network 3, the OR function has three inputs; each of these inputs may be connected to any binary operand. All operands are

4 Binary Logic Operations

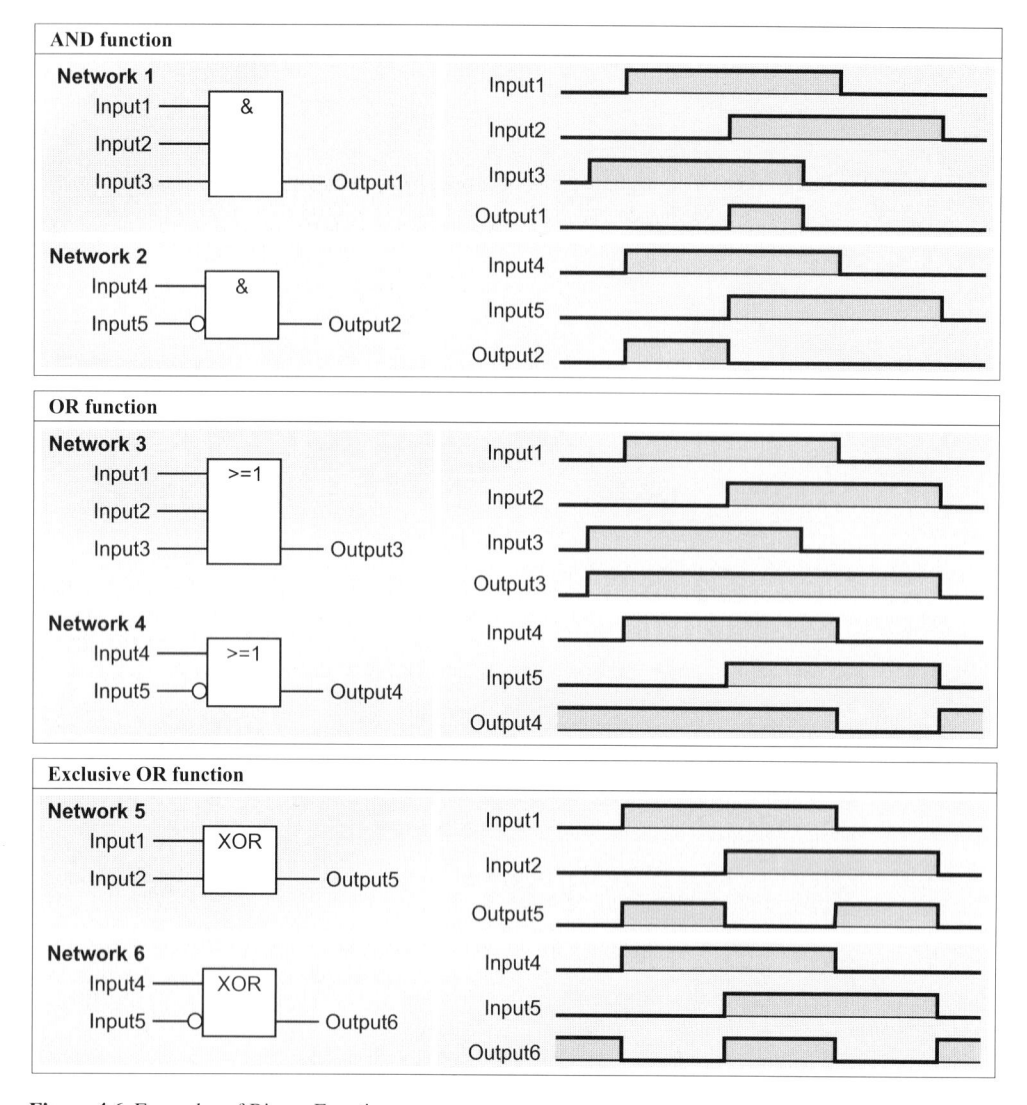

Figure 4.6 Examples of Binary Functions

scanned for signal state "1", so that the signal state of the operands is directly ORed. If one or more of the operands scanned have signal state "1", the next statement sets the operand *Output1* to "1". If all of the operands scanned have signal state "0", the OR condition is not fulfilled and operand *Output1* is reset to "0".

Network 4 shows an OR function with a negated input. Negation is represented by a circle. The scan result of a negated operand is "1" when that operand is "0", that is, the OR condition in the example is fulfilled when the operand *Input4* has a signal state of "1" or the operand *Input5* has a signal state of "0".

Exclusive OR function

The Exclusive OR function combines two binary states with one another and returns an RLO of "1" when the two states (scan results) are not the same, and RLO "0" when the two states (scan results) are identical.

Figure 4.6 shows an example of an Exclusive OR function. In Network 5, two inputs, both of which are scanned for signal state "1", lead to the Exclusive OR function. If only one of the operands scanned is "1", the Exclusive OR condition is fulfilled and the operand *Output1* is set to "1". If both operands are "1" or "0", operand *Output1* is reset to "0".

Network 6 shows an Exclusive OR function with a negated input. Negation is represented by a circle. The scan result of a negated operand is "1" when that operand is "0", that is, the Exclusive OR condition in the example is fulfilled when both input operands have the same signal state.

You can also program an Exclusive OR function with more than two inputs, in which case the Exclusive OR condition is fulfilled (in the case of a direct scan) when an uneven number of the operands scanned have a scan result of "1".

4.2.2 Combinations of Binary Logic Operations

You can easily combine binary functions with one another. For instance, you can combine several AND functions into one OR function or two OR functions into one Exclusive OR function. The number of functions per logic operation (per network) is theoretically unlimited.

The use of a "T-branch" in a logic operation gives you additional options, allowing you to program more than one output per logic operation (see Section 5.1, "Assign").

You can link the output of one binary function with the input of another binary function in order to implement complex binary logic operations. Figure 4.7 provides a number of examples.

Network 9: You are monitoring the limit switches at the ends of an X axis and a Y axis. These limit switches may not be actuated in pairs; otherwise, limit switch error will be reported.

Network 10: You can link arbitrary function inputs with binary functions, for example you can place an Exclusive OR function in front of the second input of an AND function.

Network 11: Using the negation symbol, you negate the RLO, even between binary functions, for instance you can negate the RLO of an OR function and use it as input to an AND function.

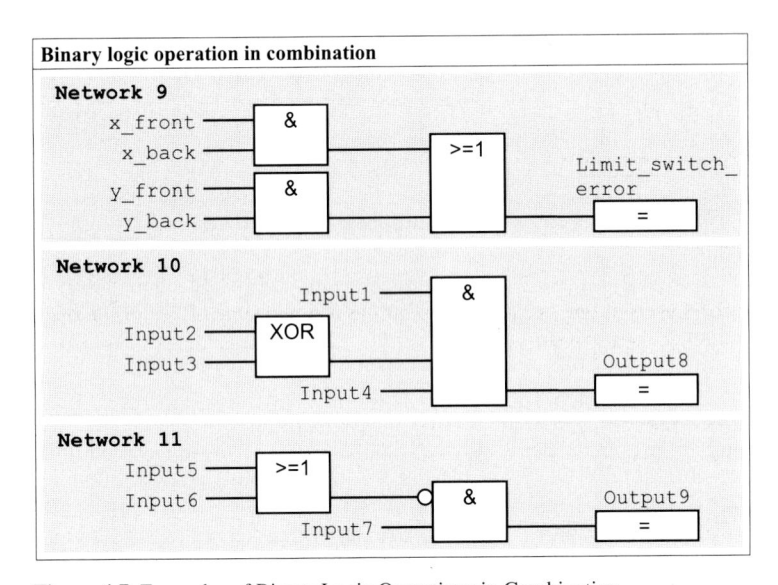

Figure 4.7 Examples of Binary Logic Operations in Combination

4 Binary Logic Operations

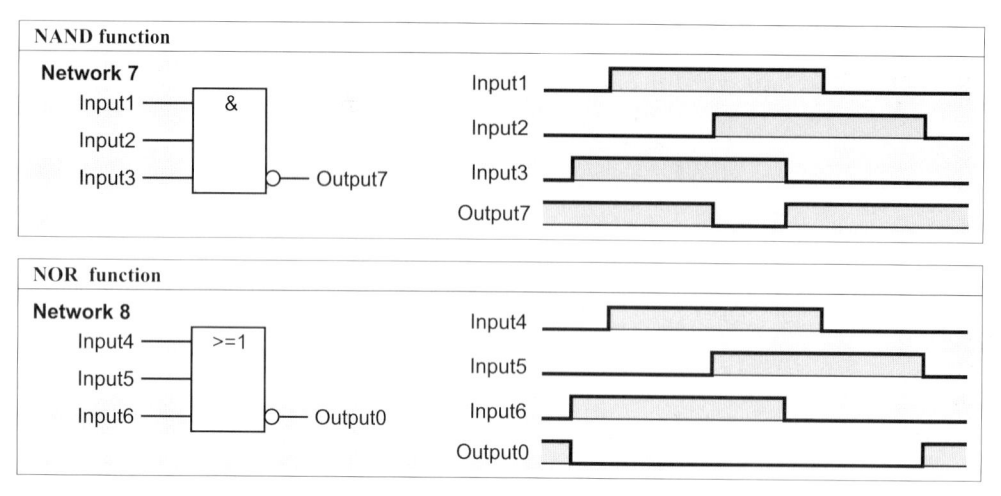

Figure 4.8 NAND and NOR Functions

4.2.3 Negating the Result of the Logic Operation

The circle at the input or output of a function symbol negates the result of the logic operation. You can use negation

▷ to scan a binary operand, which is equivalent to scanning for signal state "0" (see above)

▷ between two binary functions (which is equivalent to negating the result of the logic operation), or

▷ at the output of a binary function (for example if you want to set or reset a binary operand when the condition is not fulfilled, that is, when RLO = "0").

A negation may not immediately follow a T-branch.

Figure 4.8 shows a NAND function (an AND function with negated output) and a NOR function (an OR function with negated output). The RLO of a NAND function is "0" only when all inputs have a signal state of "1". A NOR function returns an RLO of "1" only when none of the inputs has a signal state of "1".

4.3 Taking Account of the Sensor Type

When scanning a sensor in a user program, you must take account of whether the sensor is an NO contact or an NC contact. Depending on the sensor type, there is a different signal state at the relevant input when the sensor is activated: "1" for an NO contact and "0" for an NC contact. The CPU has no means of determining whether an input is occupied by an NO contact or by an NC contact. It can only detect signal state "1" or signal state "0".

Programming with LAD

If you structure the program in such a way that you want a scan result of "1" when a sensor is activated in order to combine that scan result further, you must scan the input differently for different kinds of sensors. NO contacts and NC contacts are available to you for this purpose. An NO contact return "1" if the scanned input is also "1". An NC contact returns "1" if the scanned input is "0". In this way, you can also directly scan inputs that are to trigger activities when they are "0" (zero-active inputs) and subsequently re-gate the scan result.

4.3 Taking Account of the Sensor Type

Case 1: Both sensors are NO contacts

Case 2: One NO and one NC contact

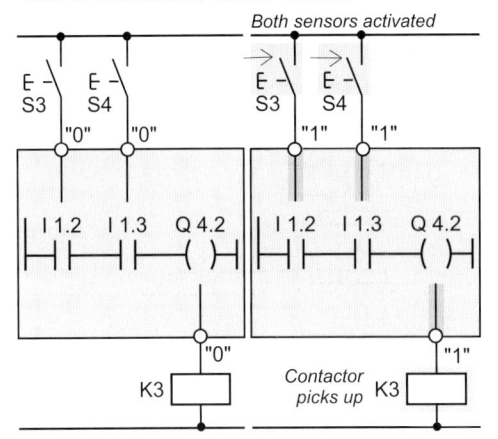

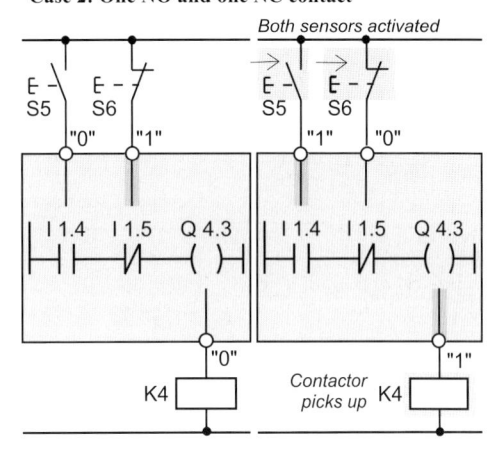

Figure 4.9 Taking Account of the Sensor Type (LAD)

The example in Figure 4.9 shows programming dependent on the sensor type. In the first case, two NO contacts are connected to the programmable controller, and in the second case one NO contact and one NC contact. In both cases, a contactor connected to an output is to pick up if both sensors are activated. If an NO contact is activated, the signal state at the input is "1", and this is scanned with an NO contact so that power can flow when the sensor is activated. If both NO contacts are activated, power flows through the rung to the coil and the contactor picks up.

If an NC contact is activated, the signal state at the input is "0". In order to have power flow in this case when the sensor is activated, the result must be scanned with an NC contact. Therefore, in the second case, an NO contact and an NC contact must be connected in series to make the contactor pick up when both sensors are activated.

Programming with FBD

If you structure the program in such a way that you want a scan result of "1" when a sensor is activated in order to combine that scan result further, you must scan the input differently for different kinds of sensors. An NO contact produces signal state "1" when activated, and is scanned directly when activation of the sensor is to produce a scan result of "1". An NC contact returns signal state "0" when activated; if you want a scan result of "1" when the NC contact is activated, it must be negated, then scanned. In this way, you can also scan inputs that are to trigger activities even when they have a signal state of "0" (zero-active inputs) and further combine the scan result.

The example in Figure 4.10 shows sensor type-dependent programming. In the first case, two NO contacts are connected to the programmable controller, and in the second case one NO contact and one NC contact. In both cases, a contactor connected to an output is to pick up if both sensors are activated. If an NO contact is activated, the signal state at the input is "1", and this is scanned directly so that the scan result is "1" when the sensor is activated. If both NO contacts are activated, and AND condition is fulfilled and the contactor picks up.

If an NC contact is activated, the signal state at the input is "0". In order to obtain a scan result of "1", the input must be negated when scanned. In the second case, you need an AND function with one direct and one negated input in order for the contactor to pick up when both sensors are activated.

111

4 Binary Logic Operations

Case 1: Both sensors are NO contacts

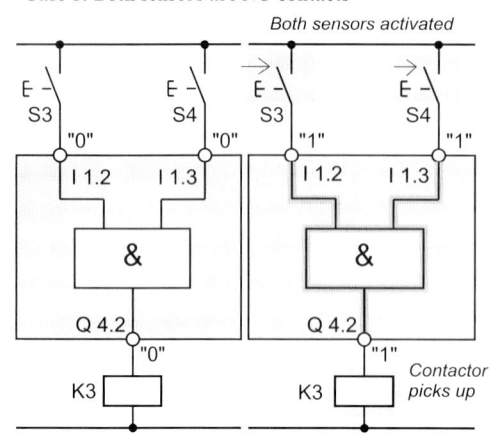

Both sensors activated

Case 2: One NO and one NC contact

Both sensors activated

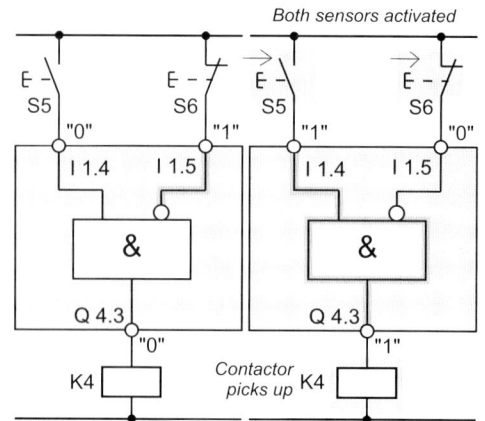

Figure 4.10 Taking Account of the Sensor Type (FBD)

5 Memory Functions

5.1 LAD Coils

In a ladder diagram (LAD), the memory functions are used in conjunction with series and parallel circuits in order to influence the signal states of the binary operands with the aid of the result of the logic operation (RLO) generated in the CPU.

The following memory functions are available:

▷ The single coil as an assignment of the RLO
▷ The coils S and R as individually programmed memory functions
▷ The boxes RS and SR as memory functions
▷ The midline outputs as intermediate buffers
▷ The coils P and N as edge evaluations of the power flow
▷ The boxes POS and NEG as edge evaluations of operands

Midline outputs and edge evaluations are discussed in detail in subsequent chapters.

You can use the memory functions described in this chapter in conjunction with all binary operands. There are restrictions when using temporary local data bits as edge memory bits.

The examples shown in this chapter are also represented on the diskette enclosed with the book in function block 105 of the "Basic Functions" program, library "LAD_Book".

For incremental programming, you will find the program elements for the memory functions in the program element catalog (with VIEW → CATALOG [Ctrl – K] or with INSERT → PROGRAM ELEMENTS) under "Bit Logic".

5.1.1 Single Coil

The single coil as terminator of a rung assigns the power flow directly to the operand located at the coil. The function of the single coil depends on the Master Control Relay (MCR): If the MCR is activated, signal state "0" is assigned to the binary operand located over the coil.

Single Coil	Binary operand —( )—

If power flows into the coil, the operand is set; if there is no power, the operand is reset (Figure 5.1, Network 1). With a NOT contact before the coil, you reverse the function (Network 2).

You can also direct the power flow into several coils simultaneously by arranging the coils in parallel with the help of a T-branch (Network 3). All operands specified over the coils respond in the same way. Up to 16 coils can be connected in parallel.

You can arrange further contacts in series and parallel circuits after the T-branch and before the coil (Network 4).

See Section 4.1, "Series and Parallel Circuits (LAD)", for further examples of the single coil.

5.1.2 Set and Reset Coil

Set and reset coils also terminate a rung. These coils only become active when power flows through them.

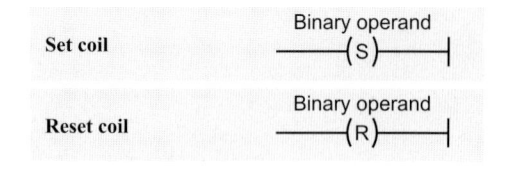

Set coil	Binary operand —(S)—
Reset coil	Binary operand —(R)—

If power flows in the set coil, the operand over the coil is set to signal state "1". If power flows in the reset coil, the operand over the coil is reset to signal state "0". If there is no power in the set or reset coil, the binary operand remains unaffected (Figure 5.1, Networks 5 and 6).

113

5 Memory Functions

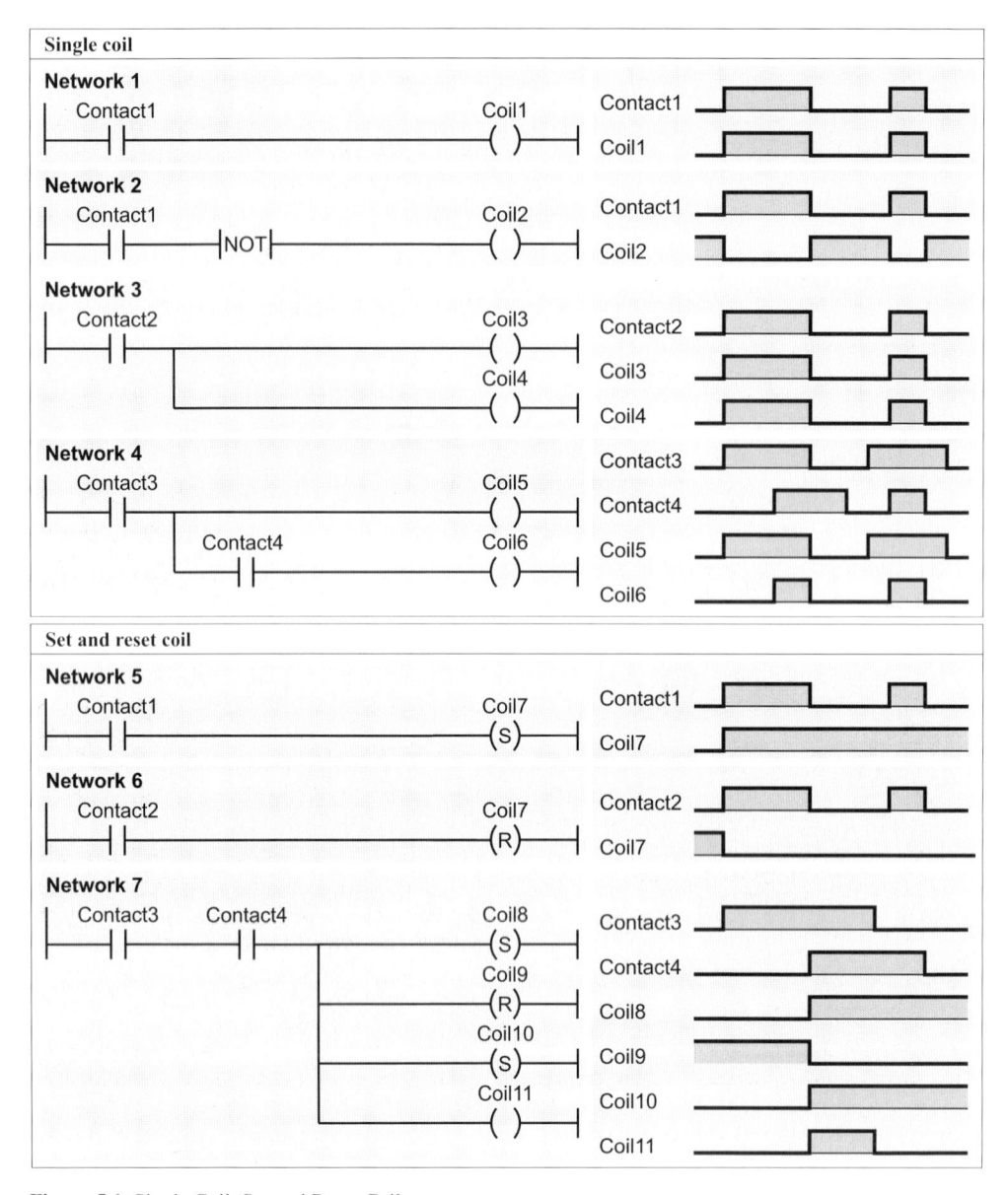

Figure 5.1 Single Coil, Set and Reset Coil

The function of the set and reset coils depends on the Master Control Relay. If the MCR is activated, the binary operand over the coil is not affected.

Please note that the operand used with a set or reset coil is usually reset at startup (complete restart). In special cases, the signal state is retained. This depends on the startup mode (for example warm restart), on the operand used (for example static local data), and on the settings in the CPU (for example retentive characteristics).

You can arrange several set and reset coils in any combination and together with single coils in the same rung (Network 7). To achieve clarity in your programming, it is advisable to group the set and reset coils affecting an operand together in pairs, and to use them only once in each case. You should also avoid additionally controlling these operands with a single coil.

As with the single coil, you can also arrange contacts after the branch and before a set and reset coil.

5.1.3 Memory Box

The functions of the set and reset coil are summarized in the box of a memory function. The common binary operand is located over the box. Input S of the box corresponds here to the set coil, input R to the reset coil. The signal state of the binary operand assigned to the memory function is at output Q of the memory function.

There are two versions of the memory function: As SR box (reset priority) and RS box (set priority). Apart from the labeling, the boxes also differ from each other in the arrangement of the S and R inputs.

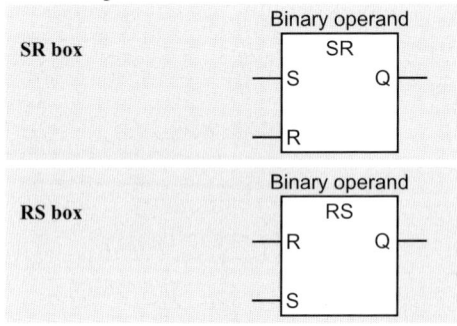

SR box

```
         Binary operand
              SR
          ┤S      Q├
          ┤R
```

RS box

```
         Binary operand
              RS
          ┤R      Q├
          ┤S
```

A memory function is set (or, more precisely, the binary operand over the memory box is set) if the set input has signal state "1" and the reset input has signal state "0". A memory function is reset if there is a "1" at the reset input and a "0" at the set input. Signal state "0" at both inputs has no effect on the memory function. If both inputs are "1" at the same time, the two memory functions respond differently: the SR memory function is reset and the RS memory function is set.

The function of the memory box depends on the Master Control Relay. If the MCR is active, the binary operand of a memory box is no longer affected.

Please note that the operand used with a memory function at startup (complete restart) is usually reset. In special cases, the signal state of a memory box is retained. This depends on the startup mode (for example warm restart), on the operand used (for example static local data), and on the settings in the CPU (for instance retentive characteristics).

SR memory function

In the SR memory box, the reset input has priority. Reset priority means that the memory function is or remains reset if power flows "simultaneously" in the set input and the reset input. The reset input has priority over the set input (Figure 5.2, Network 8).

Because the statements are executed in sequence, the CPU initially sets the memory operand because the set input is processed first, but resets it again when it processes the reset input. The memory operand remains reset while the rest of the program is processed.

If the memory operand is an output, this brief setting only takes place in the process-image output table, and the (external) output on the relevant output module remains unaffected. The CPU does not transfer the process-image output table to the output modules until the end of the program cycle.

The memory function with reset priority is the "normal" form of the memory function, since the reset state (signal state "0") is normally the safer or less hazardous state.

RS memory function

In the RS memory box, the set input has priority. Set priority means that the memory function is or remains set if power flows "simultaneously" in the set input and the reset input. The set input then has priority over the reset input (Figure 5.2, Network 9).

In accordance with the sequential execution of the instructions, the CPU resets the memory operand with the reset input first processed, but then sets it again when processing the set input. The memory operand remains set while the rest of the program is processed.

If the memory operand is an output, this brief resetting takes place only in the process-image output table, and the (external) output on the relevant output module is not affected. The

115

5 Memory Functions

Memory boxes

Network 8

Contact1 — S — Memory1 SR — Q
Contact2 — R

Network 9

Contact3 — R — Memory2 RS — Q
Contact4 — S

Memory boxes in the circuit

Network 10

Contact1 — Contact2 — S — Memory3 SR — Q — Contact3 — Coil12
Contact4 — R
Contact5
Contact6 — Contact7

Network 11

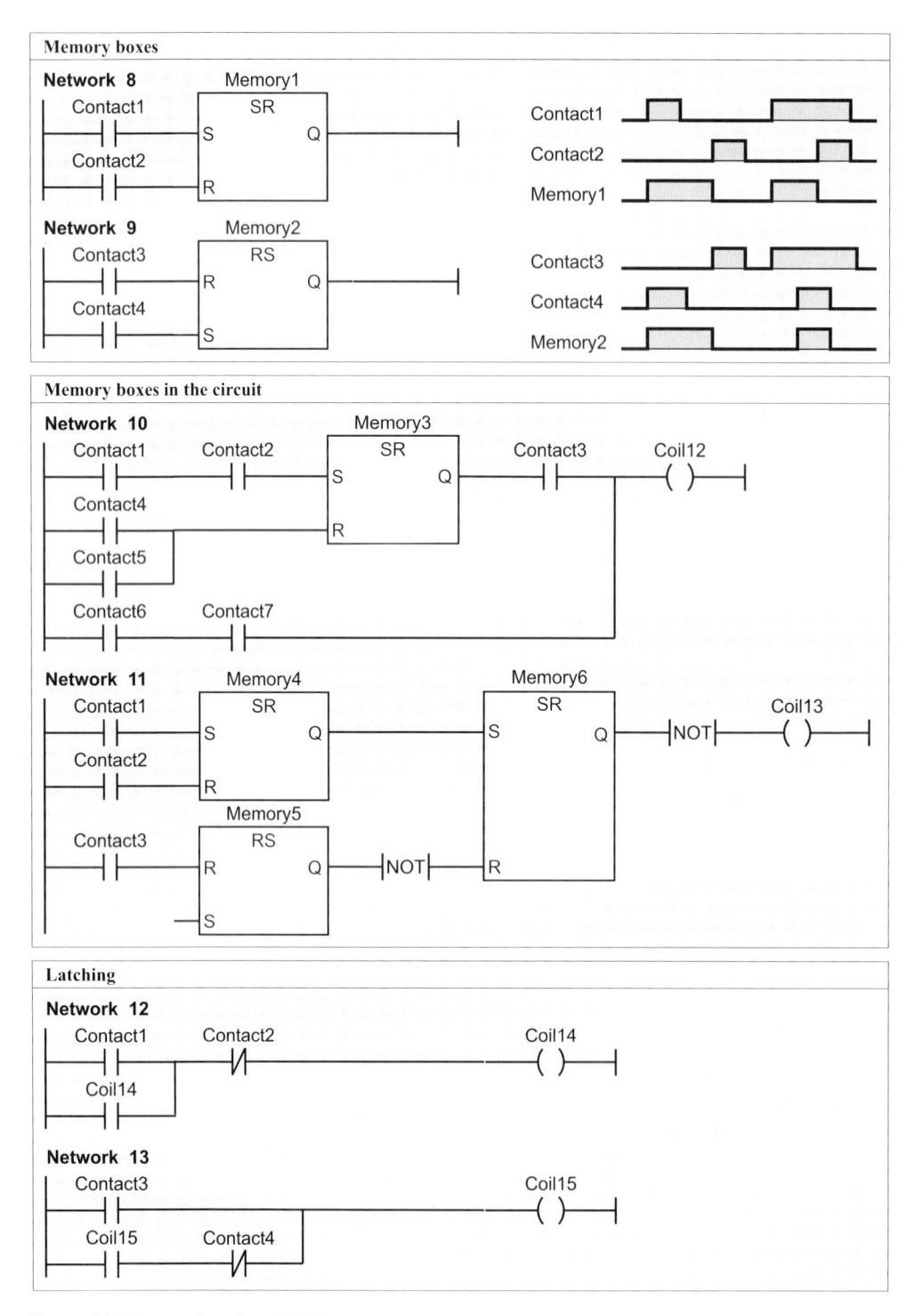

Contact1 — S — Memory4 SR — Q — S — Memory6 SR — Q — |NOT| — Coil13
Contact2 — R
Contact3 — R — Memory5 RS — Q — |NOT| — R
— S

Latching

Network 12

Contact1 — Contact2 (/) — Coil14 ( )
Coil14

Network 13

Contact3 — Coil15 ( )
Coil15 — Contact4 (/)

Figure 5.2 Memory Functions (LAD)

CPU does not transfer the process-image output table to the output modules until the end of the program cycle.

Set priority is the exception when using the memory function. It is used, for example, in the implementation of a fault message buffer if the still current fault message at the set input is to continue to set the memory function despite an acknowledgement at the reset input.

Memory function within a rung

You can also place a memory box within a rung. Contacts can be connected in series and in parallel both at the inputs and at the output (Figure 5.2, Network 10). It is also possible to leave the second input of a memory box unswitched. You can also connect several memory boxes together within one rung. You can arrange the memory boxes in series or in parallel (Network 11).

You can locate a memory function after a T-branch or in a branch that starts at the left power rail.

Memory function with latching

In a relay logic diagram, the memory function is usually implemented by latching the output to be controlled. This method can also be used when programming in ladder logic. However, it has the disadvantage, when compared with the memory box, that the memory function is not immediately recognizable.

Networks 12 and 13 in Figure 5.2 show both types of memory function, set priority and reset priority, using latching. The principle of latching is a simple one. The binary operand controlled with the coil is scanned, and this scan (the "contact of the coil") is connected in parallel to the set condition. If *Contact1* closes, *Coil14* energizes and closes the contact parallel to *Contact1*. If *Contact1* now opens again, *Coil14* remains energized. *Coil14* de-energizes if *Contact2* opens. If signal state "1" is present at both *Contact1* and *Contact2*, power does not flow into the coil (reset priority). This situation looks different in the lower network: If signal state "1" is present at both *Contact3* and *Contact4*, power flows into the coil (set priority).

5.2 FBD Boxes

In FBD, the memory boxes are used in conjunction with binary logic operations in order to influence the signal states of binary operands with the aid of the result of the logic operation (RLO) generated in the CPU.

Available memory functions are:

▷ The assign box for dynamic control

▷ The boxes set and reset as individually programmed memory functions

▷ The boxes RS and SR as full-fledged memory functions

▷ the midline output box as intermediate buffer

▷ the boxes P and N as edge evaluations of the result of the logic operation

▷ the boxes POS and NEG as edge evaluations of operands

The boxes for midline outputs and edge evaluations are discussed in detail in subsequent chapters.

You can use the memory functions described in this chapter in conjunction with all binary operands. There are restrictions when using temporary local data bits as edge memory bits.

The examples shown in this chapter are also represented on the diskette enclosed with the book in function block FB 105 of the "Basic Functions", library "FBD_Book". For incremental programming, you will find the program elements for the memory functions in the program element catalog (with VIEW → CATALOG [Ctrl – K] or with INSERT → PROGRAM ELEMENTS) under "Bit Logic".

5.2.1 Assign

The assign box as terminator of a rung assigns the result of the logic operation directly to the operand adjacent to the box. If the RLO is "1" at the input of the assign box, the binary operand is set; if the RLO is "0", the operand is reset. The function of the assign box depends on the Master Control Relay (MCR): If the MCR is activated, signal state "0" is assigned to the binary operand over the box.

5 Memory Functions

Assign Binary operand
—[=]

A number of examples in Figure 5.3 explain how the assign box works.

Network 1: The operand Output1 directly assumes the signal state of the operand Input1.

Network 2: You can use negation to reverse the function of the assign box.

Network 3: You can direct the RLO to several boxes simultaneously by inserting a T-branch and arranging the boxes with the relevant operands one below the other ("multiple output"). All operands over the boxes respond in the same way.

Network 4: You can insert binary functions between the T-branch and the terminating box, thus expanding a logic operation by additional function boxes.

You will find additional examples for the assign box in Chapter 4.2, "Binary Logic Operations (FBD)".

5.2.2 Set and Reset Box

Set and reset boxes also terminate a logic operation. These boxes are activated only when the result of the logic operation going into the box is "1".

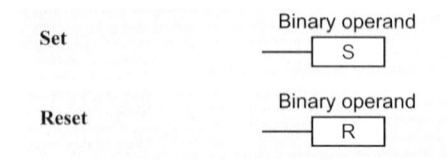

Set   Binary operand
—[S]

Reset   Binary operand
—[R]

If the RLO going into the set box is "1", the operand over the box is set to signal state "1". If the RLO going into the reset box is "1", the operand over the box is set to signal state "0". If the RLO going into the set or reset box is "0", the binary operand remains unaffected. The function of the set and reset boxes depends on the Master Control Relay. If the MCR is activated, the binary operand over the box is not affected.

Figure 5.3 shows several examples of how the set and reset boxes work.

Network 5: The operand *Output7* is set when the operand *Input1* is "1". *Output7* remains set when *Input1* returns to signal state "0".

Network 6: The operand *Output7* is reset when the operand *Input2* is "1". *Output7* remains reset when *Input2* returns to signal state "0".

Network 7: You can arrange several set and reset boxes in any combination and together with assign boxes in the same box after a T-branch. As with the assign box, you can also program binary functions after the T-branch and before a set and reset box.

To achieve clarity in your programming, it is advisable to group the set and reset boxes affecting an operand in pairs, and to use them only once in each case. You should also avoid controlling these operands with an assign box.

Please note that the operand used with a set or reset box is usually reset on startup (complete restart). In special cases, the signal state is retained. This depends on the startup mode (for instance warm restart), on the operand used (for example static local data), and on the settings in the CPU (such as retentive characteristics).

5.2.3 Memory Box

The functions of the set and reset boxes are summarized in the box of a memory function. The common binary operand is located over the box. Input S of the box corresponds here to the set box, and input R to the reset box. The signal state of the binary operand assigned to the memory function is at output Q of the memory function.

There are two versions of the memory function: As SR box (reset priority) and as RS box (set priority). Apart from the labeling, the boxes also differ from each other in the arrangement of the S and R inputs.

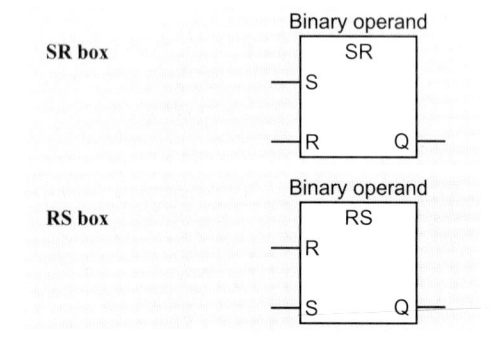

SR box   Binary operand
   SR
 —S
 —R   Q—

RS box   Binary operand
   RS
 —R
 —S   Q—

5.2 FBD Boxes

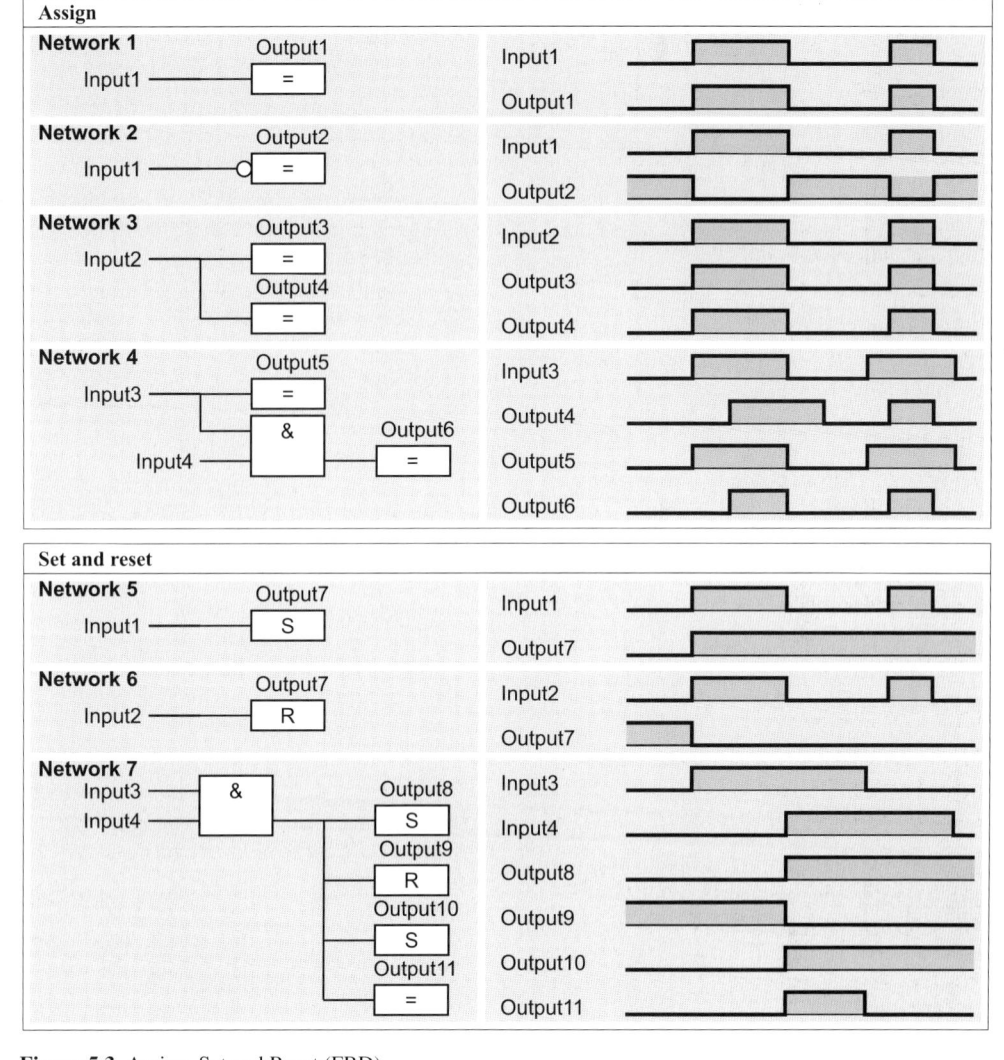

Figure 5.3 Assign, Set and Reset (FBD)

A memory function is set (or, more precisely, the binary operand over the memory box is set) when the set input is "1" and the reset input is "0". A memory function is reset when the reset input is "1" and the set input is "0". Signal state "0" at both inputs has no effect on the memory function. If both inputs have signal state "1" simultaneously, the two memory functions respond differently: the SR memory function is reset and the RS memory function is set. The function of the memory box depends on the Master Control Relay. If the MCR is active, the binary operand of a memory box is no longer affected.

Please note that the operand used with a memory function is normally reset on startup (complete restart). In special cases, the signal state of a memory box is retained. This depends on the startup mode (for instance warm restart), on the operand used (for example static local data), and on the settings in the CPU (such as retentive characteristics).

119

SR memory function

In the SR memory box, the reset input has priority. Reset priority means that the memory function is or remains reset if the RLO is "1" at the set and reset inputs "simultaneously". The reset input then has priority over the set input (Figure 5.4, Network 8).

Because the statements are executed in sequence, the CPU first sets the memory operand because the set input is processed first, but resets it again when it processes the reset input. The memory operand remains reset while the rest of the program is processed.

If the memory operand is an output, this brief setting takes place only in the process-image output table, and the (external) output on the relevant output module remains unaffected. The CPU does not transfer the process-image output table to the output modules until the end of the program cycle.

The memory function with reset priority is the "normal" form of the memory function, as the reset state (signal state "0") is usually the safer or less hazardous state.

RS memory function

In the RS memory box, the set input has priority. Set priority means that the memory function is or remains set if the RLO is "1" at the set and reset inputs "simultaneously". The set input then has priority over the reset input (Figure 5.4, Network 9).

Because the statements are executed in sequence, the CPU initially resets the memory operand because the reset input is processed first, then sets it again when the set input is processed. The memory operand remains set while the rest of the program is processed.

If the memory operand is an output, this brief resetting takes place only in the process-image output table, and the (external) output on the relevant output module is not affected. The CPU does not transfer the process-image output table to the output modules until the end of the program cycle.

Set priority is the exception rather than the rule. Set priority is used, for example, in the implementation of a fault message buffer if the still current fault message at the set input is to continue to set the memory function despite an acknowledgement at the reset input.

Memory function within a logic operation

You can also place a memory box within a logic operation. Binary functions can be programmed both at the inputs and at the output (Figure 5.4, Networks 10 and 11). It is also possible to leave the second input of a memory box unconnected. Within a logic operation, you can also interconnect several memory boxes. The memory boxes may be placed one behind the other or under one another after a T-branch.

5.3 Midline Outputs

Midline outputs are intermediate binary buffers in a ladder diagram or function block diagram. The RLO valid at the midline output is stored in the operand over the midline output. This operand can be scanned again at another point in the program, allowing you to also post-process the RLO valid at midline output elsewhere in the program.

The following binary operands are suitable for intermediate storage of binary results:

▷ You can use temporary local data bits if you only require the intermediate result within the block. All code blocks have temporary local data.

▷ Static local data bits are available only within a function block; they store the signal state until they are reused, even beyond the block boundaries.

▷ Memory bits are available globally in a fixed CPU-specific quantity; for clarity of programming, try to avoid multiple use of memory bits (the same memory bits for different tasks).

▷ Data bits in global data blocks are also available throughout the entire program, but before they are used require the relevant data block to be opened (even if implied through mass addressing).

The function of the midline output depends on the Master Control Relay. If the MCR is acti-

5.3 Midline Outputs

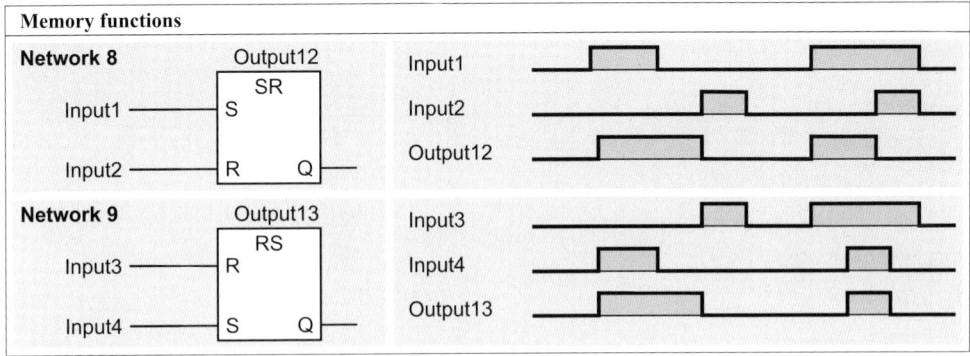

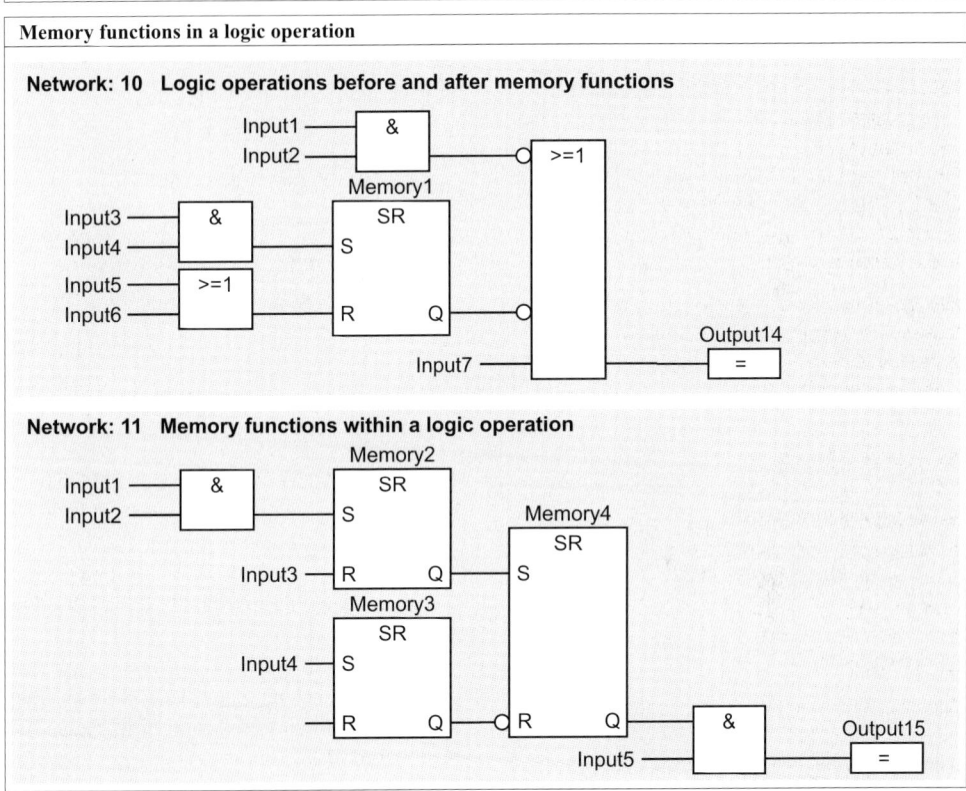

Figure 5.4 Memory Functions (FBD)

vated, the binary operand adjacent to the midline output is assigned signal state "0". The RLO is then "0" following the midline output (that is, there is no longer a "flow of power").

Note: You can replace the "scratchpad memory" with the temporary local data available in every block.

5.3.1 Midline Outputs in LAD

A midline output is a single coil within a rung. The RLO valid up to this point (the power that flows in the rung at this point) is stored in the binary operand over the midline output. The midline output itself has no effect on the power flow.

5 Memory Functions

Midline outputs (LAD)

Network: 14 Midline outputs (1)

```
   Contact1    Contact2    Midl_out1    Contact3      Coil16
  ──┤ ├───────┤ ├──────────(#)─────────┤ ├──────────( )──┤
   Contact4    Contact5
  ──┤ ├───────┤ ├──
```

Network: 15 Midline outputs (2)

```
   Midl_out1   Contact6                               Coil17
  ──┤ ├───────┤ ├──────────────────────────────────( )──┤
   Midl_out1   Contact7
  ──┤/├───────┤ ├──
```

Midline outputs (FBD)

Network: 12 Midline outputs (1)

```
  Input1 ──┐ &
  Input2 ──┘ ──┐ >=1
  Input3 ──┐ & │         Midl_out1
  Input4 ──┘ ──┘──────────  #  ──────┐ &
                                     │       Coil16
                           Input5 ───┘────────  =
```

Network: 13 Midline outputs (2)

```
  Midl_out1 ──┐ &
  Input6    ──┘ ──┐ >=1
  Midl_out1 ──○┐ & │       Coil17
  Input7    ───┘──┘─────────  =
```

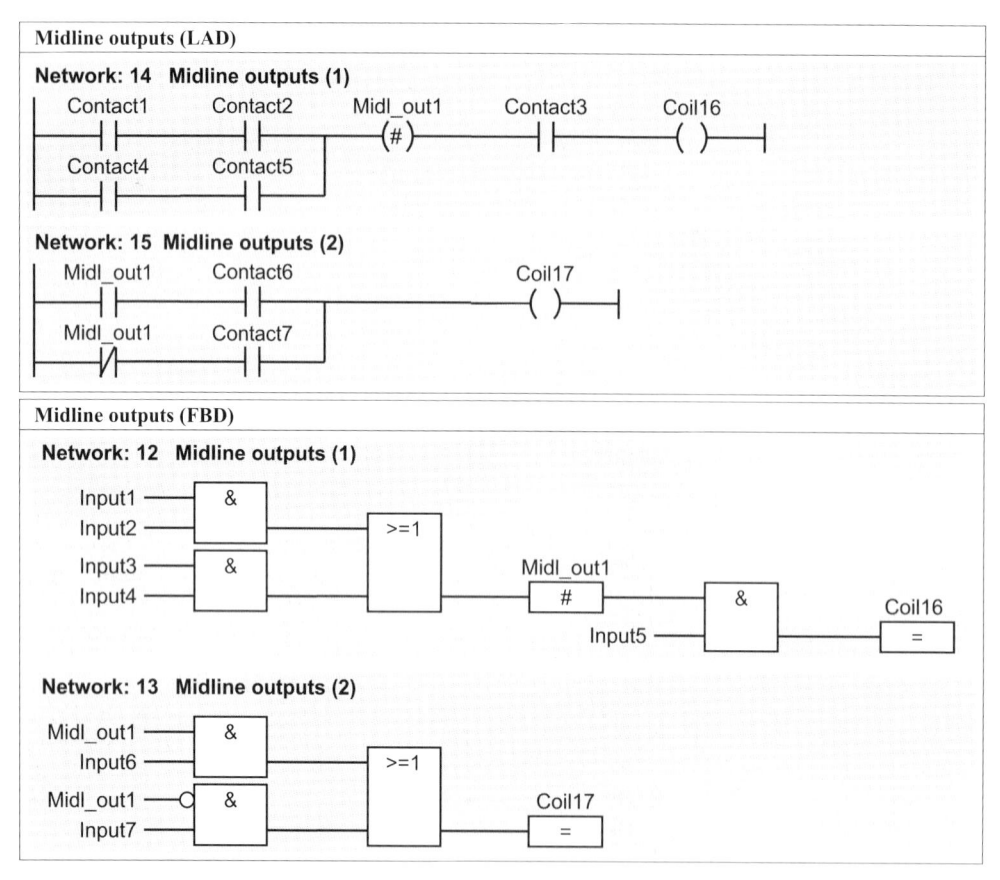

Figure 5.5 Midline Outputs

Midline output	Binary operand ───(#)───

You can scan the binary operand over the midline output at another point in the program with NO and NC contacts. Several midline outputs can be programmed in one rung.

You can place a midline output in a branch that starts at the left power rail. However, it must not be located directly at the power rail. A midline output may also follow a T-branch, but may not terminate a rung; the single coil is available for this purpose.

Figure 5.5 shows an example of how an intermediate result is stored in a midline output. The RLO from the circuit formed by *Contact1*, *Contact2*, *Contact4* and *Contact5* is stored in midline output *Midl_out1*. If the condition of the logic operation is fulfilled (power flows in the midline output) and if *Contact3* is closed, *Coil16* is energized. The RLO stored is used in two ways in the next network. On the one hand, a check is made to see if the condition of the logic operation was fulfilled and the bit logic combination made with *Contact6*, and on the other hand a check is made to see if the condition of the logic operation was not fulfilled and a bit combination made with *Contact7*.

5.3.2 Midline Outputs in FBD

A midline output is an assign box within a logic operation. The RLO valid up to this point is stored in the midline output over the midline output box.

	Binary operand
Midline output	— [#] —

You can check the binary operand over the midline output at another point in the program. Several midline outputs may be programmed in one logic operation. A midline output box must not terminate a logic operation; the assign box is available for this purpose.

Networks 12 and 13 in Figure 5.5 illustrate how an intermediate result is stored in a midline output. The RLO from the circuit formed by *Input1*, *Input2*, *Input3* and *Input4* is stored in midline output *Midl_out1*. If the condition of the logic operation is fulfilled and if the signal state of *Input5* is "1", *Output16* is activated. The stored RLO is used in two ways in the next network. On the one hand, a check is made to see if the condition of the logic operation was fulfilled and the bit logic combination made with *Input6*, and on the other hand a check is made to see if the condition of the logic operation was not fulfilled and a bit logic combination made with *Input7*.

5.4 Edge Evaluation

5.4.1 How Edge Evaluation Works

With an edge evaluation, you detect the change in a signal state, a signal edge. An edge is positive (rising) when the signal changes from "0" to "1". The opposite is referred to as a negative (falling) edge.

In a circuit diagram, the equivalent of an edge evaluation is the pulse contact element. If this pulse contact element emits a pulse when the relay is switched on, this corresponds to the rising edge. A pulse from the pulse contact element on switching off corresponds to a falling edge.

Detection of a signal edge (change in a signal state) is implemented in the program. The CPU compares the current RLO (the result of an input check, for example) with a stored RLO. If the two signal states are different, a signal edge is present.

The stored RLO is located in an "edge memory bit" (it does not necessarily have to be a memory bit). This must be an operand whose signal state must be available when the edge evaluation is again encountered (in the next program cycle), and which is not used elsewhere in the program. Memory bits, data bits in global data blocks, and static local data bits in function blocks are all suitable as operands.

The edge memory bit stored the "old" RLO with which the CPU last processed the edge evaluation. If a signal edge is now present, that is, if the current RLO differs from the signal state of the edge memory bit, the CPU corrects the signal state of the edge memory bit by assigning it the "value" of the "new" RLO. When the edge evaluation is next processed (usually in the next program cycle), the signal state of the edge memory bit is the same as that of the current RLO (if this has not change in the meantime), and the CPU no longer detects an edge.

A detected edge is indicated by the RLO after edge evaluation. If the CPU detects a signal edge, it sets the RLO to "1" after edge evaluation (power then flows). If there is no signal edge, the RLO is "0".

Signal state "1" after an edge evaluation therefore means "edge detected". Signal state "1" is present only briefly, usually only for the length of one program cycle. Since the CPU does not detect an edge in the next cycle (if the "input RLO" of the edge evaluation does not change), it sets the RLO back to "0" after edge evaluation.

Please note the performance characteristics of the edge evaluation when the CPU is switched on. If no edge is to be detected, the RLO prior to edge evaluation must be identical to the signal state of the edge memory bit when the CPU is switched on. Under certain circumstances, the edge memory bit must be reset in the start-up routine (depending on the required performance and on the operand used).

5 Memory Functions

Edge evaluations (LAD)

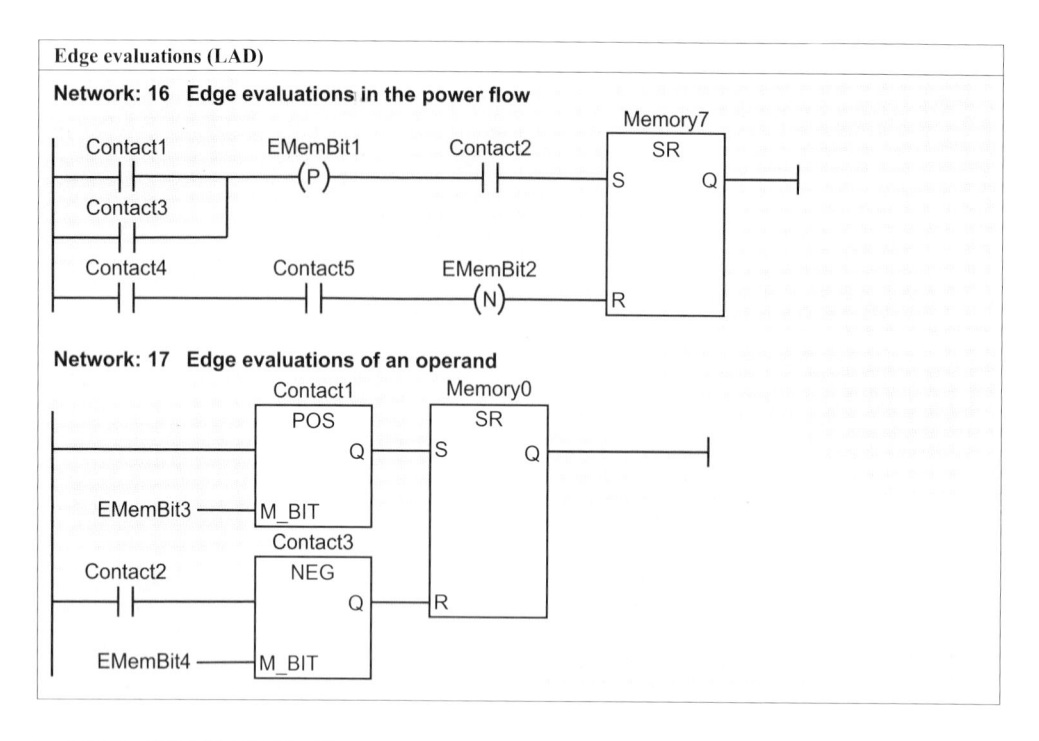

Network: 16   Edge evaluations in the power flow

```
      Contact1        EMemBit1         Contact2              Memory7
       ──┤├──┬─────────(P)───────────────┤├─────────────────┤ SR    │
           │                                                │ S    Q├──
      Contact3                                              │       │
       ──┤├──┘                                              │       │
                                                            │       │
      Contact4         Contact5         EMemBit2            │       │
       ──┤├─────────────┤├──────────────(N)────────────────┤R       │
```

Network: 17   Edge evaluations of an operand

```
                     Contact1         Memory0
                     ┌──────┐         ┌──────┐
                     │ POS  │         │  SR  │
                     │     Q├────────┤S     Q├──
      EMemBit3 ──────┤M_BIT │         │      │
                     └──────┘         │      │
                     Contact3         │      │
      Contact2       ┌──────┐         │      │
       ──┤├──────────┤ NEG  │         │      │
                     │     Q├────────┤R      │
      EMemBit4 ──────┤M_BIT │         └──────┘
                     └──────┘
```

Edge evaluations (FBD)

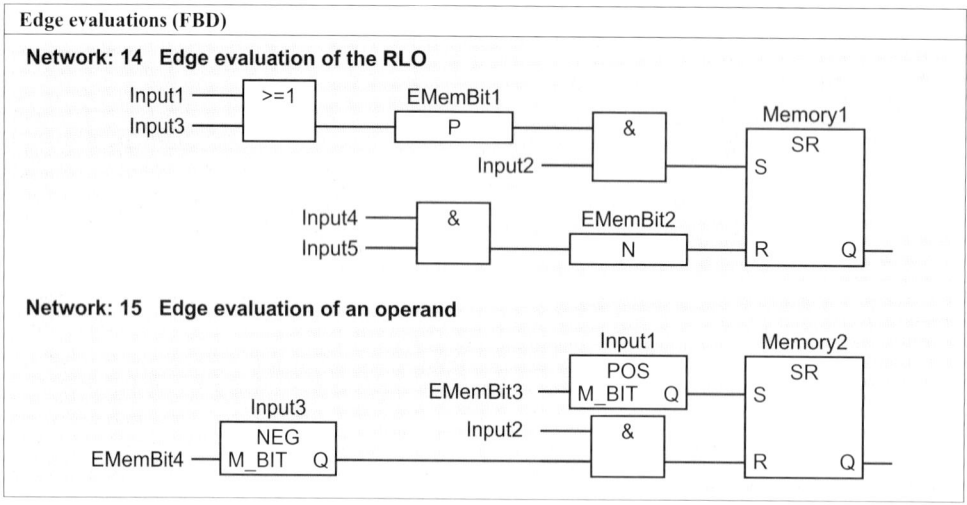

Network: 14   Edge evaluation of the RLO

```
      Input1 ──┐┌────┐    EMemBit1
               ││>=1 │    ┌────┐                    Memory1
      Input3 ──┘│    ├────┤ P  ├───┐┌───┐           ┌──────┐
                └────┘    └────┘   ││ & │           │  SR  │
                            Input2─┘│   ├──────────┤S     │
                                    └───┘           │      │
                                                    │      │
      Input4 ──┐┌───┐            EMemBit2           │      │
               ││ & ├────────────┌────┐             │      │
      Input5 ──┘└───┘            │ N  ├────────────┤R    Q├──
                                 └────┘             └──────┘
```

Network: 15   Edge evaluation of an operand

```
                                        Input1         Memory2
                                        ┌──────┐       ┌──────┐
                                        │ POS  │       │  SR  │
                      EMemBit3 ─────────┤M_BIT Q├─────┤S      │
                 Input3                  └──────┘      │      │
                 ┌──────┐        Input2 ─┐┌───┐        │      │
                 │ NEG  │                ││ & │        │      │
      EMemBit4 ─┤M_BIT Q├                └┤   ├───────┤R     Q├──
                 └──────┘                 └───┘        └──────┘
```

Figure 5.6 Edge Evaluations

124

5.4.2 Edge Evaluation in LAD

The LAD programming language provides four different elements for edge evaluation:

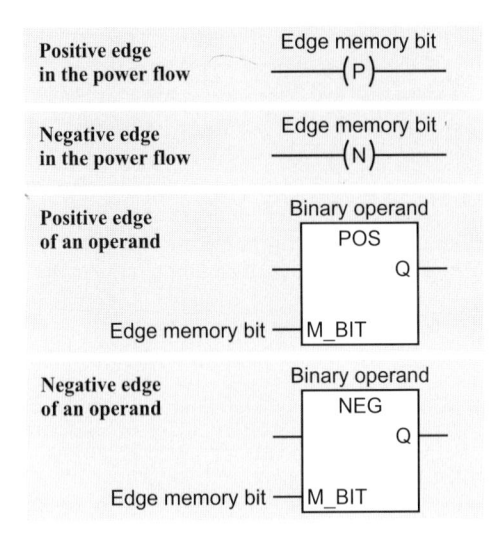

Positive edge in the power flow	Edge memory bit —(P)—
Negative edge in the power flow	Edge memory bit —(N)—
Positive edge of an operand	Binary operand / POS / Q / Edge memory bit — M_BIT
Negative edge of an operand	Binary operand / NEG / Q / Edge memory bit — M_BIT

You can process the RLO directly following an edge evaluation, that is to say, store it with a set coil, combine it with downstream contacts, or store it in a binary operand (a so-called "pulse memory bit").

You use a pulse memory bit when the RLO from the edge evaluation is to be used elsewhere in the program; it is, so to speak, the intermediate buffer for a detected edge (the pulse contact element in the circuit diagram). Operands suitable as impulse memory bit are memory bits, data bits in global data blocks, and temporary and static local data bits.

Edge evaluation in the power flow

An edge evaluation in the power flow is indicated by a coil that contains a P (for positive, rising edge) or an N (for negative, falling edge). Above the coil is an edge memory bit, a binary operand, in which the "old" RLO from the preceding edge evaluation is stored. An edge evaluation like this detects a change in the power flow from "power flowing" to "power not flowing" and vice versa.

The example in Figure 5.6 shows a positive and a negative edge evaluation in Network 16. If the parallel circuit consisting of *Contact1* and *Contact3* is fulfilled, the edge evaluation emits a brief pulse with *EmemBit1*. If *Contact2* is closed at this instant, *Memory7* is set. *Memory7* is reset again by a pulse from *EmemBit2* if the series circuit consisting of *Contact4* and *Contact5* interrupts the power flow.

You may program an edge evaluation with coil after a T-branch or in a branch that starts at the left power rail. It must not be placed directly at the left power rail.

Edge evaluation of an operand

LAD represents the edge evaluation of an operand using a box. Above the box is the operand whose signal state change is to be evaluated. The edge memory bit that stores the "old" signal state from the preceding program cycle is located at input M_BIT.

With the unlabeled input and the output Q, the edge evaluation is "inserted" in the rung instead of a contact. If power flows into the unlabeled input, output Q emits a pulse at an edge; if no power flows in this input, output Q is also always reset. You can arrange this edge evaluation in place of any contact, even in a parallel branch that does not begin at the left power rail.

Figure 5.6 shows the use of an edge evaluation of an operand in Network 17. The edge evaluation in the upper branch emits a pulse if the operand *Contact1* changes its signal state from "0" to "1" (positive edge). This pulse sets *Memory0*. The edge evaluation is always enabled by the direct connection of the unlabeled input to the left power rail. The power edge evaluation is enabled by *Contact2*. If it is enabled with "1" at this input, it emits a pulse if the binary operand *Contact3* changes its signal state from "1" to "0" (negative edge).

5.4.3 Edge Evaluation in FBD

The FBD programming language provides four different elements for edge evaluation:

5 Memory Functions

Positive edge of the RLO

```
      Edge memory bit
    ─────┤ P ├─────
```

Negative edge of the RLO

```
      Edge memory bit
    ─────┤ N ├─────
```

Positive edge of an operand

```
                Binary operand
                ┌─────────┐
                │   POS   │
Edge memory bit─┤M_BIT   Q├─
                └─────────┘
```

Negative edge of an operand

```
                Binary operand
                ┌─────────┐
                │   NEG   │
Edge memory bit─┤M_BIT   Q├─
                └─────────┘
```

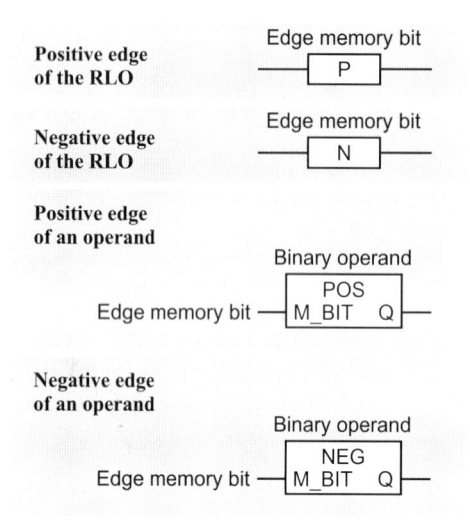

The RLO after an edge evaluation can be directly processed, for example stored with a set box, combined with subsequent binary functions, or assigned to a binary operand (a so-called "pulse memory bit"). A pulse memory bit is used when the RLO from the edge evaluation is to be processed elsewhere in the program; it is, as it were, the intermediate buffer for a detected edge. Operands suitable as pulse memory bits are memory bits, data bits in global data blocks, and temporary and static local data bits. An edge evaluation is not permitted after a T-branch.

Edge evaluation of the RLO

An edge evaluation of the RLO is indicated by a box that contains a P (for positive, rising edge) or an N (for negative, falling edge). Above the box is the edge memory bit, a binary operand containing the "old" RLO from the previous cycle. An edge evaluation like this detects a change of the RLO within a logic circuit from RLO "1" to RLO "0" and vice versa.

The example in Network 14 of Figure 5.6 shows a positive and a negative edge evaluation. When the OR function consisting of *Input1* and *Input3* is fulfilled, the edge evaluation emits a brief pulse with *EmemBit1*. If *Input2* is "1" at this instant, *Memory1* is set. *Memory1* is reset again by a pulse from *EmemBit2* when the AND function comprising *Input4* and *Input5* is no longer fulfilled.

Edge evaluation of an operand

The edge evaluation of an operand is located at the beginning of a binary logic operation. Over the box is the operand whose signal state change is to be evaluated. The edge memory bit that holds the "old" signal state from the last program cycle is located at input M_BIT. Output Q is "1" when the CPU detects a signal state change in the operand.

Network 15 in Figure 5.6 shows an edge evaluation of an operand. Upper edge evaluation POS emits a pulse when operand *Input1* goes from "0" to "1" (positive edge). This pulse sets *Memory2*. Lower edge evaluation NEG emits a pulse when binary operand *Input3* goes from "1" to "0" (negative edge). This pulse resets *Memory2* when operand *Input2* is "1".

5.5 Binary Scaler

A binary scaler has one input and one output. If the signal at the input of the binary scaler changes its state, for example from "0" to "1", the output also changes its signal state (Figure 5.7). This (new) signal state is retained until the next, in our example positive, signal state change. Only then does the signal state of the output change again. This means that half the input frequency appears at the output of the binary scaler.

5.5.1 Solution in LAD

There are many different ways of solving this task, two of which are presented below.

The first solution uses memory functions (Figure 5.8, Networks 18 and 19). If the signal state of the operand *Input_1* is "1", the operand *Output_1* is set (the operand *Memory_1* is still reset). If the signal state of the operand *Input_1* changes to "0", *Memory_1* is also set (*Output_1*

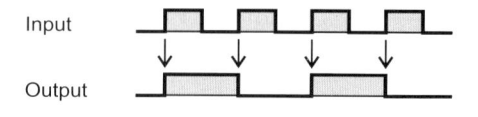

Figure 5.7 Pulse Diagram of a Binary Scaler

5.5 Binary Scaler

Binary scaler (LAD)

Network: 18 Binary scaler with memory boxes (1)

```
Input_1    Memory_1         Output_1
 ──┤ ├──────┤/├──────┬──S  SR  Q──┬──
Input_1    Memory_1  │              │
 ──┤ ├──────┤ ├──────┴──R           │
```

Network: 19 Binary scaler with memory boxes (2)

```
Input_1    Output_1         Memory_1
 ──┤/├──────┤ ├──────┬──S  SR  Q──┬──
Input_1    Output_1  │              │
 ──┤/├──────┤/├──────┴──R           │
```

Network: 20 Binary scaler with latches (1)

```
Input_2    Memory_2                    Output_2
 ──┤ ├──────┤/├────────────────────────( )──
Input_2    Output_2
 ──┤/├──────┤ ├──
```

Network: 21 Binary scaler with latches (2)

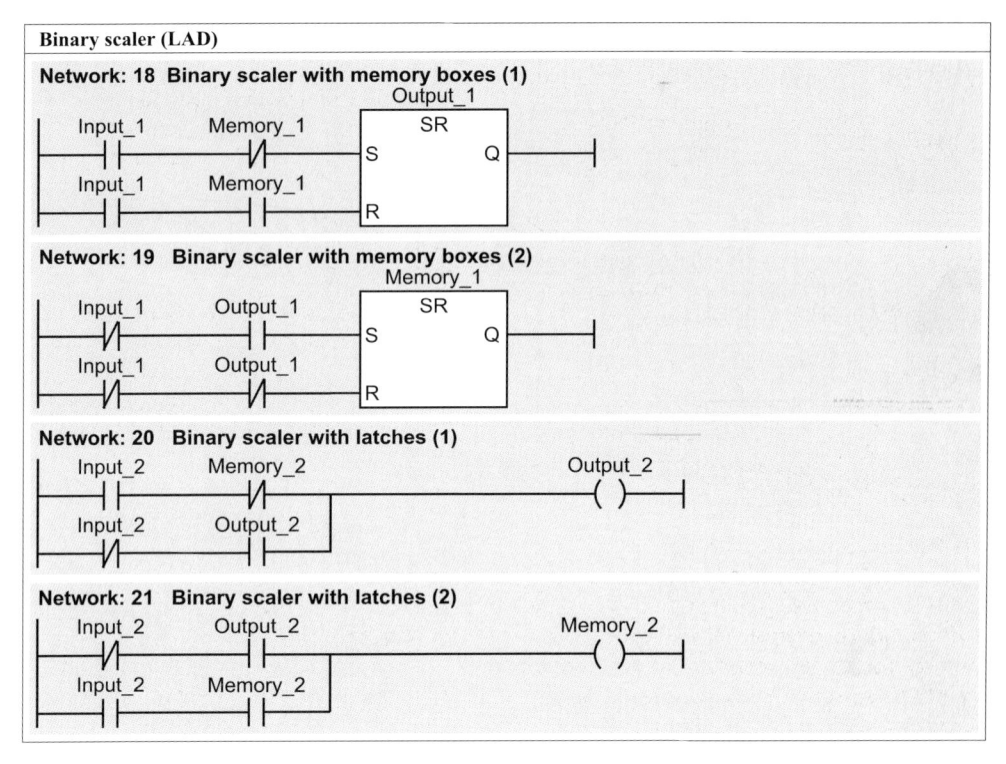

```
Input_2    Output_2                    Memory_2
 ──┤/├──────┤ ├────────────────────────( )──
Input_2    Memory_2
 ──┤ ├──────┤ ├──
```

Figure 5.8 Binary Scaler Examples (LAD)

is now "1"). If *Input_1* is "1" the next time around, *Output_1* is reset again (*Memory_1* is now "1"). If *Input_1* is once again "0", *Memory_1* is reset (since *Output_1* is now also reset). Now the basic state has been reached–changes to "0", *Memory_1* is also set (*Output_1* again after two input pulses and one output pulse.

The second solution uses the latching function (Networks 20 and 21) common in circuit diagrams. The principle is the same as in the first solution except that the reset condition – as is usual with latching – is "zero active".

5.5.2 Solution in FBD

There are many different ways of solving this task, two of which are presented below.

The first solution uses memory functions (Figure 5.9, Networks 16 and 17). If the signal state of the operand *Input* is "1", the operand *Output* is set (the operand *Memory* is still reset). If the signal state of the operand *Input* changes to "0", *Memory* is also set (*Output* is now "1"). If *Input* is "1" the next time around, *Output* is reset again (*Memory* is now "1"). If *Input* is once again "0", *Memory* is reset (since *Output* is now also reset). Now the basic state has been reached again after two input pulses and one output pulse.

The second solution uses an edge evaluation of the operand Input (Figure 5.9, Networks 18 to 20). If no edge is detected at Input, the RLO is "0" following edge evaluation and the jump instruction JCN is executed (you will find detailed descriptions of the jump operations in Chapter 16). In our example, the jump label is called "bin" and is in Network 20. It is here that the program scan is resumed if no edge is detected. The actual binary scaler is in Network 19: If *Output* is "0", it is set; if it is "1", it is reset. This network is processed only when an edge is detected at *Input*. In essence, every time an edge is detected at *Input*, *Output* changes its signal state.

5 Memory Functions

Binary scaler (FBD)

Network: 16   Binary scaler with memory boxes (1)

```
Input   —[ & ]——┐         Output
Memory  —o       │    ┌─ SR ─┐
                 └────┤ S    │
Input   —[ & ]──┐     │      │
Memory  —o      └─────┤ R   Q├─
                      └──────┘
```

Network: 17   Binary scaler with memory boxes (2)

```
Input   —o[ & ]──┐        Memory
Output  —        │    ┌─ SR ─┐
                 └────┤ S    │
Input   —o[ & ]──┐    │      │
Output  —o       └────┤ R   Q├─
                      └──────┘
```

Network: 18   Binary scaler with edge evaluation (1)

```
              Input0
           ┌── POS ──┐          bin
EMemBit0 ──┤ M_BIT  Q├──────[ JMPN ]
           └─────────┘
```

Network: 19   Binary scaler with edge evaluation (2)

```
                                Output0
Output0 ──o[ & ]──────────────[ = ]
```

Network: 20   (next network with entry point)

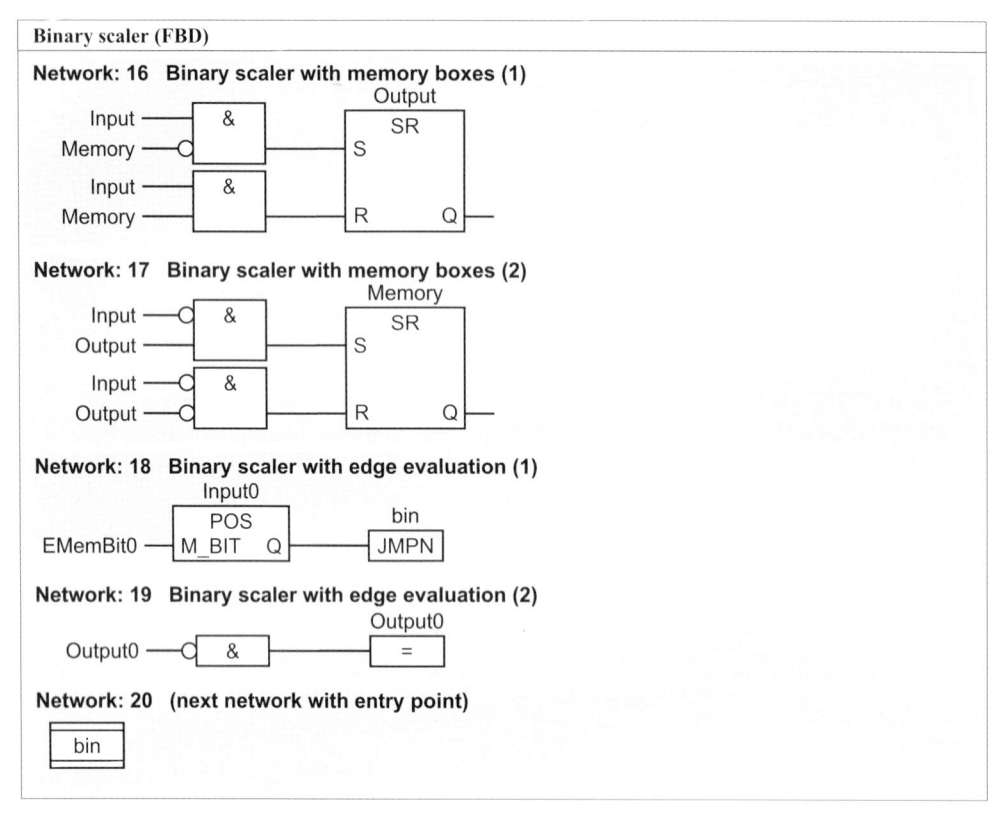

```
[ bin ]
```

Figure 5.9 Binary Scaler Examples (FBD)

5.6   Example of a Conveyor Control System

The following example of a functionally extremely simple conveyor belt control system illustrates the use of binary logic operations and memory functions in conjunction with inputs, outputs and memory bits.

Functional description

Parts are to be transported by conveyor belt, one crate or pallet per belt. The essential functions are as follows:

▷ When the belt is empty, the controller requests more parts by issuing the "ready-load" signal (ready to load).

▷ When the "Start" signal is issued, the belt starts up and transports the parts

▷ At the end of the conveyor belt, an "end-of-belt' sensor (for instance a light barrier) detects the parts, at which point the belt motor switches off and triggers the "ready_rem" signal (ready to remove)

▷ When the "continue" signal is issued, the parts are transported further until the "end-belt" (end-of-belt) sensors no longer detects them.

The example is programmed with inputs, outputs, and memory bits, and may be programmed in any block at any location. In this case, a function without function value was chosen as block.

Signals, symbols

A few additional signals supplement the functionality of the conveyor belt control system:

5.6 Example of a Conveyor Control System

Table 5.1 Symbol Table for the Example "Conveyor Belt Control System"

Symbol	Address	Data Type	Comment
Belt_control	FC 11	FC 11	Belt control system
Basic_st	I 0.0	BOOL	Set controllers to the basic state
Man_on	I 0.1	BOOL	Switch on conveyor belt motor
/Stop	I 0.2	BOOL	Stop conveyor belt motor (zero-active)
Start	I 0.3	BOOL	Start conveyor belt
Continue	I 0.4	BOOL	Acknowledgment that parts were removed
Light_barrier1	I 1.0	BOOL	(Light barrier) sensor signal "End of belt" for belt 1
/Mfault1	I 2.0	BOOL	Motor protection switch belt 1, zero-active
Readyload	Q 4.0	BOOL	Load new parts onto belt (ready to load)
Ready_rem	Q.4.1	BOOL	Remove parts from belt (ready to remove)
Belt_mot1_on	Q 5.0	BOOL	Switch on belt motor for belt 1
Load	M 2.0	BOOL	Load parts command
Remove	M 2.1	BOOL	Remove parts command
EM_Rem_N	M 2.2	BOOL	Edge memory bit for negative edge of "remove"
EM_Rem_P	M 2.3	BOOL	Edge memory bit for positive edge of "remove"
EM_Loa_N	M 2.4	BOOL	Edge memory bit for negative edge of "load"
EM_Loa_P	M 2.5	BOOL	Edge memory bit for positive edge of "load"

▷ Basic_st
Sets the controller to the basic state

▷ Man_on
Switches the belt on, regardless of conditions

▷ /Stop
Stops the conveyor as long as the "0" signal is present (an NC contact as sensor, "zero active")

▷ End_belt
The parts have reached the end of the belt

▷ /Mfault
Fault signal from the belt motor (e.g. motor protection switch); designed as "zero active" signal so that, for example, a wire break also produces a fault signal

We want symbolic addressing, that is, the operands are given names which we then use to write the program. Before entering the program, we create a symbol table (Table 5.1) containing the inputs, outputs, memory bits, and blocks.

Program for LAD

The example is located in a function block that you call in organization block OB 1 (selected from the Program Elements Catalog "FC Blocks") for processing in a CPU.

Here, the example is programmed with memory boxes. In Chapter 19, "Block Parameters", the same example is shown using latches. The program in Chapter 19 can be found in a function block with block parameters which can also be called as often as needed (for several conveyor belts).

When programming, the global symbols can also be used without quotation marks provided they do not contain any special characters. If a symbol does contain a special character (such as an umlaut or a space), it must be placed in quotation marks. In the compiled block, the editor indicates all global symbols by setting them in quotation marks.

Figure 5.10 shows the program for the conveyor control system. On the diskette supplied with the book, you can find this program in the

5 Memory Functions

"LAD_Book" library in function block FC 11 under "Conveyor Example".

Program for FBD

The example is located in a function which you call in organization block OB 1 (from the Program Elements Catalog "FC Blocks") for processing in a CPU.

In Chapter 19, "Block Parameters", the same example is shown using latches. The program in Chapter 19 can be found in a function block with block parameters which can also be called as often as needed (for several conveyor belts).

When programming, the global symbols can also be used without quotation marks provided they do not contain any special characters. If a symbol does contain a special character (such as an umlaut or a space), it must be placed in quotation marks. In the compiled block, the editor indicates all global symbols by setting them in quotation marks.

Figure 5.11 shows the program for the conveyor control system. On the diskette supplied with the book, you can find this program in the "FBD_Book" library in function block FC 11 under "Conveyor Example".

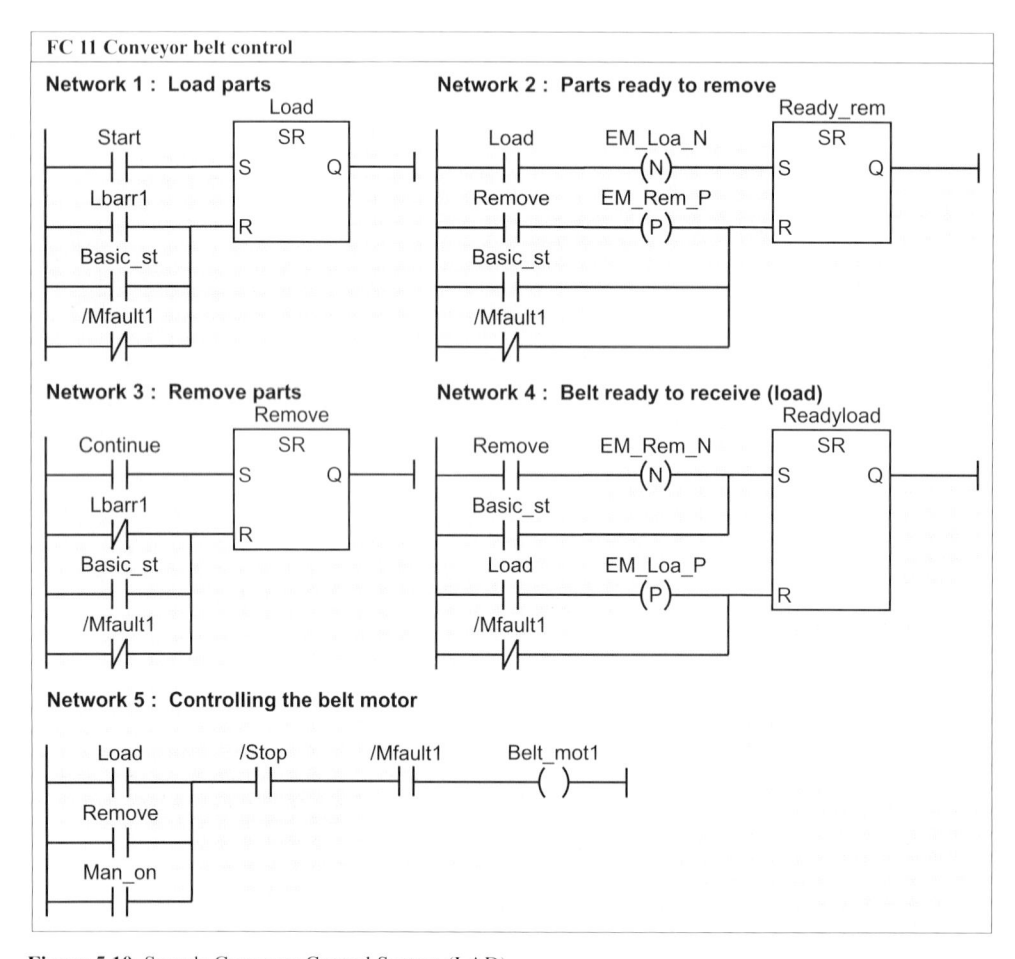

Figure 5.10 Sample Conveyor Control System (LAD)

5.6 Example of a Conveyor Control System

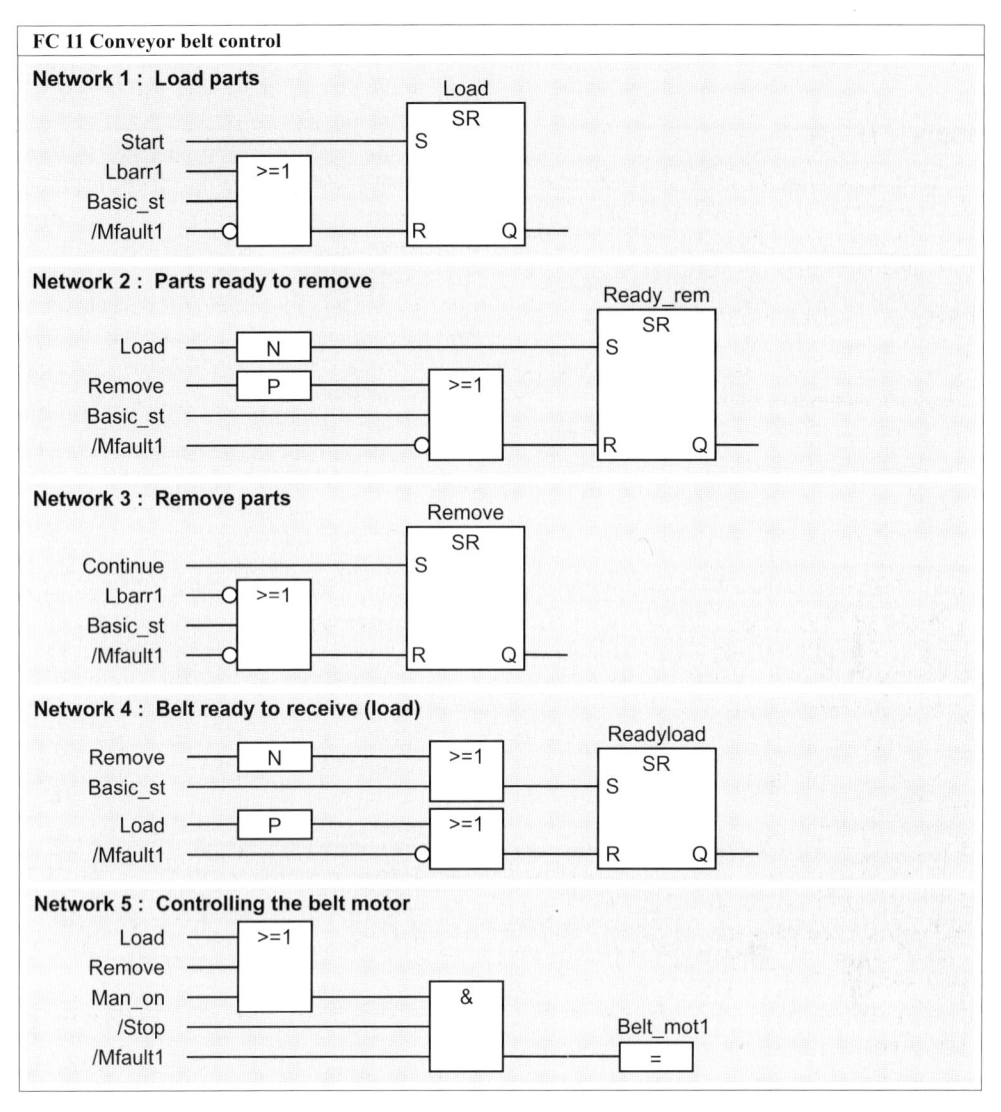

FC 11 Conveyor belt control

Network 1 : Load parts

```
Start ─────────────┐
Lbarr1 ──┐         │
         │ >=1 ├───S  Load
Basic_st ┤         │    SR
/Mfault1 ─○─┘      │
                   R    Q
```

Network 2 : Parts ready to remove

```
Load ──┤ N ├──────────────────S  Ready_rem
                                     SR
Remove ──┤ P ├──┐
                │ >=1 ├────
Basic_st ───────┤
/Mfault1 ──────○┘         R    Q
```

Network 3 : Remove parts

```
Continue ────────────────S  Remove
                              SR
Lbarr1 ──○──┐
            │ >=1 ├──
Basic_st ───┤
/Mfault1 ──○┘    R    Q
```

Network 4 : Belt ready to receive (load)

```
Remove ──┤ N ├────┐
                  │ >=1 ├──
Basic_st ─────────┤        S  Readyload
                                 SR
Load ──┤ P ├──────┐
                  │ >=1 ├──
/Mfault1 ────────○┘      R    Q
```

Network 5 : Controlling the belt motor

```
Load ────────┐
Remove ──────┤ >=1 ├──┐
Man_on ──────┤        │ & ├──── Belt_mot1
/Stop ───────────────┤             =
/Mfault1 ────────────┘
```

Figure 5.11 Sample Conveyor Control System (FBD)

6 Move Functions

The LAD and FBD programming languages provide the following move functions:

- MOVE box
  Copy operands and variables with elementary data types
- SFC 20 BLKMOV
  Copy data area
- SFC 21 FILL
  Fill data area
- SFC 81 UBLKMOV
  Uninterruptible copying of data area

The SFCs are system functions from the standard library *Standard Library* in the *System Function Blocks* program.

6.1 General

You use the move functions to copy information between the system memory, the user memory, and the user data area of the modules (Figure 6.1). Information is transferred via a CPU-internal register that functions as intermediate storage. This register is called accumulator 1. Moving information from memory to accumulator 1 is referred to as "loading" and moving from accumulator 1 to memory is called "transferring". The MOVE box contains both transfer paths. It moves information at input IN to accumulator 1 (load) and immediately following this from accumulator 1 to the operand at the output (transfer).

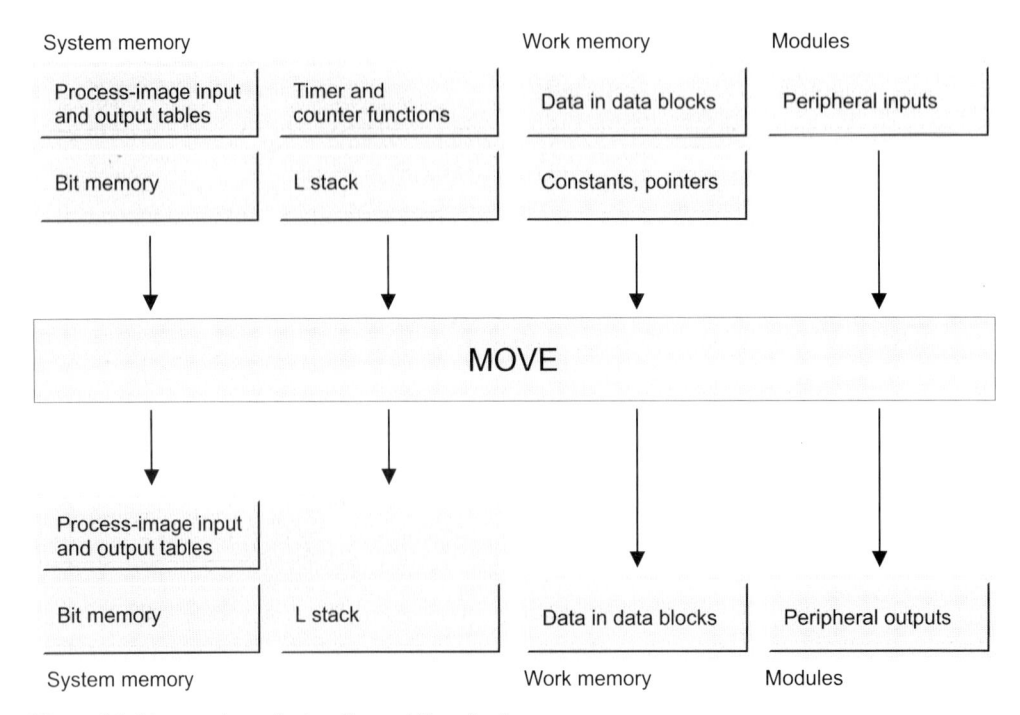

Figure 6.1 Memory Areas for Loading and Transferring

6.2 MOVE Box

6.2.1 Processing the MOVE Box

Representation

In addition to the enable input EN and the enable output ENO, the MOVE box has an input IN and an output OUT. At the input IN and the output OUT, you can apply all digital operands and digital variables of elementary data type (except BOOL). The variables at input IN and output OUT can have different data types.

MOVE box

LAD representation

```
       MOVE
    —| EN   ENO |—
     |          |
    —| IN   OUT |—
```

FBD representation

```
       MOVE
    —| EN   OUT |—
     |          |
    —| IN   ENO |—
```

The bits in the data formats are discussed in detail in Section 3.4, "Data Types".

For incremental programming, you will find the MOVE box in the Program Element Catalog (with VIEW → CATALOG [Ctrl – K] or with INSERT → PROGRAM ELEMENTS) under "Shift".

Different operand widths

The operand widths (byte, word, doubleword) at the input and the output of the MOVE box may vary. If the operand at the input is "less" than at the output, it is moved to the output operand right-justified and is padded at the left with zeroes. If the input operand is "greater" than the output operand, only that part of the input operand on the right that fits into the output operand is moved.

Figure 6.2 explains this. A byte or word at the input is loaded right-justified into accumulator 1 and the remainder is padded with zeroes. A byte or a word at output OUT is removed right-justified from accumulator 1.

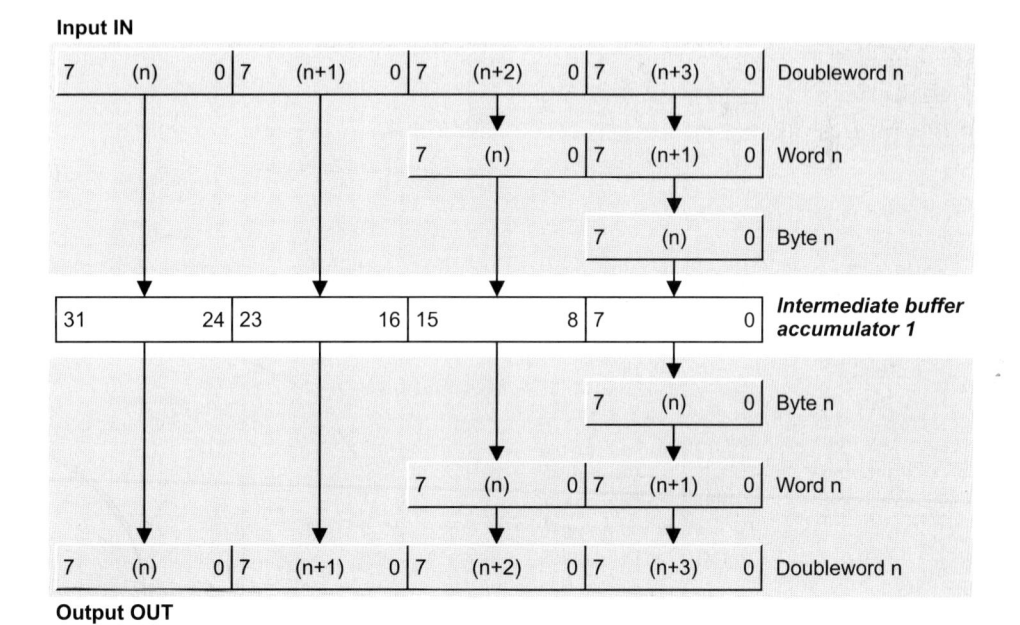

Figure 6.2 Moving Different Operand Widths

6 Move Functions

Function

The MOVE box moves the information of the operand at input IN to the operand at output OUT. The MOVE box only moves information when the enable input is "1" or is unused, and when the master control relay is deenergized.

If EN = "1" and the MCR is energized, zero is written to output OUT. With "0" at the enable input, the operand at output OUT is unaffected. MOVE does not report errors.

IF EN == "1" or not wired		
THEN		ELSE
ENO := "1"		
IF MCR enabled		
THEN	ELSE	
OUT := 0	OUT := IN	ENO := "0"

Example

The contents of input word IW 0 are moved to memory word MW 60.

LAD representation:

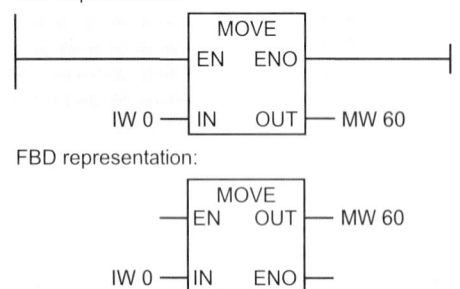

```
        MOVE
     ┤ EN   ENO ├
IW 0 ─┤ IN   OUT ├─ MW 60
```

FBD representation:

```
        MOVE
       ┤ EN   OUT ├─ MW 60
  IW 0 ─┤ IN   ENO ├
```

MOVE box in a rung (LAD)

You can arrange contacts in series and in parallel before input EN and after output ENO.

The MOVE box must only be placed in a branch that leads directly to the left power rail. This branch can also have contacts before the input EN and it need not be the top branch. With the direct connection to the left power rail you can connect MOVE boxes in parallel. When connecting boxes in parallel, you require a coil to terminate the rung. If you have not provided for error evaluation, assign a "dummy" operand to the coil, for example a temporary local data bit.

You can connect MOVE boxes in series. In doing so, the ENO output of the preceding box leads to the EN input of the following box.

If you arrange several MOVE boxes in one rung (parallel at the left power rail and then continuing in series), the boxes in the uppermost branch are processed first, from left to right, and then the boxes in the parallel branch from left to right, etc.

You can find examples of the move functions on the diskette supplied with the book (library "LAD_Book" in function block FB 106 of the "Basic Functions" program).

MOVE box in a logic circuit (FBD)

If you want to process the MOVE box in dependence on specific conditions, you can program binary logic circuits before the EN input. You can connect the ENO output with binary inputs of other functions; for instance, you can arrange MOVE boxes in series, whereby the ENO output of the preceding box leads to the EN input of the following box.

EN and ENO need not be wired.

You can find examples of the move functions on the diskette supplied with the book in the library "FBD_Book", FB 106 in the "Basic Functions" program.

6.2.2 Moving Operands

In addition to the operands mentioned in this chapter, you can also move timer and counter values (see Chapter 7, "Timers" and Chapter 8, "Counters"). Section 18.2, "Block Functions for Data Blocks", deals with using data operands.

Moving inputs

IB n   Moving an input byte

IW n   Moving an input word

ID n   Moving an input doubleword

On some CPUs, loading from or transferring to the process-image input table is permissible only for the input bytes that are also available as input module (see Section 1.5.2 "Process Image").

Moving outputs

QB n   Moving an output byte

QW n   Moving an output word

QD n   Moving an output doubleword

On some CPUs, loading from or transferring to the process-image output table is permissible only for the output bytes that are also available as output module (see Section 1.5.2 "Process Image").

Moving from the I/O

PIB n   Loading a peripheral input byte

PIW n   Loading a peripheral input word

PID n   Loading a peripheral input doubleword

PQB n   Transferring to a peripheral input word

PQW n   Transferring to a peripheral output word

PQD n   Transferring to a peripheral output doubleword

When moving in the I/O area, you can access different operands depending on the direction of the move. You specify I/O inputs (PIs) at the IN input of the MOVE box, and I/O outputs (PQs) at the OUT output.

When moving from the I/O to memory (loading), the input modules are accessed as peripheral inputs (PIs). Only the available modules may be addressed. Please note that direct loading from the I/O modules can move a different value than loading from the inputs of the module with the same address. While the signal states of the inputs corresponds to the values at the start of the program cycle (when the CPU updated the process image), the values loaded directly from the I/O modules are the current values.

The peripheral outputs (PQs) are used for transfers to the I/O. Only those addresses can be accessed that are also occupied by I/O modules. Transferring to I/O modules that have a process-image output table simultaneously updates that process-image output table, so there is no difference between identically addressed outputs and peripheral outputs.

Moving bit memory

MB n   Moving a memory byte

MW n   Moving a memory word

MD n   Moving a memory doubleword

Moving from and to the bit memory address area is always permissible, since the whole bit memory is in the CPU. Please note here the difference in bit memory area size on the various CPUs.

Moving temporary local data

LB n   Moving a local data byte

LW n   Moving a local data word

LD n   Moving a local data doubleword

Moving from and to the L stack is always allowed. Please note the information in Section 1.5.3, "Temporary Local Data".

6.2.3 Moving Constants

You may specify constant values only at the IN input of the MOVE box.

Moving constants of elementary data type

A fixed value, or constant, can be transferred to an operand. To enhance clarity, this constant can be transferred in one of several different formats. In Section 3.4.2, "Elementary Data Types", you will find an overview of all the different formats. All constants that can be moved using the MOVE box belong to the elementary data types. Examples:

B#16#F1	Moving a 2-digit hexadecimal number
–1000	Moving an INT number
5.0	Moving a REAL number
S5T#2s	Moving an S5 timer
TOD#8:30	Moving a time of day

Moving pointers

Pointers are a special form of constant used for calculating addresses in standard blocks. You can use the MOVE box to store these pointers in operands.

P#1.0	Moving an area-internal pointer
P#M2.1	Moving an area-crossing pointer

6.3 System Functions for Data Transfer

The following system functions are available for data transfer:

▷ SFC 20 BLKMOV
  Copy data area

▷ SFC 21 FILL
  Fill data area

▷ SFC 81 UBLKMOV
  Uninterruptible copying of a data area

These system functions have two parameters each of data type ANY (Table 6.1). In principle, you can assign any operand, any variable, or any absolute-addressed area to these parameters. If you use a variable of complex data type, it can only be a "complete" variable; components of a variable (individual array or structure components, for instance) are not permissible. Use the ANY pointer for specifying an absolute-addressed area.

6.3.1 ANY Pointer

You require the ANY pointer when you want to specify an absolute-addressed operand area as block parameter of type ANY. The general format of the ANY pointer is as follows:

P#[DataBlock]Operand Type Quantity

Examples:

P#M16.0 BYTE 8
  Area of 8 bytes beginning with MB 16

P#DB11.DBX30.0 INT 12
  Area of 12 words in DB 11 beginning with DBB 30

P#I18.0 WORD 1
  Input word IW 18

P#I1.0 BOOL 1
  Input I 1.0

Please note that the operand address in the ANY pointer must always be a bit address.

It makes sense to specify a constant ANY pointer when you want to access a data area for which you have not declared variables. In principle, you can assign variables or operands to an ANY parameter. For example, 'P#I1.0 BOOL 1' is identical to 'I 1.0' or the relevant symbolic address.

6.3.2 Copy Data Area

The system function **SFC 20 BLKMOV** copies the contents of a source area (parameter SLCBLK) to a destination area (parameter DSTBLK) in the direction of ascending addresses (incremental).

The following actual parameters may be assigned:

▷ Any variables from the operand areas for inputs (I), outputs (Q), bit memory (M), and data blocks (variables from global data blocks and from instance data blocks)

▷ Variables from the temporary local data (special circumstances govern the use of data type ANY)

Table 6.1 Parameters for SFC 20, 21 and 81

SFC	Parameter	Declaration	Data Type	Contents, Description
20	SRCBLK	INPUT	ANY	Source area from which data are to be copied
	RET_VAL	OUTPUT	INT	Error information
	DSTBLK	OUTPUT	ANY	Destination to which data are to be copied
21	BVAL	INPUT	ANY	Source area to be copied
	RET_VAL	OUTPUT	INT	Error information
	BLK	OUTPUT	ANY	Destination to which the source area is to be copied (including multiple copies)
81	SRCBLK	INPUT	ANY	Source area from which data are to be copied
	RET_VAL	OUTPUT	INT	Error information
	DSTBLK	OUTPUT	ANY	Destination to which data are to be copied

6.3 System Functions for Data Transfer

▷ Absolute-addressed data areas, which require specification of an ANY pointer

You cannot use SFC 20 to copy timers or counters, to copy information from or to the modules (operand area P), or to copy system data blocks (SDBs).

In the case of inputs and outputs, the specified area is copied regardless of whether or not the addresses actually reference input or output modules. You may also specify a variable or an area from a data block in load memory (a data block programmed with the keyword UNLINKED) as source area.

Source area and destination area may not overlap. If the source area and the destination area are of different lengths, the transfer is completed only up to the length of the shorter of the two areas.

Example (Figure 6.3, Network 4): Starting with memory byte MB 64, 16 bytes are to be copied to data block DB 124 starting from DBB 0.

6.3.3 Uninterruptible Copying of a Data Area

System function **SFC 81 UBLKMOV** copies the contents of a source area (parameter SRCBLK) to a destination area (parameter DSTBLK) in the direction of ascending addresses (incrementally). The copy function is uninterruptible, creating the possibility of increased response times to interrupts. A maximum of 512 bytes can be copied.

The following actual parameters may be assigned:

▷ Any variables from the operand areas for inputs (I), outputs (Q), bit memory (M), and data blocks (variables from global data blocks or from instance data blocks)

▷ Variables from the temporary local data (special circumstances govern the use of data type ANY)

▷ Absolute-addressed data areas, which require specification of an ANY pointer

SFC 81 cannot be used to copy timers or counters, to copy information from or to the modules (operand area P), or to copy system data blocks (SDBs) or data blocks in load memory (data blocks programmed with the keyword UNLINKED).

In the case of inputs and outputs, the specified area is copied regardless of whether their addresses reference input or output modules.

Source area and destination area may not overlap. If the source area and the destination area are of different lengths, the transfer is completed only up to the length of the shorter of the two areas.

6.3.4 Fill Data Area

System function **SFC 21 FILL** copies a specified value (source area) to a memory area (destination area) as often as required to fully overwrite the destination area. The transfer is made in the direction of ascending addresses (incrementally). The following actual parameters may be assigned:

▷ Any variables from the operand areas for inputs (I), outputs (Q), bit memory (M), and data blocks (variables from global data blocks and from instance data blocks)

▷ Absolute-addressed data areas, requiring specification of an ANY pointer

▷ Variables from the temporary local data of data type ANY (special circumstances apply)

SFC 21 cannot be used to copy timers or counters, to copy information from or to the modules (operand area P), or system data blocks (SDBs).

In the case of inputs and outputs, the specified area is copied regardless of whether the addresses actually reference input or output modules.

Source area and destination area may not overlap. The destination area is always fully overwritten, even when the source area is longer than the destination area or when the length of the destination area is not an integer multiple of the length of the source area.

Example (Figure 6.3, Network 5): The contents of memory byte MB 80 is to be copied 16 times to data block DB 124, beginning DBB 16.

6 Move Functions

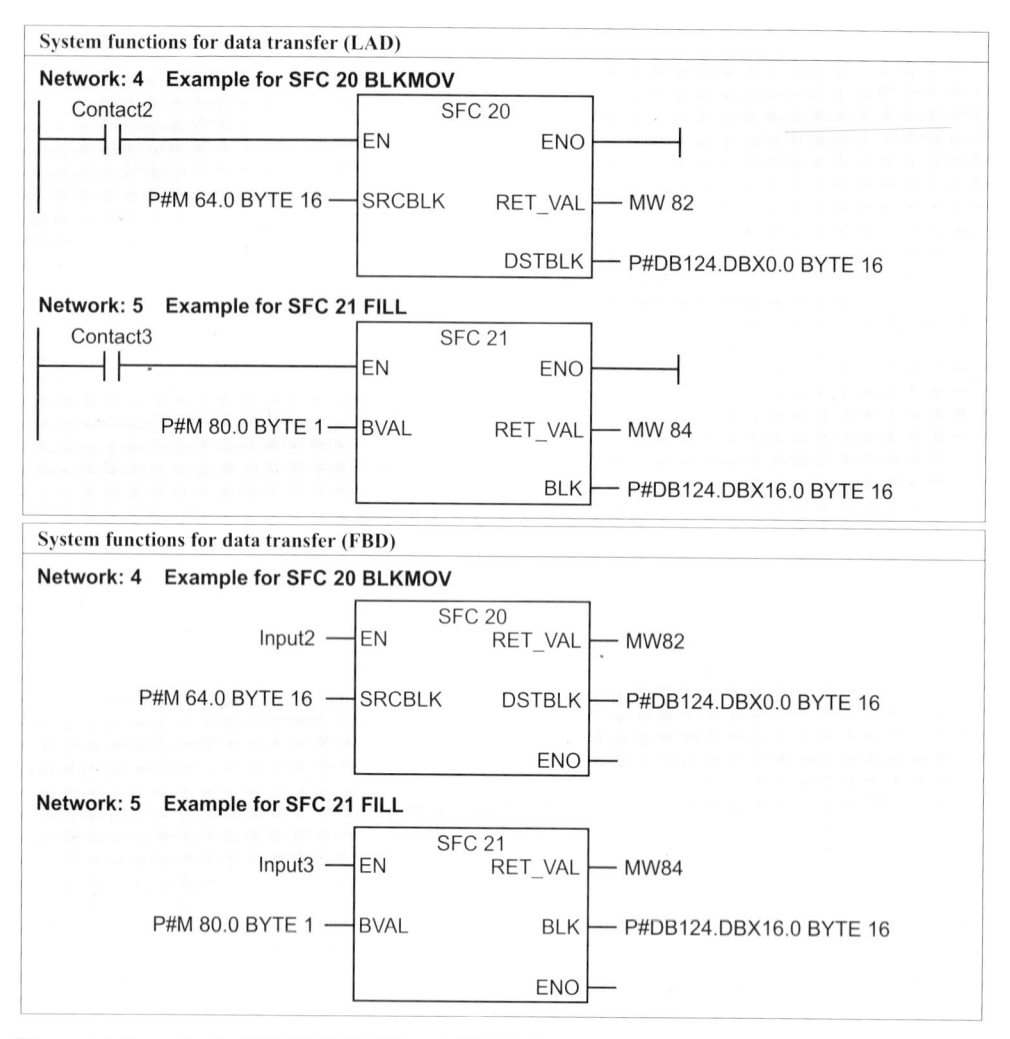

System functions for data transfer (LAD)
Network: 4 Example for SFC 20 BLKMOV
Contact2 —│ │— EN (SFC 20) ENO —
P#M 64.0 BYTE 16 — SRCBLK RET_VAL — MW 82
DSTBLK — P#DB124.DBX0.0 BYTE 16
Network: 5 Example for SFC 21 FILL
Contact3 —│ │— EN (SFC 21) ENO —
P#M 80.0 BYTE 1 — BVAL RET_VAL — MW 84
BLK — P#DB124.DBX16.0 BYTE 16

System functions for data transfer (FBD)
Network: 4 Example for SFC 20 BLKMOV
Input2 — EN (SFC 20) RET_VAL — MW82
P#M 64.0 BYTE 16 — SRCBLK DSTBLK — P#DB124.DBX0.0 BYTE 16
ENO —
Network: 5 Example for SFC 21 FILL
Input3 — EN (SFC 21) RET_VAL — MW84
P#M 80.0 BYTE 1 — BVAL BLK — P#DB124.DBX16.0 BYTE 16
ENO —

Figure 6.3 Examples for SFC 20 BLKMOV and SFC 21 FILL

7 Timers

The timers allow software implementation of timing sequences such as waiting and monitoring times, the measuring of intervals, or the generating of pulses.

The following timer types are available:

▷ Pulse timers

▷ Extended pulse timers

▷ On-delay timers

▷ Retentive on-delay timers

▷ Off-delay timers

You can program a timer complete as box or using individual program elements. When you start a timer, you specify the type of timer you want it to be and how long it should run; you can also reset a timer. A timer is checked by querying its status ("Timer running") or the current time value, which you can fetch from the timer in either binary or BCD code.

7.1 Programming a Timer

7.1.1 General Representation of a Timer

You can perform the following operations on a timer:

▷ Start a timer with specification of the time

▷ Reset a timer

▷ Check (binary) timer status

▷ Check (digital) timer value in binary

▷ Check (digital) time value in BCD

The box for a timer contains the coherent representation of all timer operations in the form of function inputs and function outputs (Figure 7.1). Above the box is the absolute or symbolic address of the timer. In the box, as a header, is the timer mode (S_PULSE means "Start pulse timer"). Assignments for the S and TW inputs are mandatory, while assignments for the other inputs and outputs are optional.

Individual program elements in LAD

You can also program a timer using individual program elements (Figure 7.2). The timer is then started via a coil. The timer mode is in the coil (SP = start pulse timer), and below the coil is the value, in S5TIME format, defining the duration. To reset a timer, use the reset coil, and use an NO or NC contact to check the status of the timer. Finally, you can store the current time value, in binary, in a word operand using the MOVE box.

Individual program elements in FBD

You can also program a timer using individual program elements (Figure 7.3). The timer is

Timer box
(in the example: pulse timer)

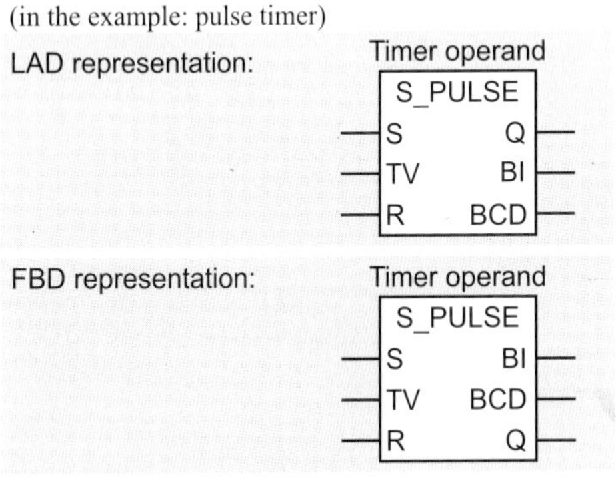

LAD representation: Timer operand
```
      S_PULSE
   ─┤S      Q├─
   ─┤TV    BI├─
   ─┤R    BCD├─
```

FBD representation: Timer operand
```
      S_PULSE
   ─┤S      BI├─
   ─┤TV   BCD├─
   ─┤R      Q├─
```

Name	Data Type	Description
S	BOOL	Start input
TV	S5TIME	Duration of time specification
R	BOOL	Reset input
BI	WORD	Current time value in binary
BCD	WORD	Current time value in BCD
Q	BOOL	Timer status

Figure 7.1 Timer in Box Representation

7 Timers

Starting a timer with time value specification (start coil with behavioral characteristics)	Timer operand —(SP)— Duration
Resetting a timer (reset coil)	Timer operand —(R)—
Checking the timer status (NO contact, NC contact)	Timer operand —│ │— Timer operand —│/│—
Reading a time value in binary (MOVE box)	MOVE / EN ENO / Timer operand —IN OUT— Digital operand

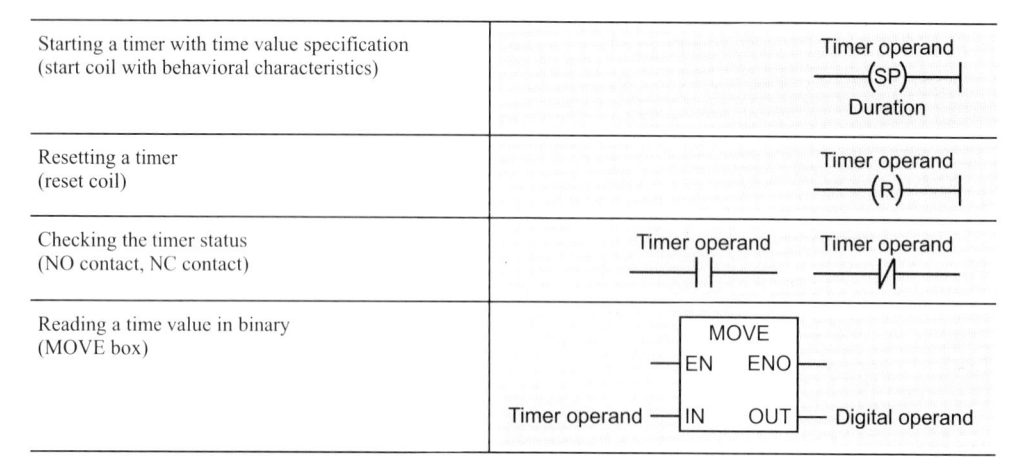

Figure 7.2 Individual Elements of a Timer (LAD)

then started via a simple box containing the timer mode (SP = start pulse timer). Below the box is the value, in S5TIME format, defining the duration. To reset a timer, use the reset box. You can scan the status of a timer directly or in negated form with any binary input. Finally, you can store the current time value, in binary, in a word operand using the MOVE box.

For incremental programming, you will find the timers in the Program Element Catalog (with VIEW → CATALOG [Ctrl – K] or INSERT → PROGRAM ELEMENTS) under "Timers".

7.1.2 Starting a Timer

A timer starts when the result of the logic operation (RLO) changes before the start input or before the start coil/box. Such a signal change is always required to start a timer. In the case of an off-delay timer, the RLO must change from "1" to "0"; all other timers start when the RLO goes from "0" to "1".

You can start a timer in one of five different modes (Figure 7.4). There is, however, no point in using any given timer in more than one mode.

Starting a timer with time value specification (start box with behavioral characteristics)	Timer operand / SP / Duration —TV
Resetting a timer (reset box)	Timer operand / R
Checking the timer status (direct or negated binary input)	Timer operand —☐ Timer operand —o☐
Reading a time value in binary (MOVE box)	MOVE / EN OUT— Digital operand / Timer operand —IN ENO

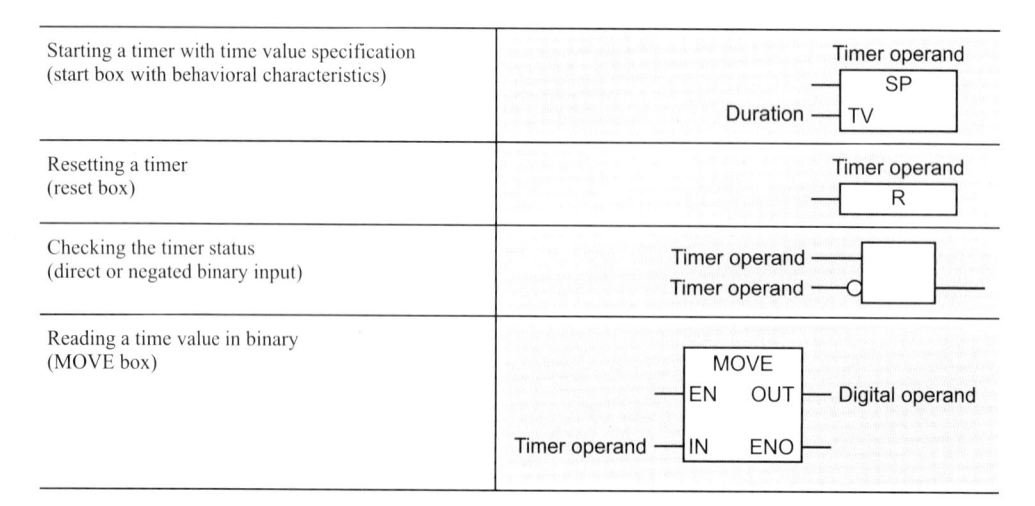

Figure 7.3 Individual Elements of a Timer (FBD)

7.1 Programming a Timer

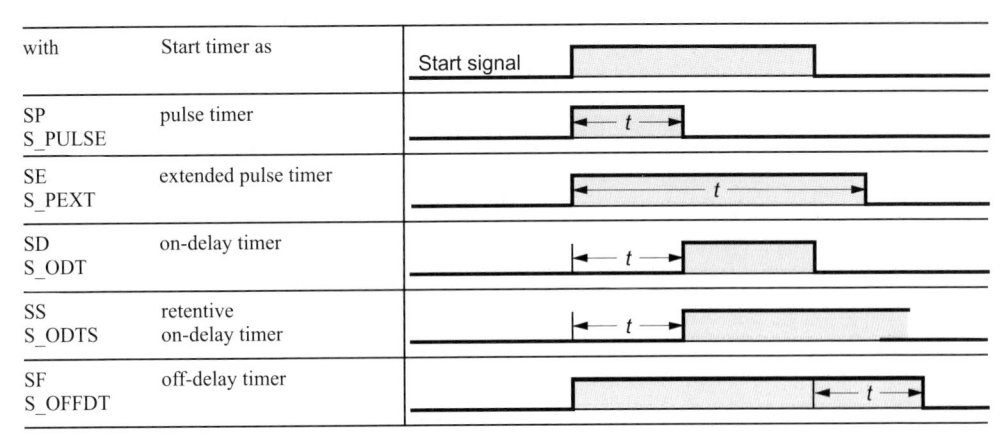

with	Start timer as	Start signal
SP S_PULSE	pulse timer	
SE S_PEXT	extended pulse timer	
SD S_ODT	on-delay timer	
SS S_ODTS	retentive on-delay timer	
SF S_OFFDT	off-delay timer	

Figure 7.4 Behavioral Characteristics of a Timer

7.1.3 Specifying the Duration of Time

The timer adopts the value below the coil/box or the value at input TV as the duration. You can specify the duration as constant, as word operand, or as variable of type S5TIME.

Specifying the duration as constant

S5TIME#10s Duration of 10 s
S5T#1m10ms Duration of 1 min + 10 ms

The duration is specified in hours, minutes, seconds and milliseconds. The range extends from S5TIME#10ms to S5TIME#2h46min30s (which corresponds to 9990 s). Intermediate values are rounded off to 10 ms. You can use S5TIME# or S5T# to identify a constant.

Specifying the duration as operand or variable

MW 20   Word operand containing the duration
"Time1"   Variable of data type S5TIME

The value in the word operand must correspond to data type S5TIME (see "Structure of the duration of time value", below).

Structure of the duration of time value

Internally, the duration is composed of the time value and the time base: duration = time value × time base. The duration is the time during which a timer is active ("timer running"). The time value represents the number of cycles for which the timer is to run. The time base defines the interval at which the CPU is to change the time value (Figure 7.5).

You can also build up a duration of time right in a word operand. The smaller the time base, the more accurate the actual duration. For example, if you want to implement a duration of one second, you can make one of three specifications:

Duration = 2001_{hex} Time base 1 s

Duration = 1010_{hex} Time base 100 ms

Duration = 0100_{hex} Time base 10 ms

The last of the these is the preferred one in this case.

15	12	11	8	7	4	3	0	Bit
		10^2		10^1		10^0		

Time value in binary-coded decimal (BCD)

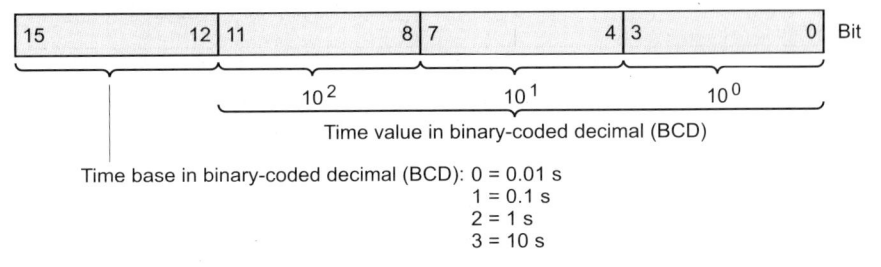

Time base in binary-coded decimal (BCD): 0 = 0.01 s
1 = 0.1 s
2 = 1 s
3 = 10 s

Figure 7.5 Description of the Bits in the Duration

When starting a timer, the CPU adopts the programmed time value. The operating system updates the timers at fixed intervals and independently of the user program, that is, it decrements the time value of all active timers as per the time base. When a timer reaches zero, it has run down. The CPU then sets the timer status (signal state "0" or "1", depending on the mode, or "type", of timer) and drops all further activities until the timer is restarted. If you specify a duration of zero (0 ms or W#16#0000) when starting a timer, the timer remains active until the CPU has processed the timer and discovered that the time has elapsed.

Timers are updated asynchronously to the program scan. As a result, it is possible that the time status at the beginning of a cycle is different than at the end of the cycle. If you use the timers at only one point in the program and in the suggested order (see below), the asynchronous updating will prevent the occurrence of malfunctions.

7.1.4 Resetting A Timer

LAD: A timer is reset when power flows in the reset input or in the reset coil (when the RLO is "1"). As long as the timer remains reset, a scan with an NO contact will return "0" and a scan with an NC contact will return "1".

FBD: A timer is reset when a "1" is present at the reset input. As long as the timer remains reset, a direct scan of the timer status will return "0" and a negated scan will return "1".

Resetting of a timer sets that timer and the time base to zero. The R input at the timer box need not be wired.

7.1.5 Checking a Timer

Checking the timer status (LAD)

The timer status is found at output Q of the timer box. You can also check the timer status with an NO contact (corresponds to output Q) or with an NC contact. The results of a check with an NO contact or with output Q differ according to the type of timer (see the description of the timer types, below). As is the case with inputs, for example, a check with an NC contact produces exactly the opposite check result as the one produced by a check with an NO contact. Output Q need not be used at the timer box.

Checking the timer status (FBD)

The timer status is available at output Q of the timer box. You can also check the timer status with a binary function input (corresponding to output Q). The results of a timer check depend on the type of timer involved (see the description of the timer types, below). Output Q need not be used at the timer box.

Checking the time value

Outputs BI and BCD provide the timer's time value in binary (BI) or binary-coded decimal (BCD). It is the value current at the time of the check (if the timer is active, the time value is counted from the set value down towards zero). The value is stored in the specified operand (transfer as with a MOVE box). You do not need to use these outputs at the timer box.

Direct checking of a time value

The time value is available in binary-coded decimal, and can be retrieved in this form from the timer. In so doing, the time base is lost and is replaced with "0". The value corresponds to a positive number in INT format. Please note: it is the time value that is checked, not the duration ! You can also program direct checking of a time value with the MOVE box.

Coded checking of a time value

You can also retrieve the binary time value in "coded" form from the timer. In this case, both the time value and the time base are available in binary-coded decimal. The BCD value is structured in the same way as for the specification of a time value (see above).

7.1.6 Sequence of Timer Operations

When you program a timer, you do not need to use all the operations available for it. You need use only the operations required to execute a particular function. Normally, these are the operations for starting a timer and for checking the timer status.

In order for a timer to behave as described in this chapter, it is advisable to observe the following order when programming with individual program elements:

▷ Start the timer

▷ Reset the timer

▷ Check the time value or the duration

▷ Check the timer status

Omit unnecessary elements when programming. If you observe the order shown above and the timer is started and reset "simultaneously", the timer will start but will be immediately reset. The subsequent timer check will fail to detect the fact that the timer was started.

7.1.7 Timer Box in a Rung (LAD)

You can connect contacts in series and in parallel before the start input and the reset input as well as after output Q.

The timer box itself may be located after a T-branch and in a branch that is directly connected to the left power rail. This branch can also have contacts before the start input and it need not be the uppermost branch.

You can find further examples for the representation and arrangement of timers in library "LAD_Book" on the diskette supplied with the book, FB 107 in the "Basic Functions" program.

7.1.8 Timer Box in a Logic Circuit (FBD)

You can program binary functions and memory functions before the start input and the reset input as well as after output Q.

The timer box and the individual elements for starting and resetting the timer may also be programmed after a T-branch.

You can find further examples for the representation and arrangement of timers in library "FBD_Book" on the diskette supplied with the book, FB 107 in the "Basic Functions" program.

7.2 Pulse Timer

Starting a pulse timer

The diagram in Figure 7.6 describes the characteristics of a timer when it is started as pulse timer and when it is reset. The description applies if you observe the order shown in Section 7.1.6 when programming with individual elements.

① When the signal state at the timer's start input changes from "0" to "1" (positive edge), the timer is started. It runs for the programmed duration as long as the signal state at the start input is "1". Output Q supplies signal state "1" as long as the timer runs.

With the start value as the starting point, the time value is counted down toward zero as per the time base.

② If the signal state at the timer's start input changes to "0" before the time has elapsed, the timer stops. Output Q then goes to "0". The time value shows how much longer the timer would have run had it not been prematurely interrupted.

Resetting a pulse timer

The resetting of a pulse time has a static effect, and takes priority over the starting of a timer (Figure 7.6).

③ Signal state "1" at the reset input of an active timer resets that timer. Output Q is then "0". The time value and the time base are also set to zero. If the signal state at the reset input goes from "1" to "0" while the signal state at the set input is still "1", the timer remains unaffected.

④ Signal state "1" at the reset input of an inactive timer has no effect.

⑤ If the signal state at the start input goes from "0" to "1" (positive edge) while the reset signal is still present, the timer starts but is immediately reset (shown by a line in the diagram). If the timer status check was programmed after the reset, the brief starting of the timer does not affect the check.

7 Timers

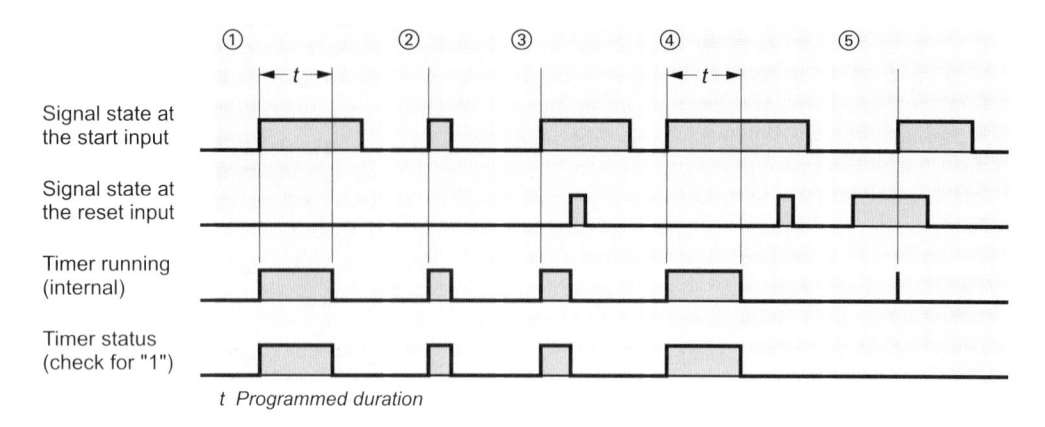

Figure 7.6 Behavioral Characteristics when Starting and Resetting a Pulse Timer

7.3 Extended Pulse Timer

Starting an extended pulse timer

The diagram in Figure 7.7 describes the behavioral characteristics of the timer after it is started and when it is reset. The description applies if you observe the order shown in Section 7.1.6 when programming with individual elements.

❶❷ When the signal state at the timer's start input goes from "0" to "1" (positive edge), the timer is started. It runs for the programmed duration, even when the signal state at the start input changes back to "0". A check for signal state "1" (timer status) returns a check result of "1" as long as the timer is running.

With the start value as starting point, the time value is counted down towards zero as per the time base.

❸ If the signal state at the start input goes from "0" to "1" (positive edge) while the timer is running, the timer is restarted with the programmed time value (the timer is "retriggered"). It can be restarted any number of times without first elapsing.

Resetting an extended pulse timer

The resetting of an extended pulse timer has a static effect, and takes priority over the starting of a timer (Figure 7.7).

❹ Signal state "1" at the timer's reset input while the timer is running resets the timer. A check for signal state "1" (timer status) returns a check result of "0" for a reset timer. The time value and the time base are also reset to zero.

❺ A "1" at the reset input of an inactive timer has no effect.

❻ If the signal state at the start input goes from "0" to "1" (positive edge) while the reset signal is present, the timer is started but is immediately reset (indicated by a line in the diagram). If the timer status check is programmed after the reset, the brief starting of the timer does not affect the timer check.

7.4 On-Delay Timer

Starting an on-delay timer

The diagram in Figure 7.8 describes the behavioral characteristics of the timer after it is started and when it is reset. The description applies

144

7.4 On-Delay Timer

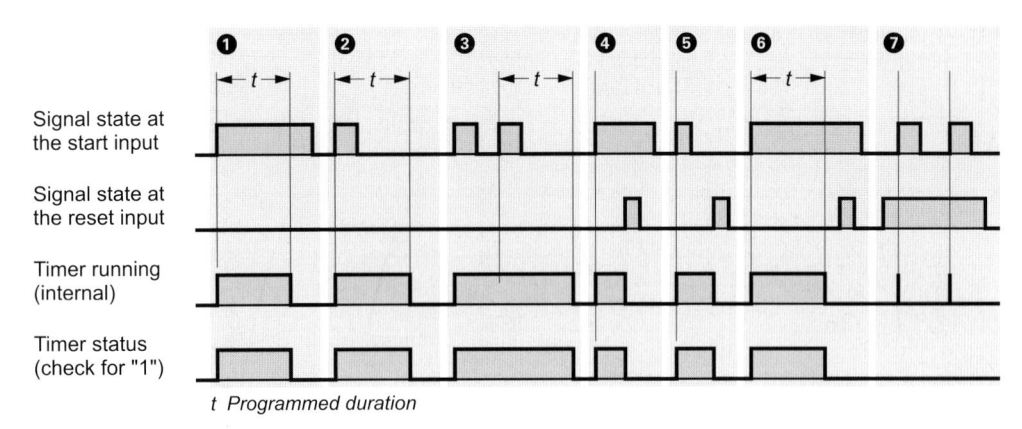

Figure 7.7 Behavioral Characteristics when Starting and Resetting an Extended Pulse Timer

if you observe the order shown in Section 7.1.6 when programming with individual elements.

① When the signal state at the timer's start input changes from "0" to "1" (positive edge), the timer is started. It runs for the programmed duration. Checks for signal state "1" return a check result of "1" when the time has duly elapsed and signal state "1" is still present at the start input (on-delay).

With the start value as starting point, the time value is counted down towards zero as per the time base.

② If the signal state at the start input changes from "1" to "0" while the timer is running, the timer stops. A check for signal state "1" (timer status) always returns a check result of "1" in such cases. The time value shows the amount of time still remaining.

Resetting an on-delay timer

The resetting of an on-delay timer has a static effect, and takes priority over the starting of the timer (Figure 7.8).

③④ Signal state "1" at the reset input resets the timer whether it is running or not. A check for signal state "1" (timer status) then returns a check result of "0", even when the timer is not running and signal state "1" is still present at the start input. Time value and time base are also set to zero.

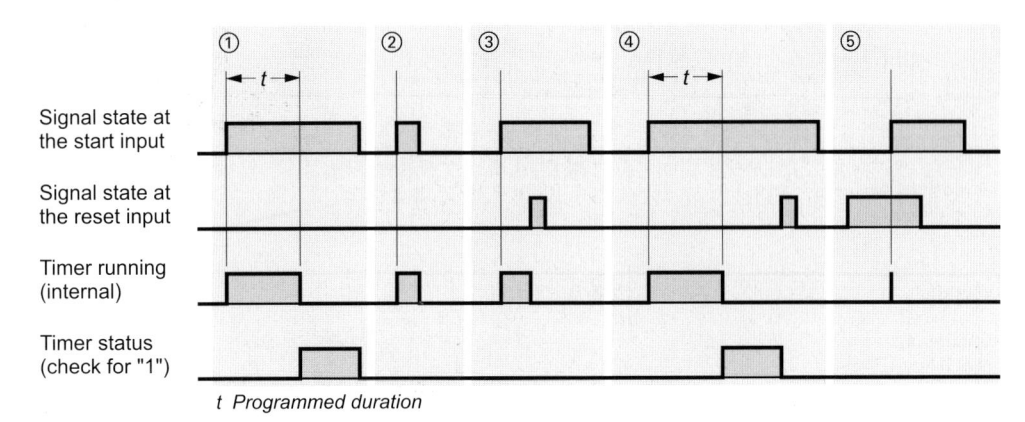

Figure 7.8 Behavioral Characteristics when Starting and Resetting an On-Delay Timer

145

7 Timers

A change in the signal state at the reset input from "1" to "0" while signal state "1" is still present at the start input has no effect on the timer.

⑤ If the signal state at the start input goes from "0" to "1" (positive edge) while the reset signal is present, the timer starts, but is immediately reset (indicated by a line in the diagram). If the timer status check is programmed after the reset, the brief starting of the timer does not affect the check.

7.5 Retentive On-Delay Timer

Starting a retentive on-delay timer

The diagram in Figure 7.9 describes the behavioral characteristics of the timer after it is started and when it is reset. The description applies if you observe the order shown in Section 7.1.6 when programming with individual elements.

❶❷ When the signal state at the timer's start input goes from "0" to "1" (positive edge), the timer is started. It runs for the programmed duration, even when the signal state at the start input changes back to "0". When the time has elapsed, a check for signal state "1" (timer status) returns a check result of "1" regardless of the signal state at the start input. A check result of "0" is not returned until the timer has been reset, regardless of the signal state at the start input. With the start value as starting point, the time value is counted down towards zero as per the time base.

❸ If the signal state at the start input changes from "0" to "1" (positive edge) while the timer is running, the timer restarts with the programmed time value (the timer is "retriggered"). It can be restarted any number of times without first having to run down.

Resetting a retentive on-delay timer

The resetting of a retentive on-delay timer has a static effect, and takes priority over the starting of the timer (Figure 7.9).

❹❺ Signal state "1" at the reset input resets the timer, regardless of the signal state at the start input. A check for signal state "1" (timer status) then returns a check result of "0". The time value and the time base are also set to zero.

❻ If the signal state at the start input goes from "0" to "1" (positive edge) while the reset signal is present, the timer starts, but is immediately reset (indicated by a line in the diagram). If the timer status check is programmed after the reset, the brief starting of the timer has no effect on the check.

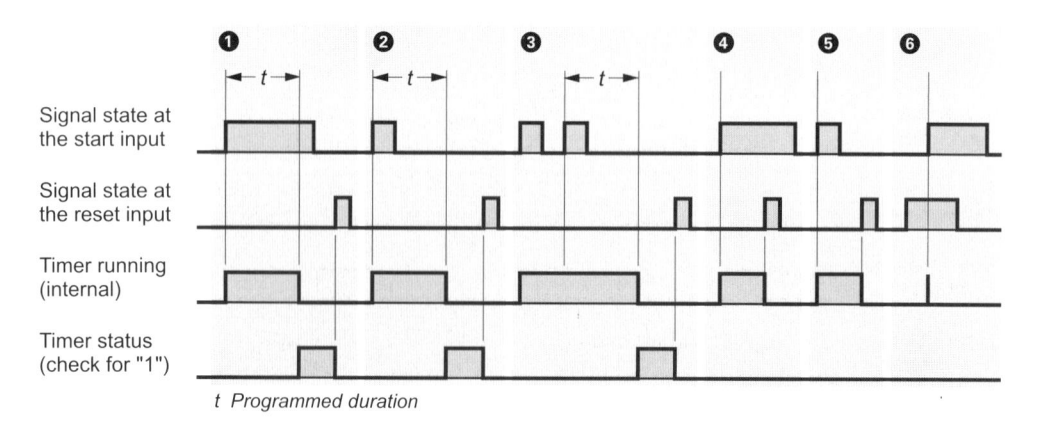

Figure 7.9 Behavioral Characteristics when Starting and Resetting an Retentive On-Delay Timer

7.6 Off-Delay Timer

Starting an off-delay timer

The diagram in Figure 7.10 describes the behavioral characteristics of the timer after it is started and when it is reset. The description applies if you observe the order shown in Section 7.1.6 when programming with individual elements.

① ③ The timer starts when the signal state at the timer's start input changes from "1" to "0" (negative edge). It runs for the programmed duration. Checks for signal state "1" (timer status) return a check result of "1" when the signal state at the start input is "1" or when the timer is running (off-delay).

With the start value as starting point, the time value is counted down towards zero as per the time base.

② If the signal state at the start input changes from "0" to "1" (positive edge) while the timer is running, the timer is reset. It is restarted only when there is a negative edge at the start input.

Resetting an off-delay timer

The resetting of an off-delay timer has a static effect, and takes priority over the starting of the timer (Figure 7.10).

④ Signal state "1" at the timer's reset input while the timer is running resets the timer. The check result of a check for signal state "1" (timer status) is then "0". Time value and time base are also set to zero.

⑤ ⑥ Signal state "1" at the start input and at the reset input resets the timer's binary output (a check for signal state "1" (timer status) then returns a check result of "0"). If the signal state at the reset input now changes back to "0", the timer's output once again goes to "1".

⑦ If the signal state at the start input goes from "1" to "0" (negative edge) while the reset signal is present, the timer starts, but is immediately reset (indicated by a line in the diagram). The check for signal state "1" (timer status) then returns a check result of "0".

7.7 IEC Timers

The IEC timers are integrated in the CPU's operating system as system function blocks (SFBs).

The following timers are available on some CPUs:

▷ SFB 3 TP
  Pulse timer

▷ SFB 4 TON
  On-delay timer

▷ SFB 5 TOF
  Off-delay timer

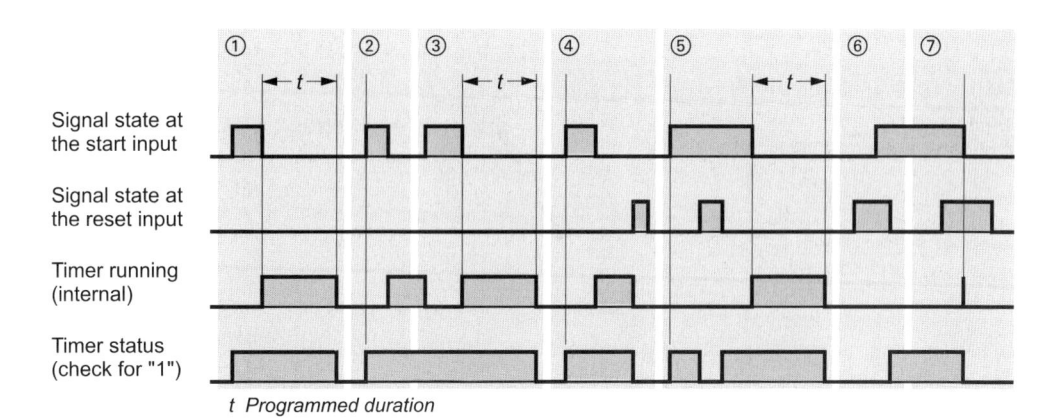

Figure 7.10 Behavioral Characteristics when Starting and Resetting an Off-Delay Timer

7 Timers

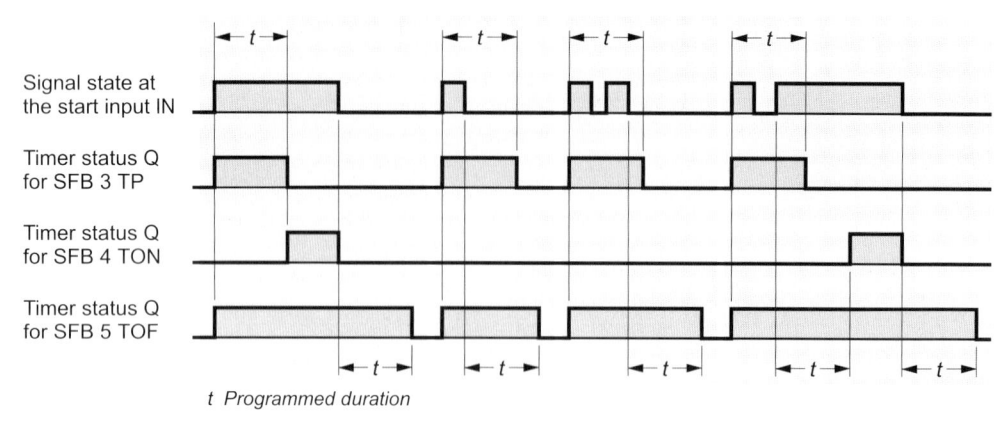

Figure 7.11 Behavioral Characteristics of the IEC Timers

Figure 7.11 shows the behavioral characteristics of these timers.

These SFBs are called with an instance data block or used as local instances in a function block. You will find the interface description for offline programming in the standard library with the name *Standard Library* under the program *Standard Function Blocks*.

You will find examples for the call on the diskette which accompanies the book on the libraries "LAD_Book" and "FBD_Book", function block FB 107 in the "Basic Functions" program.

7.7.1 Pulse Timer SFB 3 TP

IEC timer SFB 3 TP has the parameters listed in Table 7.1.

When the RLO at the timer's start input goes from "0" to "1", the timer is started. It runs for the programmed duration, regardless of any subsequent changes in the RLO at the start input. The signal state at output Q is "1" as long as the timer is running.

Output ET supplies the duration of time for output Q. This duration begins at T#0s and ends at the set duration PT. When PT has elapsed, ET remains set to the elapsed time until input IN goes back to "0". If input IN goes to "0" before PT elapses, output ET goes to T#0s the instant PT elapses.

To reinitialize the timer, simply start it with PT = T#0s.

Timer SFB 3 TP is active in START and RUN mode. It is reset (initialized) when a cold start is executed.

7.7.2 On-Delay Timer SFB 4 TON

The IEC timer SFB 4 TON has the parameters listed in Table 7.1.

The timer starts when the RLO at its start input changes from "0" to "1". It runs for the programmed duration. Output Q shows signal state "1" when the time has elapsed and as long as the signal state at the start input remains at "1". If the RLO at the start input changes from "1" to "0" before the time has run out, the timer is reset. The next positive edge restarts the timer.

Output ET supplies the duration of time for the timer. This duration begins at T#0s and ends at set duration PT. When PT has elapsed, ET remains set to the elapsed time until input IN changes back to "0". If input IN goes to "0" be-

Table 7.1 Parameters for the IEC Timers

Name	Declaration	Data Type	Description
IN	INPUT	BOOL	Start input
PT	INPUT	TIME	Pulse length or delay duration
Q	OUTPUT	BOOL	Timer status
ET	OUTPUT	TIME	Elapsed time

fore PT elapses, output ET immediately goes to T#0s.

To reinitialize the timer, simply start it with PT = T#0s.

SFB 4 TON is active in START and RUN mode. It is reset on a cold start.

7.7.3 Off-Delay Timer SFB 5 TOF

The IEC timer SFB 5 TOF has the parameters listed in Table 7.1.

The signal state at output Q is "1" when the RLO at the timer's start input changes from "0" to "1". The timer is started when the RLO at the start input changes back to "0". Output Q retains signal state "1" as long as the timer runs.

Output Q is reset when the time has elapsed. If the RLO at the start input goes back to "1" before the time has elapsed, the timer is reset and output Q remains at "1".

Output ET supplies the duration of time for the timer. This duration begins at T#0s and ends at set duration PT. When PT has elapsed, ET remains set to the elapsed time until input IN changes back to "1". If input IN goes to "1" before PT has elapsed, output ET immediately goes to T#0s.

To reinitialize the timer, simply start the timer with PT = T#0s.

SFB 5 TOF is active in START and RUN mode. It is reset on a cold start.

8 Counters

Counters allow you to use the CPU to perform counting tasks. The counters can count both up and down. The counting range extends over three decades (000 to 999). The counters are located in the CPU's system memory; the number of counters is CPU-specific.

You can program a counter complete as box or using individual program elements. You can set the count to a specific initial value or reset it, and you can count up and down. The counter is scanned by checking the counter status (zero or non-zero count value) or the current count, which you can retrieve in either binary or binary-coded decimal.

8.1 Programming a Counter

You can perform the following operations on a counter:

▷ Set counter, specifying the count value
▷ Count up
▷ Count down
▷ Reset counter
▷ Check (binary) counter status
▷ Check (digital) count in binary
▷ Check (digital) count in binary-coded decimal

Representation of a counter as box

A counter box contains the coherent representation of all counting operations in the form of function inputs and function outputs (Figure 8.1). Over the box is the absolute or symbolic address of the counter. In the box, as header, is the counter type (S_CUD stands for "up-down counter"). An assignment is mandatory for the first input (CU in the example) is mandatory;

Counter box
(in the example: up/down counter)

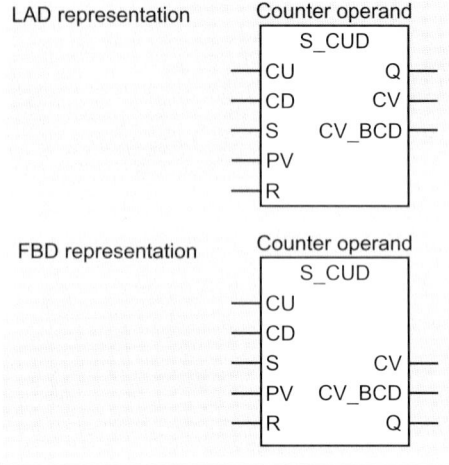

LAD representation

Name	Data Type	Description
CU	BOOL	Up Count input
CD	BOOL	Down Count input
S	BOOL	Set input
PV	WORD	Preset value
R	BOOL	Reset input
CV	WORD	Current value in binary
CV_BCD	WORD	Current value in BCD
Q	BOOL	Counter status

Figure 8.1 Counter in Box Representation

assignments for all other inputs and outputs are optional.

Counter boxes are available in three versions: up-down counter (S_CUD), up counter only (S_CU), and down counter only (S_CD). The differences in functionality are explained below.

For incremental programming, you can find the counters in the Program Element Catalog (with VIEW → CATALOG [Ctrl – K] or INSERT → PROGRAM ELEMENTS) under "Counters".

8.1 Programming a Counter

Up counter (up count coil)	Counter operand —(CU)—|
Down counter (down count coil)	Counter operand —(CD)—|
Set counter, specifying the count value (set counter coil with count value)	Counter operand —(SC)—| Count value
Reset counter (Reset coil)	Counter operand —(R)—|
Check counter status (NO contact, NC contact)	Counter operand —| |— Counter operand —|/|—
Read count value in binary (MOVE box)	MOVE / EN ENO / Counter operand —IN OUT— Digital operand

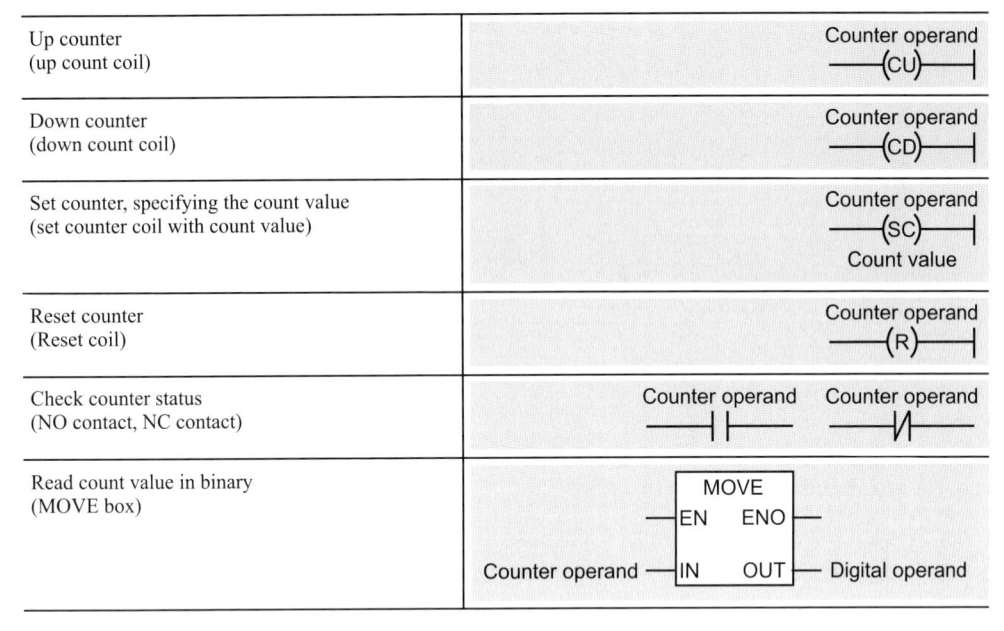

Figure 8.2 Individual Elements of a Counter (LAD)

Representation of a counter using individual elements (LAD)

You can also program a counter using individual elements (Figure 8.2).

Setting and counting are then done via coils. The set counter coil contains the counting operation (SC = Set Counter); below the coil, in WORD format, is the count value to be used to set the counter.

In the coils for counting, CU stands for count up and CD stands for count down. Use the reset coil to reset a counter and an NO or NC contact to check the status of a counter.

Finally, you can transfer the current count, in binary, with the MOVE box.

Counter box in a rung (LAD)

You can arrange contacts in series and in parallel before the counter inputs, the start input and the reset input as well as after output Q.

The counter box may be placed after a T-branch or in a branch that is directly connected to the left power rail. This branch may also have contacts before the inputs and need not be the uppermost branch.

You can find further examples for the representation and arrangement of counters on the diskette supplied with the book in library "LAD_Book", function block FB 108 of the "Basic Functions" program.

Representation of a counter using individual elements (FBD)

You can also program a counter using individual elements (Figure 8.3).

Setting and counting are then done via simple boxes. The set counter box contains the counting operation (SC = Set Counter); at input PV is the count value, in WORD format, is the count value to be used to set the counter.

In the boxes for counting, CU stands for count up and CD stands for count down. Use the reset box to reset a counter and a direct or negated binary function input to check the status of a counter.

Finally, you can transfer the current count, in binary, with the MOVE box.

8 Counters

Up counter (up count box)	Counter operand — CU
Down counter (down count box)	Counter operand — CD
Set counter, specifying the count value (set counter box with count value)	Counter operand — SC, Count value — PV
Reset counter (Reset box)	Counter operand — R
Check counter status (direct or negated binary input)	Counter operand, Counter operand
Read count value in binary (MOVE box)	MOVE: EN, IN ← Counter operand, OUT → Digital operand, ENO

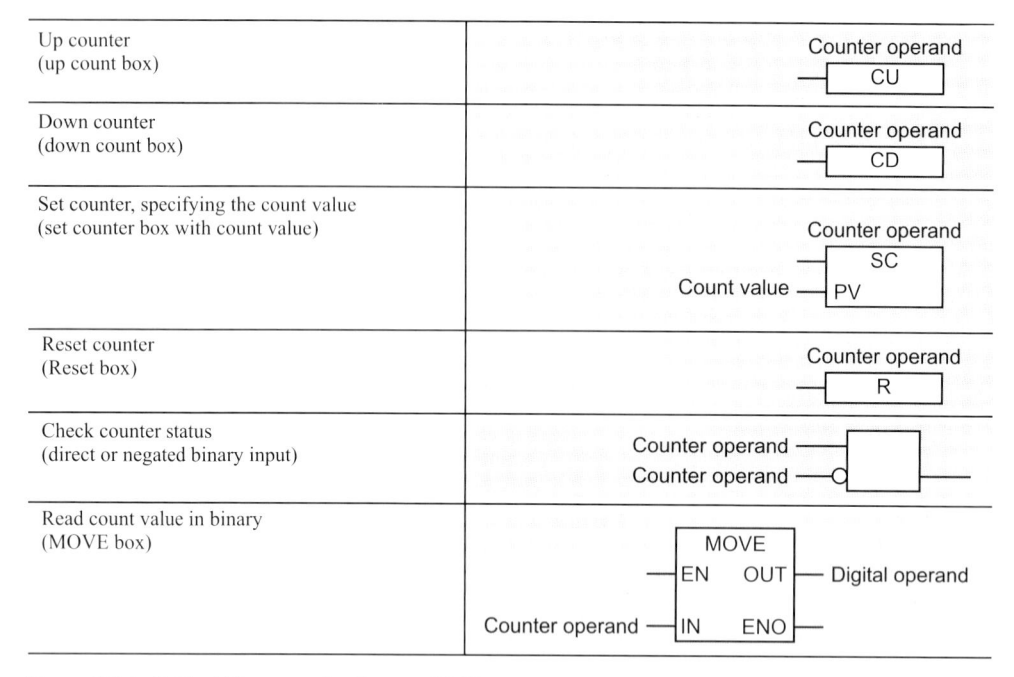

Figure 8.3 Individual Elements of a Counter (FBD)

Counter box in a logic circuit (FBD)

You can arrange binary functions and memory functions before the counter inputs, the start input and the reset input as well as after output Q.

The counter box and the individual elements for counting, setting a counter and resetting a counter may also be placed after a T-branch.

You can find further examples for the representation and arrangement of counters on the diskette supplied with the book in library "FBD_Book", function block FB 108 of the "Basic Functions" program.

Sequence of counting operations

When programming a counter, you do not need to use all the operations available for that counter. The operations required to carry out the desired functions are enough.

For example, to program a down counter, you need only programs the operations for setting the counter to its initial count, down counting, and checking the counter status.

In order that a counter's behavioral characteristics be as described in this chapter, it is advisable to observe the following order when programming with individual program elements:

▷ Count (up or down in any order)
▷ Set counter
▷ Reset counter
▷ Check count
▷ Check counter status

Omit any individual elements that are not required. If counting, setting, and resetting of the counter take place "simultaneously" when the operations are programmed in the order shown, the count will first be changed accordingly before being reset by the reset operation which follows. The subsequent check will therefore show a count of zero and counter status "0".

If counting and setting take place "simultaneously" when the operations are programmed in the order shown, the count will first be changed accordingly before being set to the programmed count value, which it will retain for the remainder of the cycle.

The order of the operations for up and down counting is not significant.

8.2 Setting and Resetting Counters

Setting counters

A counter is set when the RLO changes from "0" to "1" before set input S or before the set coil or set box. A positive edge is always required to set a counter.

"Set counter" means that the counter is set to a starting value. The value may have a range of from 0 to 999.

Specifying the count value

When a counter is set, it assumes the value at input PV or the value below the set coil or set box as count value. You may specify the count value as constant, word operand, or variable of type WORD.

Specifying the count value as constant

C#100   Count value 100
W#16#0100   Count value 100

The count value comprises three decades in the range 000 to 999. Only positive BCD values are permissible; the counter cannot process negative values. You can use C# or W#16# to identify a constant (in conjunction with decimal digits only).

Specifying the count value as operand or variable

MW 56   Word operand containing the count value
"Count value"   Variable of type WORD

Resetting counters (LAD)

A counter is reset when power flows in the reset input or in the reset coil (when RLO "1" is present). When this is the case, checking the counter with an NO contact will result in a check result of "0", and checking with an NC contact will return a check result of "1". Resetting a counter sets its count to "zero". The counter box's R input need not be connected.

Resetting counters (FBD)

A counter is reset when a "1" is present at the reset input. Then a direct scan of the counter status will return "0" and a negated scan will return "1". Resetting a counter sets its count to "zero". The counter box's R input need not be connected.

8.3 Counting

The counting frequency of a counter is determined by the execution time of your program! To be able to count, the CPU must detect a change in the state of the input pulse, that is, an input pulse (or a space) must be present for at least one program cycle. Thus, the longer the program execution time, the lower the counting frequency.

Up counting

A counter is counted up when the RLO changes from "0" to "1" before the up count input CU or at the up count coil or box. A positive edge is always required for up counting.

In up counting, each positive edge increments the count by one unit until it reaches the upper range limit of 999. Each additional positive edge for up counting then has no further effect. There is no carry.

Down counting

A counter is counted down when the RLO changes from "0" to "1" before down count input CD or at the down count coil or box. A positive edge is always required for down counting.

In down counting, each positive edge decrements the count by one unit until it reaches the lower range limit of 0. Each subsequent positive edge for down counting then has no further effect. There are no negative counts.

Different counter boxes

The Editor provides three different counter boxes:

S_CUD   Up/down counter
S_CU   Up counter
S_CD   Down counter

These counter boxes differ only in the type and number of counter inputs. Whereas S_CUD has inputs for both counting directions, S_CU has only the up count input and S_CD only the down count input.

You must always connect the first input of a counter box. If you do not connect the second input (S_CD) on S_CUD, this box will take on the same characteristics as S_CU.

8.4 Checking a Counter

Checking the counter status (LAD)

The counter status is at output Q of the counter box. You can also check the counter status with an NO contact (corresponding to output Q) or an NC contact.

Output Q is "1" (power flows from the output) when the current count is greater than zero. Output Q is "0" if the current count is equal to zero. Output Q does not need to be connected at the counter box.

Checking the counter status (FBD)

The counter status is at output Q of the counter box. You can also check the counter status directly (corresponds to output Q) with a binary function input, or in negated form.

Output Q is "1" when the current count is greater than zero. Output Q is "0" when the current count is equal to zero. Output Q need not be connected at the counter box.

Checking the count value (LAD and FBD)

Outputs CV and CV_BCD make the counter's current count available in binary (CV) or in binary-coded decimal (BCD). The count value made available by this operation is the one which is current at the time the check is made.

The value is stored in the specified operand (transfer as with a MOVE box). You need not switch these outputs at the counter box.

Direct checking of the count

The count is available in binary, and can be fetched from the counter in this form. The value corresponds to a positive number in INT format. Direct checking of a count can also be programmed with a MOVE box.

Coded checking of the count

You can also fetch the binary count from the counter in "coded" form. The binary-coded decimal (BCD) value is structured in the same way as for specifying the count (see above).

8.5 IEC Counters

The IEC counters are integrated in the CPU operating system as system function blocks (SFBs). The following counters are available in the appropriate CPUs:

▷ SFB 0 CTU
  Up counter

▷ SFB 1 CTD
  Down counter

▷ SFB 2 CTUD
  Up/down counter

You can call these SFBs with an instance data block or use them as local instances in a function block.

You will find the interface description for offline programming on standard library *Standard Library* under the *System Function Blocks* program.

You will find sample calls on the diskette supplied with the book on the libraries "LAD_Book" and "FBD_Book" in function block FB 108 of the "Basic Functions" program.

8.5.1 Up Counter SFB 0 CTU

IEC counter SFB 0 CTU has the parameters listed in Table 8.1.

When the signal state at up count input CU changes from "0" to "1" (positive edge), the current count is incremented by 1 and shown at output CV. On the first call (with signal state "0" at reset input R), the count corresponds to the default value at input PV. When the count reaches the upper limit of 32767, it is no longer incremented, and CU has no effect.

The count value is reset to zero when the signal state at reset input R is "1". As long as input R is "1", a positive edge at CU has no effect. Output Q is "1" when the value at CV is greater than or equal to the value at PV.

SFB 0 CTU executes in START and RUN mode. It is reset on a cold start.

Table 8.1 Parameters for the IEC Counters

Name	Present in SFB			Declaration	Data Type	Description
CU	0	-	2	INPUT	BOOL	Up count input
CD	-	1	2	INPUT	BOOL	Down Count input
R	0	-	2	INPUT	BOOL	Reset input
LOAD	-	1	2	INPUT	BOOL	Load input
PV	0	1	2	INPUT	INT	Preset value
Q	0	1	-	OUTPUT	BOOL	Counter status
QU	-	-	2	OUTPUT	BOOL	Up counter status
QD	-	-	2	OUTPUT	BOOL	Down counter status
CV	0	1	2	OUTPUT	INT	Current count value

8.5.2 Down Counter SFB 1 CTD

IEC counter SFB 1 CTD has the parameters listed in Table 8.1.

When the signal state at down count input CD goes from "0" to "1" (positive edge), the current count is decremented by 1 and shown at output CV. On the first call (with signal state "0" at load input LOAD), the count corresponds to the default value at input PV. When the current count reaches the lower limit of –32768, it is no longer decremented and CD has no effect.

The count is set to default value PV when load input LOAD is "1". As long as input LOAD is "1", a positive edge at input CD has no effect.

Output Q is "1" when the value at CV is less than or equal to zero.

SFB 1 CTD executes in START and RUN mode. It is reset on a cold start.

8.5.3 Up/down Counter SFB 2 CTUD

IEC counter SFB 2 CTUD has the parameters listed in Table 8.1.

When the signal state at up count input CU changes from "0" to "1" (positive edge), the count is incremented by 1 and shown at output CV. If the signal state at down count input CD changes from "0" to "1" (positive edge), the count is decremented by 1 and shown at output CV. If both inputs show a positive edge, the current count remains unchanged.

If the current count reaches the upper limit of 32767, it is no longer incremented when there is a positive edge at count up input CU. CU then has no further effect. If the current count reaches the lower limit of –32768, it is no longer decremented when there is a positive edge at down count input CD. CD then has no effect.

The count is set to default value PV when load input LOAD is "1". As long as load input LOAD is "1", positive signal edges at the two count inputs have no effect.

The count is reset to zero when reset input R is "1". As long as input R is "1", positive signal edges at the two count inputs and signal state "1" at load input LOAD have no effect.

Output QU is "1" when the value at CV is greater than or equal to the value at PV.

Output QD is "1" when the value at CV is less than or equal to zero.

SFB 2 CTUD executes in START and RUN mode. It is reset on a cold start.

8.6 Parts Counter Example

The examples illustrates the use of timers and counters. It is programmed with inputs, outputs and memory bits so that it can be programmed at any point in any block. At this point, a function without block parameters is used; the timers and counters are represented by complete boxes. You will find the same example programmed as a function block with block parameters and with individual elements in Chapter 19, "Block Parameters".

8 Counters

Function description

Parts are transported on a conveyor belt. A light barrier detects and counts the parts. After a set number, the counter sends the "Finished" signal. The counter is equipped with a monitoring circuit. If the signal state of the light barrier does not change within a specified amount of time, the monitor sends a signal.

The "Set" input passes the starting value (the number to be counted) to the counter. A positive edge at the light barrier decrements the counter by one unit. When a value of zero is reached, the counters sends the "Finished" signal. Prerequisite is that the parts be arranged singly (at intervals) on the belt.

The "Set" input also sets the "Active" signal. The controller monitors a signal state change at the light barrier in the active state only. When counting is finished and the last counted item has exited the light barrier, "Active" is reset.

In the active state, a positive edge at the light barrier starts the timer with the time value "Duration1" ("Dura1") as retentive pulse timer. If the timer's start input is processed with "0" in the next cycle, it still continues to run. A new positive edge "retriggers" the timer, that is, restarts it. The next positive edge to restart the timer is generated when the light barrier signals a negative edge. The timer is then started with the time value "Duration2" ("Dura2"). If the light barrier is now covered for a period of time exceeding "Dura1" or free for a period of time exceeding "Dura2", the timer runs down and signals "Fault". The first time it is activated, the timer is started with the time value "Dura2".

Signals, symbols

The "Set" signal activates the counter and the monitor. The light barrier controls the counter, the "active" state, selection of the time value, and the starting (retriggering) of the monitoring time via positive and negative edges.

Evaluation of the positive and negative edge of the light barrier is required often, and temporary local data are suitable as "scratchpad" memory. Temporary local data are block-local variables; they are declared in the blocks (not in the symbol table). In the example, the pulse memory bits used for edge evaluation are stored in temporary local data. (The edge memory bits require their signal states in the next cycle as well, and must therefore not be temporary local data).

Table 8.2 Symbol Table for the Parts Counter Example

Symbol	Address	Data Type	Comment
Counter_control	FC 12	FC 12	Counter and monitor control for parts
Acknowl	I 0.6	BOOL	Acknowledge fault
Set	I 0.7	BOOL	Set counter, activate monitor
Lbarr1	I 1.0	BOOL	"End_of_belt" sensor signal conveyor belt 1
Finished	Q 4.2	BOOL	Number of parts reached
Fault	Q 4.3	BOOL	Monitor responded
Active	M 3.0	BOOL	Counter and monitor active
EM_LB_P	M 3.1	BOOL	Edge memory bit for positive edge of light barrier
EM_LB_N	M 3.2	BOOL	Edge memory bit for negative edge of light barrier
EM_Ac_P	M 3.3	BOOL	Edge memory bit for positive edge of "Monitor active"
EM_ST_P	M 3.4	BOOL	Edge memory bit for positive edge of "Set"
Quantity	MW 4	WORD	Number of parts
Dura1	MW 6	S5TIME	Monitoring time for light barrier covered
Dura2	MW 8	S5TIME	Monitoring time for light barrier not covered
Count	C 1	COUNTER	Counter function for parts
Monitor	T 1	TIMER	Timer function for monitor

We want symbolic addressing, that is, the operands are assigned names which are then used for programming. Before entering the program, we create a symbol table (Table 8.2) containing the inputs, outputs, memory bits, timers, counters, and blocks.

Program

The program is located in a function that you call in the CPU in organization block OB 1 (selected from the Program Elements Catalog under "FC Blocks").

During programming, the global symbols can also be used without quotation marks provided they do not contain any special characters. If a symbol contains a special character (an Umlaut or a space, for instance), it must be placed in quotation marks. In the compiled block, the Editor shows all global symbols with quotation marks.

Figure 8.4 and Figure 8.5 shows the program for the parts counter. You can find this program on the diskette supplied with the book on libraries "LAD_Book" and "FBD_Book" in function FC 12 under the "Conveyor Example".

8 Counters

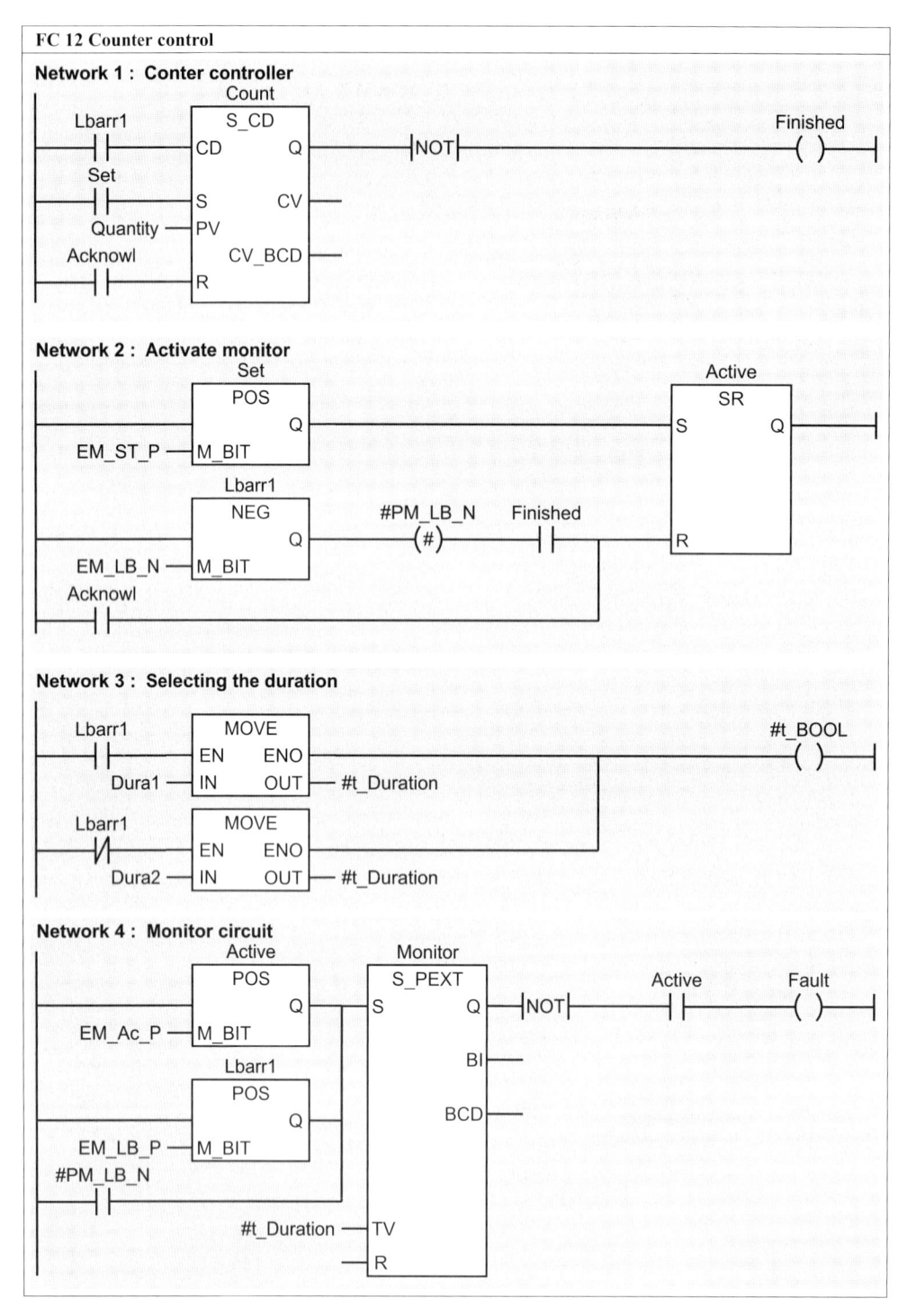

FC 12 Counter control

Network 1 : Conter controller

```
                Count
  Lbarr1        S_CD                                    Finished
  ──┤ ├──────CD        Q ──────┤NOT├──────────────────────( )──
     Set
  ──┤ ├──────S         CV ──
   Quantity ──PV
   Acknowl           CV_BCD ──
  ──┤ ├──────R
```

Network 2 : Activate monitor

```
                Set                                     Active
                POS                                      SR
                    Q ──────────────────────────────S        Q ──
  EM_ST_P ──M_BIT
                Lbarr1
                NEG           #PM_LB_N  Finished
                    Q ──────────(#)─────┤ ├────────R
  EM_LB_N ──M_BIT
   Acknowl
  ──┤ ├──
```

Network 3 : Selecting the duration

```
   Lbarr1       MOVE                                    #t_BOOL
  ──┤ ├──────EN     ENO ──────────────────────────────────( )──
    Dura1 ──IN    OUT ── #t_Duration

   Lbarr1       MOVE
  ──┤/├──────EN     ENO ──
    Dura2 ──IN    OUT ── #t_Duration
```

Network 4 : Monitor circuit

```
                Active         Monitor
                POS            S_PEXT         Active      Fault
                    Q ──────S        Q ──┤NOT├──┤ ├──────( )──
  EM_Ac_P ──M_BIT
                Lbarr1                  BI ──
                POS
                    Q ──────            BCD ──
  EM_LB_P ──M_BIT
  #PM_LB_N
  ──┤ ├──
              #t_Duration ──TV
                            R
```

Figure 8.4 Programming Example for a Parts Counter (LAD)

8.6 Parts Counter Example

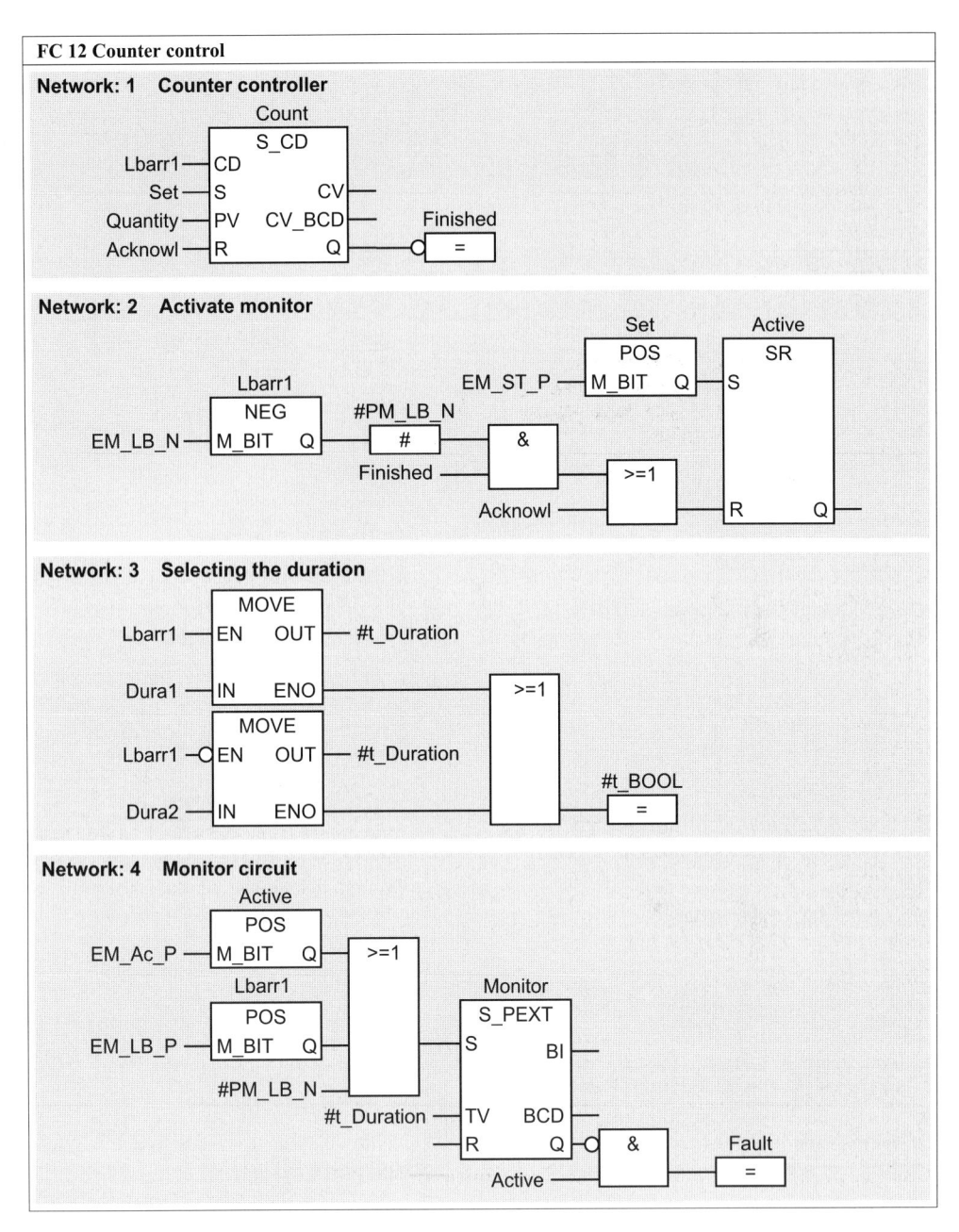

FC 12 Counter control

Network: 1   Counter controller

Network: 2   Activate monitor

Network: 3   Selecting the duration

Network: 4   Monitor circuit

Figure 8.5 Programming Example for a Parts Counter (FBD)

Digital Functions

The digital functions process digital values predominantly with the data types INT, DINT and REAL, and thus extend the functionality of the PLC.

The **comparison functions** form a binary result from the comparison of two values. They take account of the data types INT, DINT and REAL.

You use the **arithmetic functions** to make calculations in your program. All the basic arithmetic functions in data types INT, DINT and REAL are available.

The **mathematical functions** extend the calculation possibilities beyond the basic arithmetic functions to include, for example, trigonometric functions.

Before and after performing calculations, you adapt the digital values to the desired data type using the **conversion functions**.

The **shift functions** allow justification of the contents of a variable by shifting to the right or left.

With **word logic**, you mask digital values by targeting individual bits and setting them to "1" or "0".

The digital logic operations work mainly with values stored in data blocks. These can be global data blocks or instance data blocks if static local data are used. Section 18.2, "Block Functions for Data Blocks", shows how to handle data blocks and gives the methods of addressing data operands.

09   Comparison Functions

Comparison for equal to, not equal to, greater than, greater than or equal to, less than, and less than or equal to

10   Arithmetic Functions

Basic arithmetic functions with data types INT, DINT and REAL

11   Mathematical Functions

Trigonometric functions; inverse trigonometric functions; squaring, square-root extraction, exponentiation, and logarithms

12   Conversion Functions

Conversion from INT/DINT to BCD and vice versa; conversion from DINT to REAL and vice versa with different forms of rounding; one's complement, negation, and absolute-value generation

13   Shift Functions

Shifting to left and right, by word and doubleword, shifting with correct sign; rotating to left and right

14   Word Logic

AND, OR, exclusive OR; word and doubleword combinations

9 Comparison Functions

The comparison functions compare two digital variables of data type INT, DINT or REAL for equal to, not equal to, greater than greater than or equal to, les than, or less than or equal to. The comparison result is then available as binary value (Table 9.1).

9.1 Processing a Comparison Function

Comparison box
(in the example:
comparison for equal to between INT numbers)

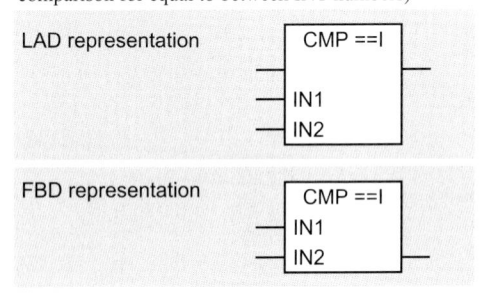

LAD representation — CMP ==I / IN1 / IN2

FBD representation — CMP ==I / IN1 / IN2

Representation (LAD)

In addition to the (unlabeled) binary input, the box for a comparison function has two inputs, IN1 and IN2, and an (unlabeled) binary output. The "header" in the box identifies a comparison operation (CMP for compare) and the type of comparison performed (CMP ==I, for instance, stands for the comparison of two INT numbers for equal to).

You can arrange a comparator in a rung in place of a contact. The unlabeled input and the unlabeled output establish the connection to the other (binary) program elements.

The values to be compared are at inputs IN1 and IN2 and the comparison result is the the output. A successful comparison is equivalent to a closed contact ("power" flows through the comparator). If the comparison is not successful, the contact is open. The comparator's output must always be interconnected.

Representation (FBD)

The box for a comparison has two inputs, IN1 and IN2, and an unlabeled binary output. The "header" in the box identifies the comparison performed (CMP ==I, for example, stands for the comparison of two INT numbers for equal to).

The values to be compared are at inputs IN1 and IN2 and the result of the comparison is at the output. If the comparison is successful, the comparator output shows signal state "1"; otherwise, it is "0". It must always be interconnected.

Table 9.1 Overview of the Comparison Functions

Comparison Function	Comparison According to Data Type		
	INT	DINT	REAL
Comparison for equal to	CMP ==I	CMP ==D	CMP ==R
Comparison for not equal to	CMP <>I	CMP <>D	CMP <>R
Comparison for greater than	CMP >I	CMP >D	CMP >R
Comparison for greater than or equal to	CMP >=I	CMP >=D	CMP >=R
Comparison for less than	CMP <I	CMP <D	CMP <R
Comparison for less than or equal to	CMP <=I	CMP <=D	CMP <=R

9 Comparison Functions

Data types

The data type of the inputs in a comparison function depends on that function. For example, the inputs are of type REAL in the comparison function CMP >R (compare REAL numbers for greater than). Variables must be of the same data type as the inputs. When using operands with absolute addresses, the operand widths must accord with the data types. For example, you can use a word operand for data type INT.

You can find the bit assignments for the data formats in Section 3.4, "Data Types".

A comparison between REAL numbers is not true if one or both REAL numbers are invalid. In addition, status bits OS and OV are set. You can find out how the comparison functions set the remaining status bits in Chapter 15, "Status Bits".

Examples

Figure 9.1 provides an example for each of the data types. A comparison function carries out a comparison according to the characteristics specified even when no data types are declared when using operands with absolute addresses.

In the case of incremental programming, you will find the comparison functions in the Program Elements Catalog (with VIEW → CATALOG [Ctrl – K] or INSERT → PROGRAM ELEMENTS) under "Comparator".

Comparison function in a rung (LAD)

You can use the comparison function in a rung in place of a contact.

You can connect contacts before and after the comparison function in series and in parallel. The comparison boxes themselves can also be

---

Comparison according to INT Memory bit M 99.0 is reset if the value in memory word MW 92 equals 120; otherwise it is not affected.

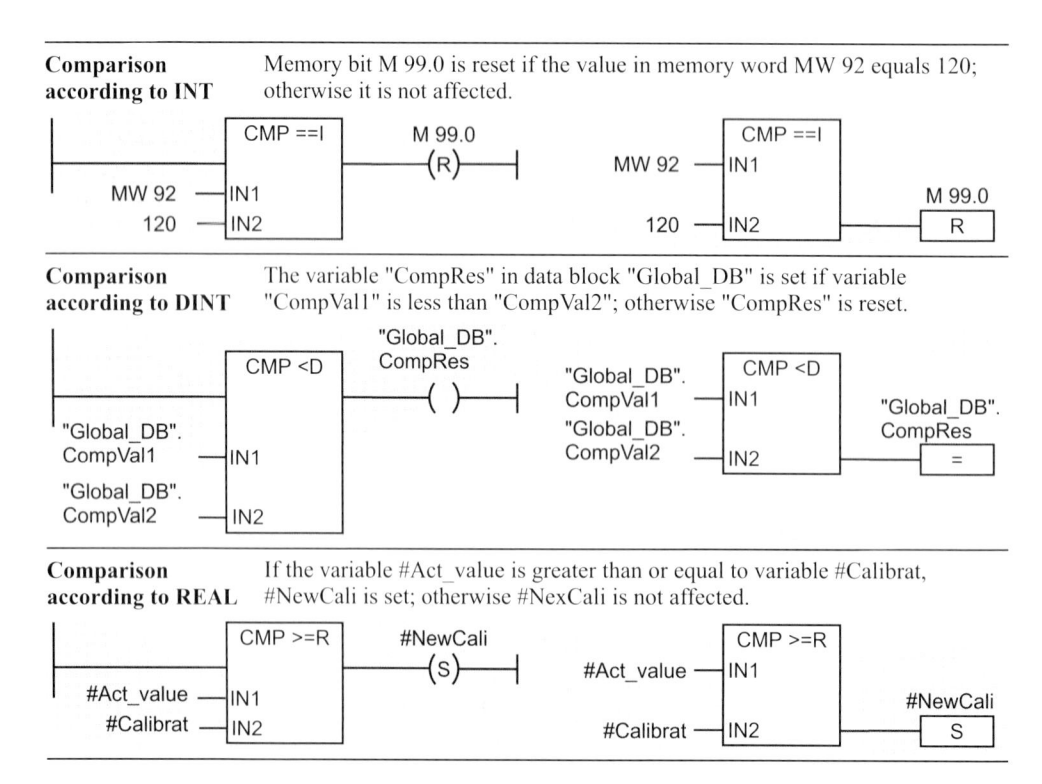

Comparison according to DINT The variable "CompRes" in data block "Global_DB" is set if variable "CompVal1" is less than "CompVal2"; otherwise "CompRes" is reset.

Comparison according to REAL If the variable #Act_value is greater than or equal to variable #Calibrat, #NewCali is set; otherwise #NexCali is not affected.

Figure 9.1 Comparison Function Examples

connected in series or in parallel. In the case of comparison functions connected in series, both comparisons must be successful for power to flow in the rung. In the case of comparators connected in parallel, only one compare condition need be fulfilled for power to flow in the parallel circuit.

The "LAD_Book" library on the diskette supplied with the book contains further examples of representation and arrangement of comparison functions (FB 109 in the "Digital Functions" program).

Comparison function in a logic circuit (FBD)

You can position the comparison function at any binary input of a program element. The result of the comparison can be subsequently combined with binary functions.

The "FBD_Book" library on the diskette supplied with the book contains further examples of representation and arrangement of comparison functions (FB 109 in the "Digital Functions" program).

9.2 Description of the Comparison Functions

Comparison for equal to

The "comparison for equal to" interprets the contents of the input variables in accordance with the data type specified in the comparison function and checks to see if the two values are equal. The compare condition is fulfilled ("power" flows through the comparator output or the RLO is "1") when the two variables have the same value.

If, in the case of a REAL comparison, one or both input variables are invalid, the comparison is not successful. Status bits OV and OS are also set.

Comparison for not equal to

The "comparison for not equal to" interprets the contents of the input variables in accordance with the data type specified in the comparison function and checks to see if the two values differ. The comparison is successful ("power" flows through the comparator output or the RLO is "1") when the two variables have different values.

If, in the case of a REAL comparison, one or both input variables are invalid, the comparison is not successful. In addition, status bits OV and OS are set.

Comparison for greater than

The "comparison for greater than" interprets the contents of the input variables in accordance with the data type specified in the comparison function and checks to see if the value at input IN1 is greater than the value at IN2. If this is the case, the comparison is successful ("power" flows through the comparator output or the RLO is "1").

If, in the case of a REAL comparison, one or both input variables are invalid, the comparison is not successful. In addition, status bits OV and OS are set.

Comparison for greater than or equal to

The "comparison for greater than or equal to" interprets the contents of the input variables in accordance with the data type specified in the comparison function and checks to see if the value at input IN1 is greater than or equal to the value at input IN2. If this is the case, the comparison is successful ("power" flows at the comparator output or the RLO is "1").

If, in the case of a REAL comparison, one or both input variables are invalid, the comparison is not successful. In addition, status bits OV and OS are set.

Comparison for less than

The "comparison for less than" interprets the contents of the input variables in accordance with the data type specified in the comparison function and checks to see if the value at input IN1 is less than the value at input IN2. If this is the case, the comparison is successful ("power" flows at the comparator output or the RLO is "1").

If, in the case of a REAL comparison, one or both input variables are invalid, the comparison is not successful. In addition, status bits OV and OS are set.

Comparison for less than or equal to

The "comparison for less than or equal to" interprets the contents of the input variables in accordance with the data type specified in the comparison function and checks to see if the value at input IN1 is less than or equal to the value at input IN2. If this is the case, the comparison is successful ("power" flows at the comparator output or the RLO is "1").

If, in the case of a REAL comparison, one or both input variables are invalid, the comparison is not successful. In addition, status bits OV and OS are set.

10 Arithmetic Functions

The arithmetic functions combine two values in accordance with the basic arithmetical operations of addition, subtraction, multiplication, and division. You can use the arithmetic functions on variables of type INT, DINT, and REAL (Table 10.1).

10.1 Processing an Arithmetic Function

Representation

Arithmetic box
(in the example: Addition according to INT)

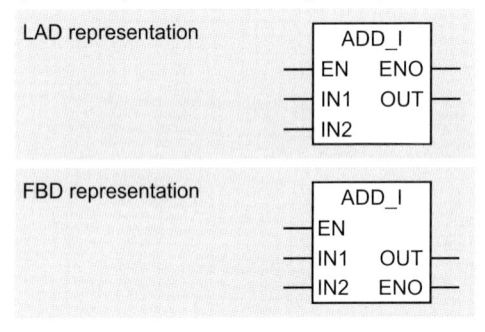

LAD representation

```
   ADD_I
──┤EN  ENO├──
──┤IN1 OUT├──
──┤IN2       │
```

FBD representation

```
   ADD_I
──┤EN        │
──┤IN1 OUT├──
──┤IN2 ENO├──
```

In addition to enable the input EN and the enable output ENO, a box for an arithmetic function has two inputs, IN1 and IN2, and an output, OUT. The "header" in the box identifies the arithmetic function executed (ADD_1, for instance, stands for the addition of INT numbers).

The values to be combined are at inputs IN1 and IN2, and the result of the calculation is at output OUT. The inputs and the output have different data types, depending on the arithmetic function. For example, in the case of the arithmetic function ADD_R (addition of REAL numbers), the inputs and the output are of data type REAL. The variables applied must be of the same data type as the inputs or the output. If you use absolute addresses for the operands, the operand widths must be matched to the data types. For example, you can use a word operand for data type INT.

You can find a description of the individual bits in each data format in Section 3.4, "Data Types".

Function

The arithmetic function is executed if "1" is present at the enable input ("power" flows in input EN). If an error occurs during the calculation, the enable output is set to "0"; otherwise, it is set to "1". If execution of the function is not enabled (EN = "0"), the calculation does not take place, and ENO is also "0".

If the Master Control Relay (MCR) is activated, output OUT is set to zero when the arithmetic function is processed (EN = "1"). The MCR does not affect the ENO output.

Table 10.1 Overview of Arithmetic Functions

Arithmetic function	With data type		
	INT	DINT	REAL
Addition	ADD_I	ADD_DI	ADD_R
Subtraction	SUB_I	SUB_DI	SUB_R
Multiplication	MUL_I	MUL_DI	MUL_R
Division with quotient as result	DIV_I	DIV_DI	DIV_R
Division with remainder as result	-	MOD_DI	-

10 Arithmetic Functions

IF EN == "1" or not wired		
THEN	ELSE	
OUT := IN1 Cfct IN2		
IF error occured		
THEN	ELSE	
ENO := "0"	ENO := "1"	ENO := "0"

with Cfct as calculation function

The following errors can occur during execution of an arithmetic function:

▷ Range violation (overflow) in INT and DINT calculations

▷ Underflow and overflow in a REAL calculation

▷ Invalid REAL number in a REAL calculation

See Chapter 15, "Status Bits", to find out how the arithmetic functions set the various status bits.

Examples

Figure 10.1 shows an example for each data type. An arithmetic function executes a calculation in accordance with the characteristic specified, even if no data types have been declared when using operands with absolute addresses.

In the case of incremental programming, you will find the arithmetic functions in the Program Element Catalog (with VIEW → CATALOG [Ctrl – K] or INSERT → PROGRAM ELEMENTS) under "Integer Math" (INT and DINT calculations) and under "Floating-Point Math" (REAL calculations).

Arithmetic function in a rung (LAD)

You can connect contacts in series and in parallel before the EN input and after the ENO output.

The arithmetic box itself may be placed after a T-branch and in a branch that leads directly to the left power rail. This branch can also have

---

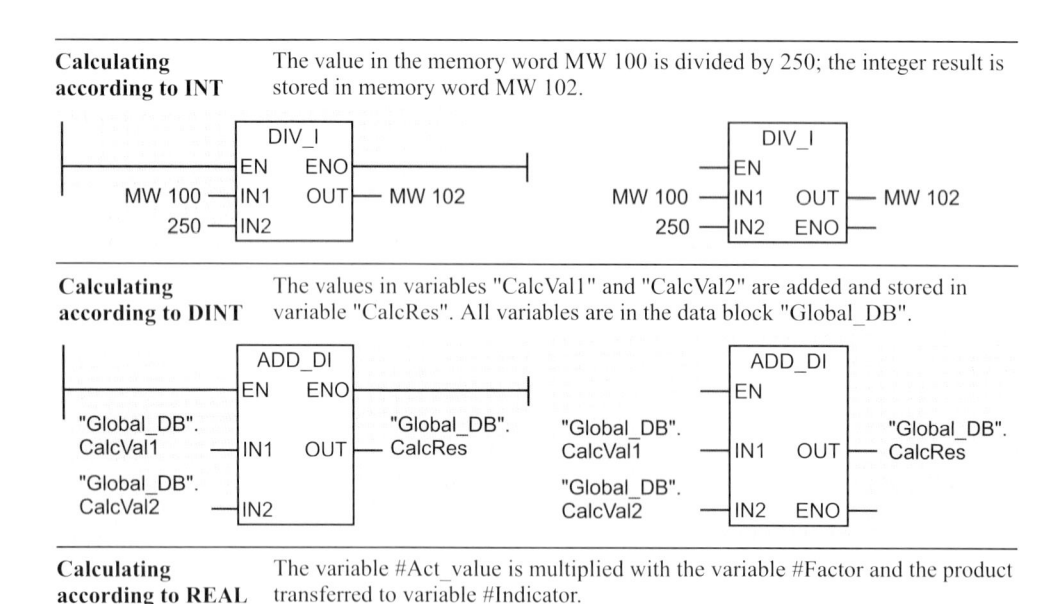

Calculating according to INT The value in the memory word MW 100 is divided by 250; the integer result is stored in memory word MW 102.

```
        DIV_I                          DIV_I
      ┤EN  ENO├                      ─┤EN
MW 100┤IN1  OUT├─ MW 102      MW 100 ┤IN1  OUT├─ MW 102
   250┤IN2                       250 ┤IN2  ENO├
```

Calculating according to DINT The values in variables "CalcVal1" and "CalcVal2" are added and stored in variable "CalcRes". All variables are in the data block "Global_DB".

```
            ADD_DI                              ADD_DI
          ┤EN  ENO├                           ─┤EN
"Global_DB".                    "Global_DB".              "Global_DB".
CalcVal1  ┤IN1  OUT├─ "Global_DB".  CalcVal1  ┤IN1  OUT├─ CalcRes
"Global_DB".         CalcRes    "Global_DB".
CalcVal2  ┤IN2                   CalcVal2    ┤IN2  ENO├
```

Calculating according to REAL The variable #Act_value is multiplied with the variable #Factor and the product transferred to variable #Indicator.

```
           MUL_R                              MUL_R
         ┤EN  ENO├                          ─┤EN
#Act_value┤IN1  OUT├─ #Indicator   #Act_value┤IN1  OUT├─ #Indicator
#Factor   ┤IN2                     #Factor   ┤IN2  ENO├
```

Figure 10.1 Examples of Arithmetic Functions

contacts before the EN input and it need not be the uppermost branch.

Direct connection to the left power rail means that you can connect arithmetic boxes in parallel. When you connect boxes in parallel, you need a coil to terminate the rung. If you have not provided any error evaluation, assign a "dummy" operand to the coil, for example a temporary local data bit.

You can connect arithmetic boxes in series. If the ENO output of the preceding box leads to the EN input of the subsequent box, the subsequent box is processed only if the preceding box has been completed without errors. If you want to use the result from the preceding box as input value for the next box, variables from the temporary local data area make convenient intermediate buffers.

If you arrange several arithmetic boxes in one rung (parallel to the left power rail, then further in series), the boxes in the uppermost branch are processed from left to right, followed by the boxes in the second branch from left to right, and so on.

The "LAD_Book" library on the diskette supplied with the book contains further examples for the representation and arrangement of arithmetic functions (FB 110 in the "Digital Functions" program).

Arithmetic function in a logic circuit (FBD)

EN and ENO need not be wired. If you want to process the arithmetic box in dependence on specific conditions, you can arrange binary logic operations before the EN input. You can interconnect the ENO output with binary inputs of other functions; for example, you can arrange arithmetic boxes in series, whereby the ENO output of the preceding box leads to the EN input of the following box. If you want to use the calculation result from the preceding box as input value for a subsequent box, variables from the temporary local data area make convenient intermediate buffers.

The "FBD_Book" library on the diskette supplied with the book contains further examples for the representation and arrangement of arithmetic functions (FB 110 in the "Digital Functions" program).

10.2 Calculating with Data Type INT

INT addition

The function ADD_I interprets the values at inputs IN1 and IN2 as numbers of data type INT. It adds the two numbers and stores the sum in output OUT.

After execution of the calculation, status bits CC0 and CC1 indicate whether the sum is negative, zero, or positive. Status bits OV and OS indicate any range violations.

INT subtraction

The function SUB_I interprets the values at inputs IN1 and IN2 as numbers of data type INT. It subtracts the value at IN2 from the value at IN1 and stores the difference in output OUT.

After execution of the calculation, status bits CC0 and CC1 indicate whether the difference is negative, zero, or positive. Status bits OV and OS indicate any range violations.

INT multiplication

The function MUL_I interprets the values at inputs IN1 and IN2 as numbers of data type INT. It multiplies the two numbers and stores the product in output OUT.

After execution of the calculation, status bits CC0 and CC1 indicate whether the product is negative, zero, or positive. Status bits OV and OS indicate any INT range violations.

INT division

The function DIV_I interprets the values at inputs IN1 and IN2 as numbers of data type INT. It divides the values at input IN1 (dividend) by the value at input IN2 (divisor) and supplies the quotient at output OUT. It is the integer result of the division. The quotient is zero if the dividend is equal to zero and the divisor is not equal to zero or if the absolute value of the dividend is less than the absolute value of the divisor. The quotient is negative if the divisor is negative.

After execution of the calculation, status bits CC0 and CC1 indicate whether the quotient is negative, zero, or positive. Status bits OV and OS indicate any range violations. Division by zero produces zero as quotient and sets status bits CC0, CC1, OV and OS to "1".

10.3 Calculating with Data Type DINT

DINT addition

The function ADD_DI interprets the values at inputs IN1 and IN2 as numbers of data type DINT. It adds the two numbers and stores the sum in output OUT.

After execution of the calculation, status bits CC0 and CC1 indicate whether the sum is negative, zero, or positive. Status bits OV and OS indicate any range violations.

DINT subtraction

The function SUB_DI interprets the values at inputs IN1 and IN2 as numbers of data type DINT. It subtracts the value at input IN2 from the value at input IN1 and stores the difference in output OUT.

After execution of the calculation, status bits CC0 and CC1 indicate whether the difference is negative, zero, or positive. Status bits OV and OS indicate any range violations.

DINT multiplication

The function MUL_DI interprets the values at inputs IN1 and IN2 as numbers of data type DINT. It multiplies the two numbers and stores the product in output OUT.

After execution of the calculation, status bits CC0 and CC1 indicate whether the product is negative, zero, or positive. Status bits OV and OS indicate any range violations.

DINT division with quotient as result

The function DIV_DI interprets the values at inputs IN1 and IN2 as numbers of data type DINT. It divides the value at input IN1 (dividend) by the value at input IN2 (divisor) and stores the quotient in output OUT. It is the integer result of the division. The quotient is zero if the dividend is equal to zero and the divisor is not equal to zero or if the absolute value of the dividend is less than the absolute value of the divisor. The quotient is negative if the divisor is negative.

After execution of the calculation, status bits CC0 and CC1 indicate whether the quotient is negative, zero, or positive. Status bits OV and OS indicate any range violations. Division by zero produces zero as quotient and sets status bits CC0, CC1, OV and OS to "1".

DINT division with remainder as result

The function MOD_DI interprets the values at inputs IN1 and IN2 as numbers of data type DINT. It divides the value at input IN1 (dividend) by the value at input IN2 (divisor) and stores the remainder of the division in output OUT. The remainder is what is left over from the division; it does not correspond to the decimal places. If the dividend is negative, the remainder is also negative.

After execution of the calculation, status bits CC0 and CC1 indicate whether the remainder is negative, zero, or positive. Status bits OV and OS indicate any range violations. Division by zero produces zero as remainder and sets status bits CC0, CC1, OV and OS to "1".

10.4 Calculating with Data Type REAL

REAL numbers are represented internally as floating-point numbers with two number ranges: One range with full accuracy ("normalized" floating-point numbers) and one range with limited accuracy ("denormalized" floating-point numbers; also see Section 3.4, "Data Types"). S7-400 CPUs calculate in both ranges, S7-300 CPUs only in the full-accuracy range. If an S7-300 CPU carries out a calculation whose result is in the limited-accuracy range, zero is returned as result and a range violation reported.

REAL addition

The function ADD_R interprets the values at inputs IN1 and IN2 as numbers of data type REAL. It adds the two numbers and stores the sum in output OUT.

After execution of the calculation, status bits CC0 and CC1 indicate whether the sum is negative, zero, or positive. Status bits OV and OS indicate any range violations.

In the case of an illegal calculation (one of the input values is an invalid REAL number or you attempt to add +∞ and –∞), ADD_R returns an invalid value at output OUT and sets status bits CC0, CC1, OV and OS to "1".

REAL subtraction

The function SUB_R interprets the values at inputs IN1 and IN2 as numbers of data type REAL. It subtracts the number at input IN2 from the number at input IN1 and stores the difference in output OUT.

After execution of the calculation, status bits CC0 and CC1 indicate whether the difference is negative, zero, or positive. Status bits OV and OS indicate any range violations.

In the case of an illegal calculation (one of the input values is an invalid REAL number or you attempt to subtract +∞ from +∞), SUB_R returns an invalid value at output OUT and sets status bits CC0, CC1, OV and OS to "1".

REAL multiplication

The function MUL_R interprets the values at inputs IN1 and IN2 as numbers of data type REAL. It multiplies the two numbers and stores the product in output OUT.

After execution of the calculation, status bits CC0 and CC1 indicate whether the product is negative, zero, or positive. Status bits OV and OS indicate any range violations.

In the case of an illegal calculation (one of the input values is an invalid REAL number or you attempt to multiply ∞ and 0), MUL_R returns an invalid value at output OUT and sets status bits CC0, CC1, OV and OS to "1".

REAL division

The function DIV_R interprets the values at inputs IN1 and IN2 as numbers of data type REAL. It divides the number at input IN1 (dividend) by the number at input IN2 (divisor) and stores the quotient in output OUT.

After execution of the calculation, status bits CC0 and CC1 indicate whether the quotient is negative, zero, or positive. Status bits CC0 and CC1 indicate any range violations.

In the case of an illegal calculation (one of the input values is an invalid REAL number or you attempt to divide ∞ by ∞ or 0 by 0), DIV_R returns an invalid value at output OUT and sets status bits CC0, CC1, OV and OS to "1".

11 Mathematical Functions

The following mathematical functions are available in LAD and FBD:

- Sine, cosine, tangent
- Arc sine, arc cosine, arc tangent
- Squaring, square-root extraction
- Exponential function to base e, natural logarithm

All mathematical functions process REAL numbers.

11.1 Processing a Mathematical Function

Representation

The box for a mathematical function has an input IN and an output OUT in addition to the enable input EN and the enable output ENO. The "header" in the box identifies the mathematical function executed (for example, SIN stands for sine).

Mathematics box
(in the example: sine)

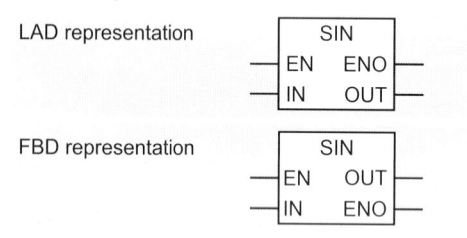

LAD representation	SIN
	EN ENO
	IN OUT

FBD representation	SIN
	EN OUT
	IN ENO

The input value is at input IN and the result of the mathematical function is at output OUT. Input and output are of data type REAL. Operands referenced with absolute addresses must be doubleword operands.

See Section 3.4, "Data Types", for a description of the bits in REAL format.

Function

The mathematical function is executed if "1" is present at the enable input or if "power" flows in input EN. If an error occurs in the calculation, the enable input is set to "0"; otherwise it is set to "1". If execution of the function is not enabled (EN = "0"), the calculation does not take place and ENO is also "0".

IF EN == "1" or not wired	
THEN	ELSE
OUT := Mfct (IN)	
IF error occured	
THEN	ELSE	
ENO := "0"	ENO := "1"	ENO := "0"

with Mfct as mathematical function

If the Master Control Relay (MCR) is active, output OUT is set to zero when the mathematical function is processed (EN = "1"). The MCR does not affect the ENO.

The following errors can occur in a mathematical function:

- Range violation (underflow and overflow)
- No valid REAL number as input value

Chapter 15, "Status Bits", explains how the mathematical functions set the status bits.

Examples

Figure 11.1 shows three examples of mathematical functions. A mathematical function performs the calculation in accordance with REAL even if no data types have been declared when using operands with absolute addresses.

In the case of incremental programming, you will find the mathematical functions in the Program Elements Catalog (with VIEW → CATALOG [Ctrl – K] or INSERT → PROGRAM ELEMENS) under "Floating-Point Math".

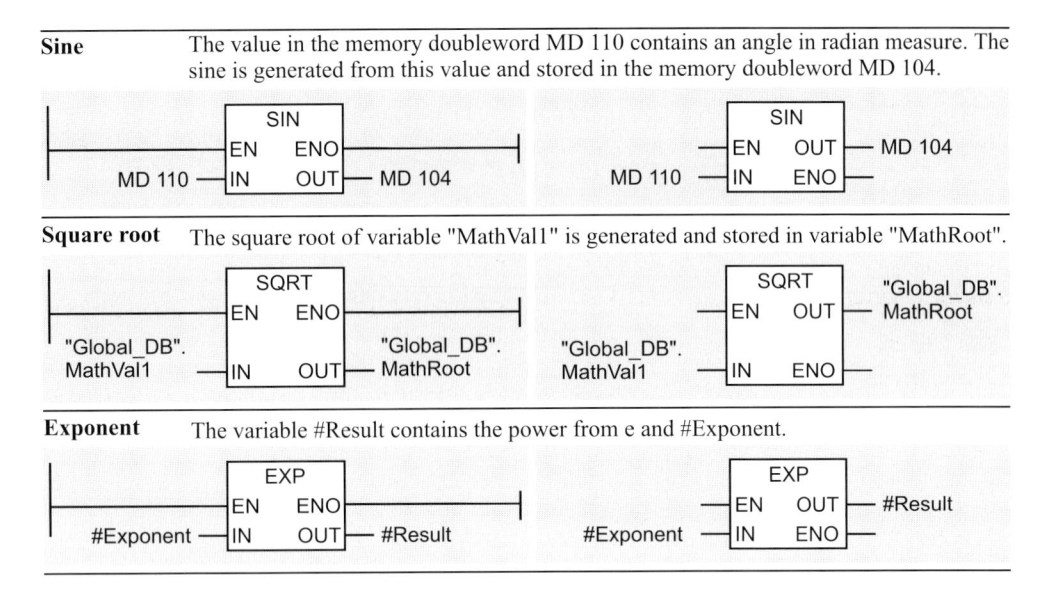

Sine	The value in the memory doubleword MD 110 contains an angle in radian measure. The sine is generated from this value and stored in the memory doubleword MD 104.

```
         SIN                              SIN
       EN  ENO                         EN  OUT — MD 104
MD 110—IN  OUT— MD 104         MD 110—IN  ENO
```

Square root	The square root of variable "MathVal1" is generated and stored in variable "MathRoot".

```
              SQRT                                SQRT      "Global_DB".
            EN  ENO                             EN  OUT  — MathRoot
"Global_DB".              "Global_DB".   "Global_DB".
MathVal1   —IN  OUT   — MathRoot         MathVal1  —IN  ENO
```

Exponent	The variable #Result contains the power from e and #Exponent.

```
            EXP                              EXP
          EN  ENO                          EN  OUT — #Result
#Exponent—IN  OUT— #Result      #Exponent—IN  ENO
```

Figure 11.1 Examples of Mathematical Functions

Mathematical function in a rung (LAD)

You can connect contacts in series and in parallel before input EN and after output ENO.

The mathematics box itself may be placed after a T-branch or in a branch that leads directly to the left power rail. This branch can also have contacts before input EN and need not be the uppermost branch.

The direct connection to the left power rail allows you to connect mathematics boxes in parallel. When connecting boxes in parallel, you require a coil to terminate the rung. If you have not provided error evaluation, assign a "dummy" operand to the coil, for example a temporary local data bit.

You can connect mathematics boxes in series. If the ENO output of the preceding box leads to the EN input of the subsequent box, the subsequent box is processed only if the preceding box has been completed without errors. If you want to use the result from the preceding box as the input value for a subsequent box, variables from the temporary local data area make convenient intermediate buffers.

If you arrange several mathematics boxes in one rung (parallel to the left power rail and then further in series), the boxes in the uppermost branch are the first to be processed from left to right, followed by the boxes in the second branch from left to right, and so on.

The library "LAD_Book" on the diskette supplied with the book contains further examples for the representation and arrangement of mathematical functions (FB 111 in the "Digital Functions" program).

Mathematical function in a logic circuit (FBD)

EN and ENO need not be wired. If you want to have a mathematics box processed in dependence on specific conditions, you can program binary logic operations before the EN input. You can connect the ENO output with binary inputs from other functions. For example, you can arrange the mathematics boxes in series, whereby the ENO output of the preceding box leads to the EN input of the following box. If you want to use the calculation result of the preceding box as input value to a subsequent box, variables from the temporary local data area make good intermediate buffers.

The library "FBD_Book" on the diskette supplied with the book contains further examples for the representation and arrangement of mathematical functions (FB 111 in the "Digital Functions" program).

11.2 Trigonometric Functions

The trigonometric functions

SIN   Sine,

COS   Cosine and

TAN   Tangent

assume an angle in radian measure as a REAL number at the input.

Two units are conventionally used for giving the size of an angle: degrees from 0° to 360° and radian measure from 0 to 2π (where π = +3.141593e+00). Both can be converted proportionally. For example, the radian measure for a 90° angle is $\pi/2$, or +1.570796e+00. With values greater than 2π (+6.283185e+00), 2π or a multiple of 2π is subtracted until the input value for the trigonometric function is less than 2π.

Example (Figure 11.2 or Figure 11.3, Network 4): Calculating the idle power $Ps = U \cdot I \cdot sine(\varphi)$.

11.3 Arc Functions

The arc functions (inverse trigonometric functions)

ASIN   Arc sine,

ACOS   Arc cosine and

ATAN   Arc tangent

Are the inverse functions of the corresponding trigonometric functions. They assume a REAL number in a specific range at input IN and return an angle in the radian measure (Table 11.1).

If the permissible value range is exceeded at input IN, the arc function returns an invalid REAL number and ENO = "0" and sets status bits CC0, CC1, OV and OS to "1".

Table 11.1 Range of arc functions

Function	Permissible Range	Value Returned
ASIN	–1 to +1	$-\pi/2$ to $+\pi/2$
ACOS	–1 to +1	0 to π
ATAN	Full range	$-\pi/2$ to $+\pi/2$

11.4 Miscellaneous Mathematical Functions

The following mathematical functions are also available:

SQR   Compute the square of a number

SQRT   Compute the square root of a number

EXP   Compute the exponent to base e

LN   Find the natural logarithm (logarithm to base e)

Computing the square

The SQR function squares the value at input IN and stores the result in output OUT.

Example: See "Computing the square root".

Computing the square root

The SQRT function extracts the square root of the value at input IN and stores the result in output OUT. If the value at input IN is less than zero, SQRT sets status bits CC0, CC1, OV and OS to "1" and returns an invalid REAL number. If the value at input IN is –0 (minus zero), –0 is returned.

Example: $c = \sqrt{a^2 + b^2}$

Figure 11.2 or Figure 11.3, Network 5: First, the squares of variables a and b are found and then added. Finally, the square root is extracted from the sum. Temporary local data are used as intermediate memory.

(If you have declared b or c as a local variable, you must precede it with # so that the Editor recognizes it as a local variable; if b or c is a global variable, it must be placed in quotation marks.)

Computing the exponent to base e

The EXP function computes the exponential value to base e (= 2.718282e+00) and the value at input IN (e^{IN}) and stores the result in output OUT.

You can calculate any exponential value using the formula $a^b = e^{b \ln a}$.

11.4 Miscellaneous Mathematical Functions

Examples of mathematical functions

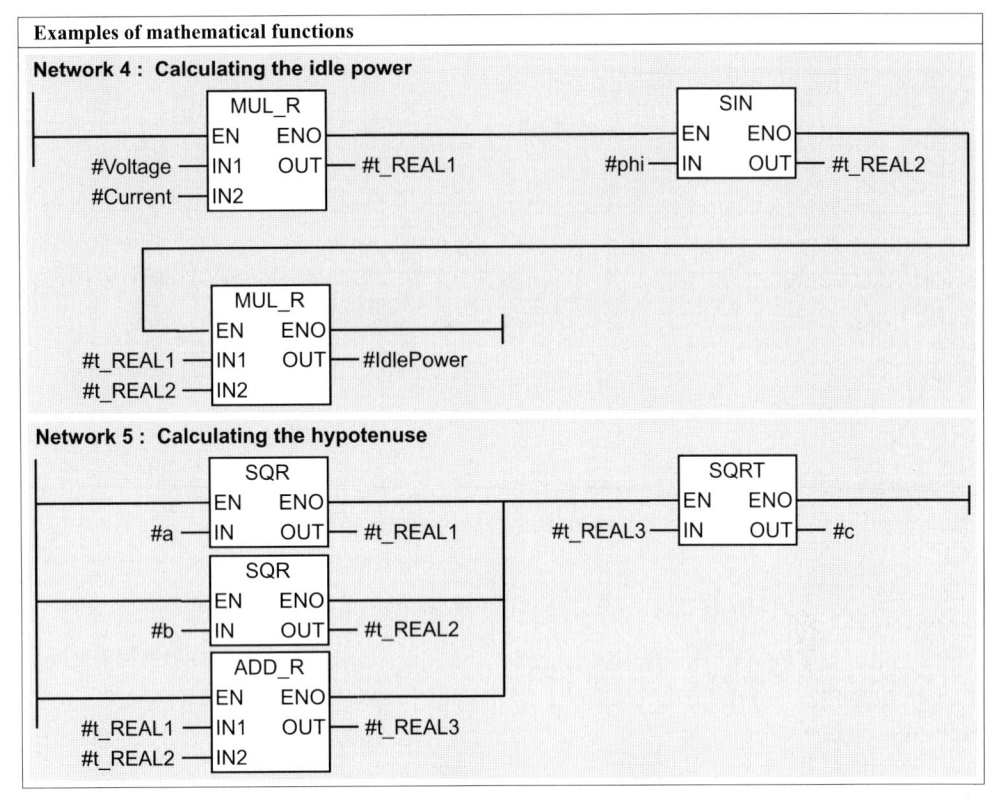

Network 4 : Calculating the idle power

```
        MUL_R                              SIN
     EN    ENO                          EN   ENO
#Voltage─IN1   OUT─#t_REAL1      #phi─IN    OUT─#t_REAL2
#Current─IN2

        MUL_R
     EN    ENO
#t_REAL1─IN1  OUT─#IdlePower
#t_REAL2─IN2
```

Network 5 : Calculating the hypotenuse

```
        SQR                               SQRT
     EN    ENO                         EN    ENO
  #a─IN    OUT─#t_REAL1     #t_REAL3─IN    OUT─#c
        SQR
     EN    ENO
  #b─IN    OUT─#t_REAL2
        ADD_R
     EN    ENO
#t_REAL1─IN1  OUT─#t_REAL3
#t_REAL2─IN2
```

Figure 11.2 Examples of Mathematical Functions (LAD)

Finding the natural logarithm

The LN function finds the natural logarithm to base e (= 2.718282e+00) from the number at input N and stores it in output OUT. If the value at input IN is less than or equal to zero, LN sets status bits CC0, CC1, OV and OS to "1" and returns an invalid REAL number.

The natural logarithm is the inverse function of the exponential function: If $y = e^x$ then $x = \ln y$.

To find any logarithm, use the formula

$$\log_b a = \frac{\log_n a}{\log_n b}$$

where b or n is any base. If you make $n = e$, you can find a logarithm to any base using the natural logarithm:

$$\log_b a = \frac{\ln a}{\ln b}$$

In the special case for base 10, the formula is as follows:

$$\lg a = \frac{\ln a}{\ln 10} = 0.4342945 \cdot \ln a$$

11 Mathematical Functions

Examples of mathematical functions

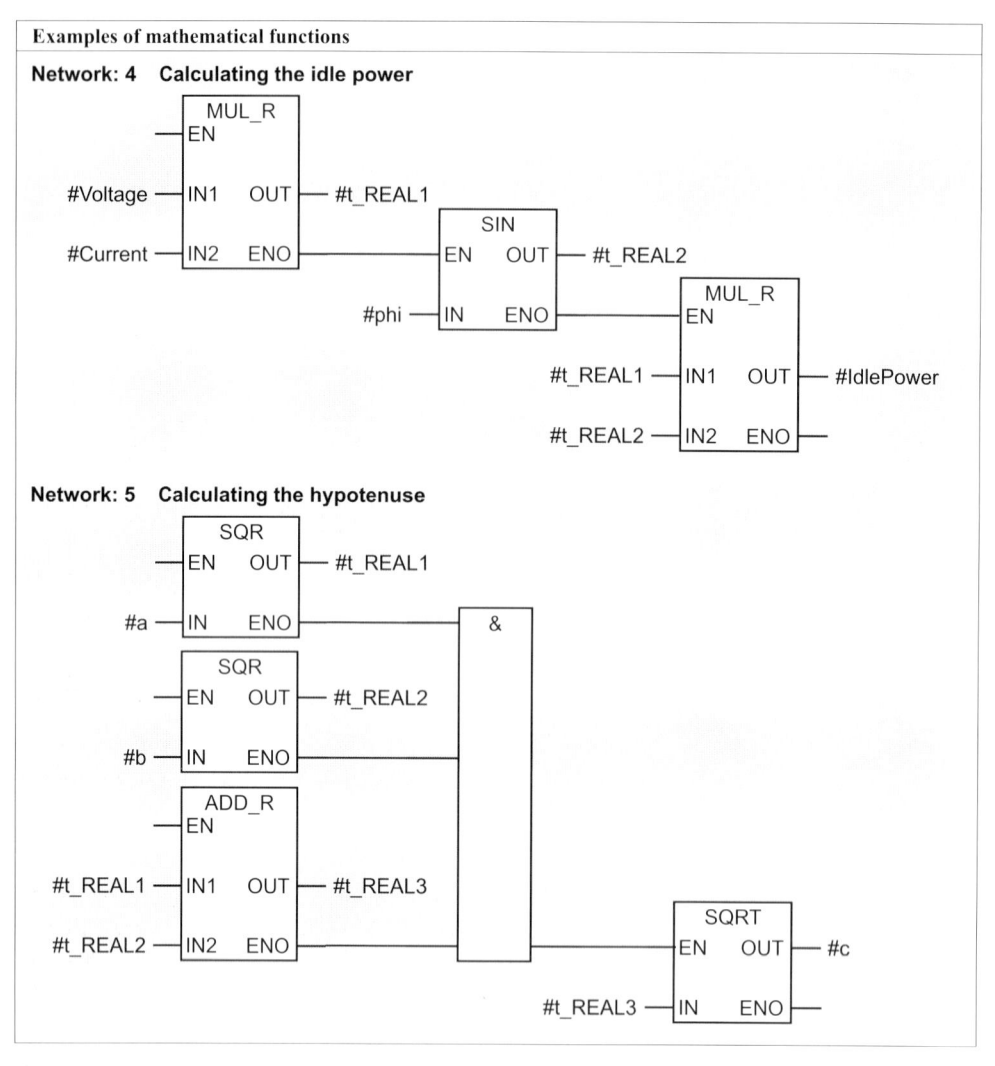

Network: 4   Calculating the idle power

```
        MUL_R
        EN
#Voltage—IN1  OUT—#t_REAL1
                              SIN
#Current—IN2  ENO—     EN   OUT—#t_REAL2
                                              MUL_R
                       #phi—IN  ENO           EN
                                      #t_REAL1—IN1  OUT—#IdlePower
                                      #t_REAL2—IN2  ENO
```

Network: 5   Calculating the hypotenuse

```
         SQR
         EN   OUT—#t_REAL1
  #a—IN  ENO—————————————┐
                          │
         SQR              │
         EN   OUT—#t_REAL2│
  #b—IN  ENO—————————————┤ &
                          │
        ADD_R             │
        EN                │
#t_REAL1—IN1  OUT—#t_REAL3│
                          │      SQRT
#t_REAL2—IN2  ENO         │      EN   OUT—#c
                          └#t_REAL3—IN  ENO
```

Figure 11.3 Examples of Mathematical Functions (FBD)

12 Conversion Functions

The conversion functions convert the data type of a variable. Figure 12.1 provides an overview of the data type conversions described in this chapter.

12.1 Processing a Conversion Function

Representation

In addition to the enable input EN and the enable output ENO, the box for a conversion function has an input IN and an output OUT. The "header" in the box identifies the conversion function executed (for example, I_BCD stands for the conversion of INT to BCD).

Conversion box
(in the example: INT to BCD)

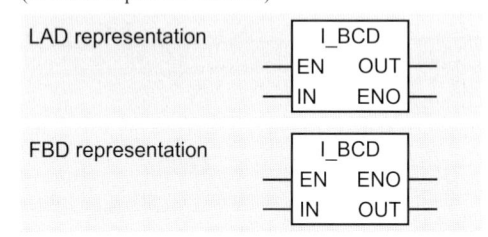

LAD representation	I_BCD
	EN OUT
	IN ENO

FBD representation	I_BCD
	EN ENO
	IN OUT

The value to be converted is at input IN and the result of the conversion is at output OUT. The data type of the input and that of the output depend on the conversion function. In the conversion function DI_R (DINT to REAL), for instance, the input is of type DINT and the output of type REAL. The variables applied must be of the same data type as the input or the output. If you use operands with absolute addresses, the sizes of the operands must be matched to the data types; for example, you can use a word operand for data type INT.

You can find the bit descriptions for the data formats in Chapter 3.4, "Data Types".

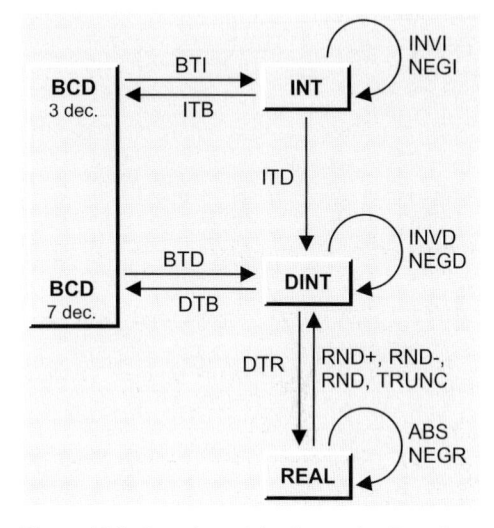

Figure 12.1 Overview of the Conversion Functions

Function

The conversion function is executed if "1" is present at the enable input (if current flows in input EN). If an error occurs during conversion, the enable output ENO is set to "0"; otherwise, it is set to "1". If execution of the function is not enabled (EN = "0"), the conversion does not take place and ENO is also "0".

IF EN == "1" or not wired		
THEN		ELSE
OUT := Confct (IN)		
IF error occured		
THEN	ELSE	
ENO := "0"	ENO := "1"	ENO := "0"

with Confct as conversion function

If the Master Control Relay (MCR) is active, output OUT is set to zero when the conversion function is processed (EN = "1"). The MCR has no effect on output ENO.

12 Conversion Functions

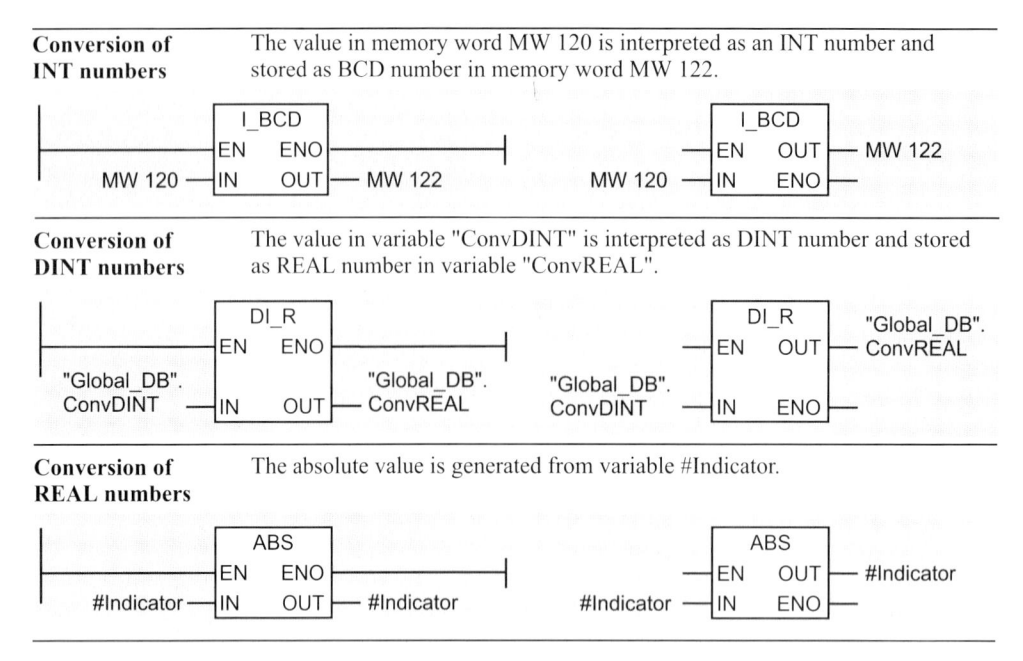

Conversion of INT numbers
The value in memory word MW 120 is interpreted as an INT number and stored as BCD number in memory word MW 122.

```
        I_BCD                               I_BCD
    ┤EN    ENO├                          ┤EN    OUT├─ MW 122
MW 120 ┤IN   OUT├─ MW 122        MW 120 ┤IN    ENO├
```

Conversion of DINT numbers
The value in variable "ConvDINT" is interpreted as DINT number and stored as REAL number in variable "ConvREAL".

```
              DI_R                                DI_R        "Global_DB".
         ┤EN    ENO├                          ┤EN    OUT├─ ConvREAL
"Global_DB".      "Global_DB".    "Global_DB".
ConvDINT ┤IN   OUT├─ ConvREAL     ConvDINT ┤IN    ENO├
```

Conversion of REAL numbers
The absolute value is generated from variable #Indicator.

```
              ABS                                 ABS
         ┤EN    ENO├                          ┤EN    OUT├─ #Indicator
#Indicator ┤IN   OUT├─ #Indicator  #Indicator ┤IN    ENO├
```

Figure 12.2 Examples of Conversion Functions

Not all conversion functions report errors. An error occurs only if the permissible number range is exceeded (I_BCD, DI_BCD) or an invalid REAL number if specified (FLOOR, CEIL, ROUND, TRUNC).

If the input value for a BCD_I or BCD_DI conversion contains a pseudo tetrad, program execution is interrupted and error organization block OB 121 (programming errors) called.

Chapter 15, "Status Bits", explains how the conversion functions set the status bits.

Figure 12.2 shows one example for each data type. A conversion function converts according to the specific characteristic even if no data types have been declared when using operands with absolute addresses.

In the case of incremental programming, you will find the conversion functions in the Program Elements Catalog (with VIEW → CATALOG [Ctrl – K] or INSERT → PROGRAM ELEMENTS) under "Convert".

Conversion function in a rung (LAD)

You can connect contacts in series and in parallel before the EN input and after the ENO output.

The conversion box itself may be placed after a T-branch or in a branch that leads directly to the left power rail. This branch can also have contacts before input EN and it need not be the uppermost branch.

With the direct connection to the left power rail you can thus connect conversion boxes in parallel. When connecting boxes in parallel, you require a coil to terminate the rung. If you have not provided error evaluation, assign a "dummy" operand to the coil, for instance a temporary local data bit.

You can connect conversion boxes in series. If the ENO output of the preceding box leads to the EN output of the subsequent box, the subsequent box is processed only if the preceding box has been completed without errors. If you want to use the result from the preceding box as the input value for a subsequent box, variables from the temporary local data area make convenient intermediate buffers.

If you arrange several conversion boxes in one rung (parallel to the left power rail and then further in series), the boxes in the uppermost branch are processed first from left to right, followed by the boxes in the second branch from left to right, and so on.

The "LAD_Book" library on the diskette supplied with the book contains further examples for the representation and arrangement of the conversion functions (FB 112 in the "Digital Functions" program).

Conversion function in a logic circuit (FBD)

EN and ENO need not be wired. If you want to process the conversion box in dependence on specific conditions, you can arrange binary logic operations before the EN input. You can connect the ENO output with binary inputs of other functions; for example, you can arrange conversion boxes in series, whereby the ENO output of the preceding box leads to the EN input of the subsequent box. If you want to use the result from the preceding box as input value for a subsequent box, variables from the temporary local data area make good temporary buffers.

The "FBD_Book" library on the diskette supplied with the book contains further examples for the representation and arrangement of the conversion functions (FB 112 in the "Digital Functions" program).

12.2 Conversion of INT and DINT Numbers

Table 12.1 shows the conversion functions for INT and DINT numbers. Variables of the specified data type or absolute-addressed operands of the relevant size must be applied to the inputs and outputs of the boxes (for example a word operand for data type INT).

Conversion from INT to DINT

The function I_DI interprets the value at input IN as a number of data type INT and transfers it to the low-order word of output OUT. The sig-

Table 12.1 Conversion of INT and DINT Numbers

Data Type Conversion	Box	Data Type for Parameter	
		IN	OUT
INT to DINT	I_DI	INT	DINT
INT to BCD	I_BCD	INT	WORD
DINT to BCD	DI_BCD	DINT	DWORD
DINT to REAL	DI_R	DINT	REAL

nal state of bit 15 (the sign) of the input is transferred to bits 16 to 31 of the high-order word of output OUT.

The conversion from INT to DINT reports no errors.

Conversion from INT to BCD

The function I_BCD interprets the value at input IN as a number of data type INT and converts it to a 3-decade BCD number at output OUT. The three right-justified decades represent the absolute value of the decimal number. The sign is in bits 12 to 15. If all bits are "0", the sign is positive; if all bits are "1", the sign is negative.

If the INT number is too large to be converted into BCD (> 999), the I_BCD function sets status bits OV and OS. The conversion does not take place.

Conversion from DINT to BCD

The function DI_BCD function interprets the value at input IN as a number of data type DINT and converts it to a 7-decade BCD number at output OUT. The seven right-justified decades represent the absolute value of the decimal number. The sign is in bits 28 to 31. If all bits are "0", the sign is positive; if all bits are "1", the sign is negative.

If the DINT number is too large to be converted to a BCD number (> 9 999 999), status bits OV and OS are set. The conversion does not take place.

Conversion from DINT to REAL

The function DI_R interprets the value at input IN as a number of data type DINT and converts it to a REAL number at output OUT.

Since a number in DINT format has a higher accuracy than a number in REAL format, rounding make take place during the conversion. The REAL number is then rounded to the next whole number (in accordance with the ROUND function).

The DI_R function does not report errors.

12.3 Conversion of BCD Numbers

Table 12.2 shows the functions for converting BCD numbers. Variables of the specified type of absolute-addressed operands of the relevant size must be applied to the input and outputs of the boxes (for example a word operand for data type INT).

Conversion from BCD to INT

The function BCD_I interprets the value at input IN as a 3-decade BCD number and converts it to an INT number at output OUT. The three right-justified decades represent the absolute value of the decimal number. The sign is in bits 12 to 15. If these bits are "0", the sign is positive; if they are "1", the sign is negative. Only bit 15 is taken into account in the conversion.

If the BCD number contains a pseudo tetrad (numerical value 10 to 15 or A to F in hexadecimal), the CPU signals a programming error and calls organization block OB 121 (synchronization error). If this block is not available, the CPU goes to STOP.

The function BCD_I sets no status bits.

Conversion from BCD to DINT

The function BCD_DI interprets the value at input IN as a 7-decade BCD number and converts it to an INT number at output OUT. The seven right-justified decades represent the absolute value of the decimal number. The sign is in bits 28 to 31. If these bits are "0", the sign is positive; if they are "1", the sign is negative. Only bit 31 is taken into account in the conversion.

If the BCD number contains a pseudo tetrad (numerical value 10 to 15 or A to F in hexadecimal), the CPU signals a programming error and calls organization block OB 121 (synchronization error). If this block is not available, the CPU goes to STOP.

The function BCD_I sets no status bits.

12.4 Conversion of REAL Numbers

There are several functions for converting a number in REAL format to DINT format (conversion of a fractional value to a whole number) (Table 12.3). They differ as regards rounding. Variables of the specified data type or absolute-addressed doubleword operands must be applied to the inputs and outputs of the boxes.

Rounding to the next higher integer number

The function CEIL interprets the value at input IN as a number in REAL format and converts it to a number in DINT format at output OUT. CEIL returns an integer that is greater than or equal to the number to be converted.

If the value at input IN exceeds or falls short of the range permissible for a number in DINT format or if it does not correspond to a number in REAL format, CEIL sets status bits OV and OS. Conversion does not take place.

Table 12.3
Conversion of REAL Numbers to DINT Numbers

Data Type Conversion with Rounding	Box	Data Type for Parameter	
		IN	OUT
To next higher integer number	CEIL	REAL	DINT
To next lower integer number	FLOOR	REAL	DINT
To next integer number	ROUND	REAL	DINT
Without rounding	TRUNC	REAL	DINT

Table 12.2 Conversion of BCD Numbers

Data Type Conversion	Box	Data Type for Parameter	
		IN	OUT
BCD to INT	BCD_I	WORD	INT
BCD to DINT	BCD_DI	DWORD	DINT

12.4 Conversion of REAL Numbers

Table 12.4 Rounding Modes for the Conversion of REAL Numbers

Eingangswert REAL	DW#16#	Ergebnis ROUND	CEIL	FLOOR	TRUNC
1.0000001	3F80 0001	1	2	1	1
1.00000000	3F80 0000	1	1	1	1
0.99999995	3F7F FFFF	1	1	0	0
0.50000005	3F00 0001	1	1	0	0
0.50000000	3F00 0000	0	1	0	0
0.49999996	3EFF FFFF	0	1	0	0
5.877476E–39	0080 0000	0	1	0	0
0.0	0000 0000	0	0	0	0
–5.877476E–39	8080 0000	0	0	–1	0
–0.49999996	BEFF FFFF	0	0	–1	0
–0.50000000	BF00 0000	0	0	–1	0
–0.50000005	BF00 0001	–1	0	–1	0
–0.99999995	BF7F FFFF	–1	0	–1	0
–1.00000000	BF80 0000	–1	–1	–1	–1
–1.0000001	BF80 0001	–1	–1	–2	–1

Rounding to the next lower integer number

The function FLOOR interprets the value at input IN as a number in REAL format and converts it to a number in DINT format at output OUT. FLOOR returns an integer that is less than or equal to the number to be converted.

If the value at input IN exceeds or falls short of the range permissible for a number in DINT format or if it does not correspond to a number in REAL format, FLOOR sets status bits OV and OS. Conversion does not take place.

Rounding to the next whole number

The function ROUND interprets the value at input IN as a number in REAL format and converts it to a number in DINT format at output OUT. ROUND returns the next whole number. If the result lies exactly between an odd and an even number, the even number is given preference.

If the value at input IN exceeds or falls short of the range permissible for a number in DINT format or if it does not correspond to a number in REAL format, ROUND sets status bits OV and OS. Conversion does not take place.

No rounding

The function TRUNC interprets the value at input IN as a number in REAL format and converts it to a number in DINT format at output OUT. TRUNC returns the integer component of the number to be converted; the fractional component is "truncated".

If the value at input IN exceeds or falls short of the range permissible for a number in DINT format or if it does not correspond to a number in REAL format, TRUNC sets status bits OV and OS. Conversion does not take place.

Summary of conversions from REAL to DINT

Table 12.4 shows the different effects of the functions for converting from REAL to DINT. The range –1 to +1 has been chosen as example.

12.5 Miscellaneous Conversion Functions

Other available conversion functions are one's complement generation, negation, and absolute-value generation of a REAL number (Table 12.5). Variables of the specified data type or absolute-addressed operands of the relevant size must be applied to the inputs and outputs of the boxes (for example a doubleword operand for data type DINT).

One's complement INT

The function INV_I negates the value at input IN bit for bit and writes it to output OUT. INV_I replaces the zeroes with ones and vice versa. The function INV_I does not signal errors.

One's complement DINT

The function INV_DI negates the value at input IN bit for bit and writes it to output OUT. INV_DI replaces the zeroes with ones and vice versa. The function INV_DI does not signal errors.

Negation INT

The function NEG_I interprets the value at input IN as an INT number, changes the sign through two's complement generation, and writes the converted value to output OUT. NEG_I is identical to multiplication with –1. The function NEG_I sets status bits CC0, CC1, OV and OS.

Negation DINT

The function NEG_DI interprets the value at input IN as a DINT number, changes the sign through two's complement generation, and writes the converted value to output OUT. NEG_DI is identical to multiplication by –1. The function NEG_DI sets status bits CC0, CC1, OV and OS.

Negation REAL

The function NEG_R interprets the value at input IN as a REAL number, multiplies this number by –1, and writes it to output OUT. NEG_R changes the sign of the mantissa even on an invalid REAL number. NEG_R does not signal errors.

Absolute-value generation REAL

The ABS function interprets the value at input IN as a REAL number, generates the absolute value from this number, and writes it to output OUT. ABS sets the sign of the mantissa to "0" even on an invalid REAL number. ABS does not signal errors.

Table 12.5 Miscellaneous Conversion Functions

Conversion	Box	Data Type for Parameter IN	Data Type for Parameter OUT
One's complement INT	INV_I	INT	INT
One's complement DINT	INV_DI	DINT	DINT
Negation of an INT number	NEG_I	INT	INT
Negation of a DINT number	NEG_DI	DINT	DINT
Negation of a REAL number	NEG_R	REAL	REAL
Absolute value of a REAL number	ABS	REAL	REAL

13 Shift Functions

The shift functions shift the contents of a variable bit by bit to the left or right. The bits shifted out are either lost or are used to pad the other side of the variable. Table 13.1 provides an overview of the shift functions.

13.1 Processing a Shift Function

13.1.1 Representation

In addition to an enable input EN and an enable output ENO, the box for a shift function has an input IN, an input N, and an output OUT. The "header" in the box identifies the shift function executed (for example, SHL_W stands for shifting a word variable to the left).

Shift box
(in the example: Shift word variable to the left)

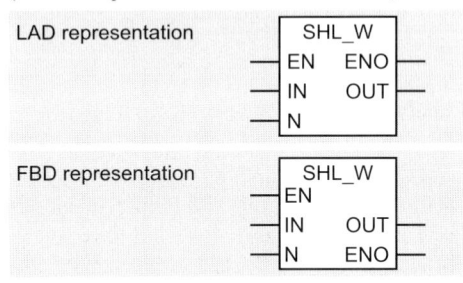

LAD representation

```
      SHL_W
   ─┤EN   ENO├─
   ─┤IN   OUT├─
   ─┤N
```

FBD representation

```
      SHL_W
   ─┤EN
   ─┤IN   OUT├─
   ─┤N    ENO├─
```

The value to be shifted is at input IN, the number of places to be shifted is at input N, and the result is at output OUT. The input and the output have different data types depending on the shift function. For example, input and output are of type DWORD for the shift function SHR_DW (shift a doubleword variable to the right). The variables applied must be of the same data type as the input or output. If you use operands with absolute addresses, the operand sizes must accord with the data types; for instance, you can use a word operand for data

Table 13.1 Overview of Shift Functions

Shift Functions	Word Variable	Doubleword Variable
Shift left	SHL_W	SHL_DW
Shift right	SHR_W	SHR_DW
Shift with sign	SHR_I	SHR_DI
Rotate left	-	ROL_DW
Rotate right	-	ROR_DW

type INT. Input N has data type WORD for every shift function.

See Chapter 3.4, "Data Types", for a description of the bits in each data format.

Function

The shift function is executed if "1" is present at the enable input or when "power" flows through input EN. ENO is then "1". If execution of the function is not enabled (EN = "0"), the shift does not take place and ENO is also "0".

IF EN == "1" or not wired	
THEN	ELSE
OUT := Sfct (IN, N)	
ENO := "1"	ENO := "0"

with Sfct as shift function

When the Master Control Relay (MCR) is activated, output OUT is set to zero when the shift function is processed (EN = "1"). The MCR does not affect output ENO.

Chapter 15, "Status Bits", explains how the shift functions set the status bits.

Examples

Figure 13.1 gives one example each for various shift functions.

13 Shift Functions

Shifting word variables

The value in memory word MW 130 is shifted 4 positions to the left and stored in memory word MW 132.

```
     SHL_W                              SHL_W
    ┤EN  ENO├                          ┤EN
MW 130┤IN  OUT├─ MW 132         MW 130 ┤IN  OUT├─ MW 132
W#16#4┤N                        W#16#4 ┤N   ENO├
```

Shifting double-word variables

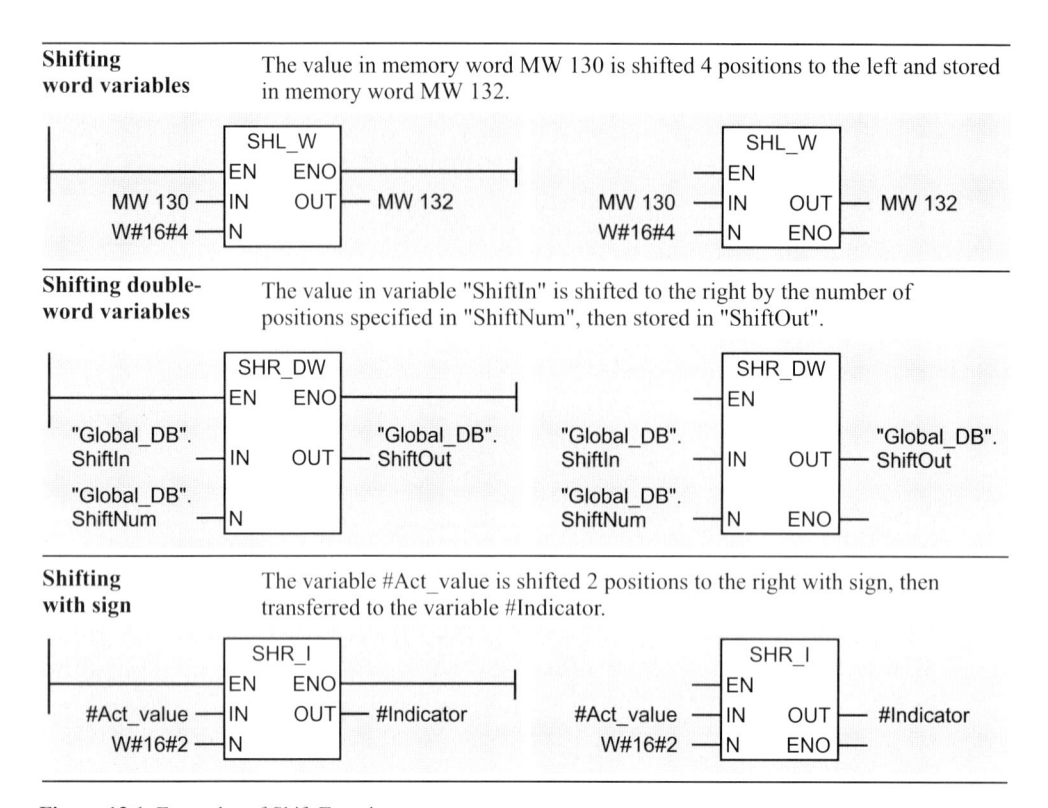

The value in variable "ShiftIn" is shifted to the right by the number of positions specified in "ShiftNum", then stored in "ShiftOut".

```
            SHR_DW                               SHR_DW
           ┤EN  ENO├                            ┤EN
"Global_DB".                    "Global_DB".
 ShiftIn   ┤IN  OUT├─ "Global_DB".  ShiftIn    ┤IN  OUT├─ "Global_DB".
                      ShiftOut                              ShiftOut
"Global_DB".                    "Global_DB".
 ShiftNum  ┤N                    ShiftNum      ┤N   ENO├
```

Shifting with sign

The variable #Act_value is shifted 2 positions to the right with sign, then transferred to the variable #Indicator.

```
             SHR_I                               SHR_I
            ┤EN  ENO├                           ┤EN
#Act_value  ┤IN  OUT├─ #Indicator    #Act_value┤IN  OUT├─ #Indicator
W#16#2      ┤N                       W#16#2    ┤N   ENO├
```

Figure 13.1 Examples of Shift Functions

Shift function in a rung (LAD)

You can connect contacts in series and in parallel before input EN and after output ENO.

The shift box itself may be placed after a T-branch or in a branch that leads directly to the left power rail. This branch can also have contacts before input EN and it need not be the uppermost branch.

The direct connection to the left power rail allows you to connect shift boxes in parallel. When connecting boxes in parallel, you require a coil to terminate the rung. If you have not provided error evaluation, assign a "dummy" operand to the coil, for example a temporary local data bit.

You can connect shift boxes in series. If the ENO output from the preceding box leads to the EN input of the subsequent box, the subsequent is always processed. If you want to use the result from the preceding box as input value for a subsequent box, variables from the temporary local data area make a convenient intermediate buffer.

If you arrange several boxes in one rung (parallel to the left power rail and then further in series), the boxes in the uppermost branch are processed first from left to right, followed by the boxes in the second branch from left to right, and so on.

The "LAD_Book" library on the diskette supplied with the book contains further examples for the representation and arrangement of the shift functions (FB 113 in the "Digital Functions" program).

Shift function in a rung (FBD)

EN and ENO need not be wired. If you want the shift box processed in dependence on specific conditions, you can arrange binary logic operations before the EN input. You can connect the ENO output with binary inputs of other functions; for instance, you can arrange shift boxes in series, whereby the EBO output of the preceding box leads to the EN output of the subsequent box. If you want to use the result from the preceding box as input value to a subsequent box, variables from the temporary local data area make convenient intermediate buffers.

The "FBD_Book" library on the diskette supplied with the book contains further examples for the representation and arrangement of the shift functions (FB 113 in the "Digital Functions" program).

13.2 Shift

Shift word variable to the left

The shift function SHL_W shifts the contents of the WORD variable at input IN bit by bit to the left the number of positions specified by the shift number at input N. The bit positions freed up by the shift are padded with zeroes. The WORD variable at output OUT contains the result.

The shift number at input N specifies the number of bit positions by which the contents are to be shifted. It can be a constant or a variable. If the shift number is 0, the function is not executed; if it is greater than 15, the output variable contains zero following execution of the SHL_W function.

Shift doubleword variable to the left

The shift function SHL_DW shifts the contents of the DWORD variable at input IN bit by bit to the left the number of positions specified by the shift number at input N. The bit positions freed up by the shift are padded with zeroes. The DWORD variable at output OUT contains the result.

The shift number at input N specifies the number of bit positions by which the contents are to be shifted. It can be a constant or a variable. If the shift number is 0, the function is not executed; if it is greater than 31, the output variable contains zero following execution of the SHL_DW function.

Shift word variable to the right

The shift function SHR_W shifts the contents of the WORD variable at input IN bit by bit to the right the number of positions specified by the shift number at input N. The bit positions freed up by the shift are padded with zeroes. The WORD variable at output OUT contains the result.

The shift number at input N specifies the number of bit positions by which the contents are to be shifted. It can be a constant or a variable. If the shift number is 0, the function is not executed; if it is greater than 15, the output variable contains zero following execution of the SHR_W function.

Shift doubleword variable to the right

The shift function SHR_DW shifts the contents of the DWORD variable at input IN bit by bit to the right the number of positions specified by the shift number at input N. The bit positions freed up by the shift are filled with zeroes. The DWORD variable at output OUT contains the result.

The shift number at input N specifies the number of bit positions by which the contents are to be shifted. It can be a constant or a variable. If the shift number is 0, the function is not executed; if it is greater than 31, the output variable contains zero following execution of the SHR_DW function.

Shift word variable with sign

The shift function SHR_I shifts the contents of the INT variable at input IN bit by bit to the right the number of positions specified by the shift number at input N. The bit positions freed up by the shift are filled with the signal state of bit 31 (which is the sign of the INT number), that is, with "0" if the number is positive and with "1" if the number is negative. The variable at output OUT contains the re

The shift number at input N specifies the number of bit positions by which the contents are to be shifted. It can be a constant or a variable. If the shift number is 0, the function is not executed; if it is greater than 15, all bit positions in the output variable contain the sign following execution of the SHR_I function.

For a number in data format INT, shifting to the right is equivalent to division with an exponential number to base 2. The exponent is the shift number. The result of such a division corresponds to the integer number rounded down.

Shift doubleword variable with sign

The shift function SHR_DI shifts the contents of the DINT variable at input IN bit by bit to the right the number of positions specified by the shift number at input N. The bit positions freed up by the shift are padded with the signal state of bit 15 (which is the sign of the DINT number), that is, with "0" if the number is positive and with "1" if the number is negative. The DINT variable at output OUT contains the result.

The shift number at input N specifies the number of bit positions by which the contents are to be shifted. It can be a constant or a variable. If the shift number is 0, the function is not executed; if it is greater than 31, all bit positions in the output variable contain the sign following execution of the SHR_DI function.

For a number in data format DINT, shifting to the right is equivalent to division with an exponential number to base 2. The exponent is the shift number. The result of such a division corresponds to the integer number rounded down.

13.3 Rotate

Rotate doubleword variable to the left

The shift function ROL_DW shifts the contents of the DWORD variable at input IN bit by bit to the left the number of positions specified in the shift number at input N. The bit positions freed up by the shift are padded with the bit positions that were shifted out. The DWORD variable at output OUT contains the result.

The shift number at input N specifies the number of bit positions by which the contents are to be shifted. It can be a constant or a variable. If the shift number is 0, the function is not executed; if it is 32, the contents of the input variable are retained and the status bits are set. If the shift number is 33, a shift of one position is executed; if it is 34, a shift of two positions is executed, and so on (shifting is carried out modulo 32).

Rotate doubleword variable to the right

The shift function ROR_DW shifts the contents of the DWORD variable at input IN bit by bit to the right the number of positions specified by the shift number at input N. The bit positions freed up by the shift are padded with the bit positions that were shifted out. The DWORD variable at output OUT contains the result.

The shift number at input N specifies the number of bit positions by which the contents are to be shifted. It can be a constant or a variable. If the shift number is 0, the function is not executed; if it is 32, the contents of the input variable are retained and the status bits are set. If the shift number is 33, a shift of one position is executed; if the shift number is 34, a shift of two positions is executed, and so on (shifting is executed modulo 32).

14 Word Logic

Word logic combines the values of two variables bit by bit according to AND, OR, or Exclusive OR. The logic operation may be applied to words or doublewords. The available word logic operations are listed in Table 14.1.

14.1 Processing a Word Logic Operation

Representation

In addition to the enable input EN and the enable output ENO, the box for a word logic operation has two inputs, IN1 and IN2, and one output, OUT. The "header" in the box identifies the word logic operation executed (for example, WAND_W stands for word ANDing).

Word logic box
(in the example: ANDing by word)

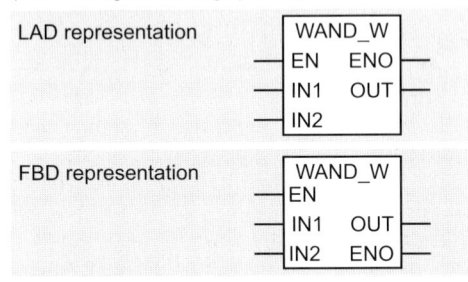

LAD representation	WAND_W – EN ENO – – IN1 OUT – – IN2

FBD representation	WAND_W – EN – IN1 OUT – – IN2 ENO –

The values to be combined are at inputs IN1 and IN2, and the result of the operation is at output OUT. The inputs and the output have different data types, depending on the operation: WORD for word (16-bit) operations, and DWORD for doubleword (32-bit) operations. The applied variables must be of the same data type as the inputs or the output.

See Section 3.4, "Data Types", for a description of the bits in the various data formats.

Table 14.1 Overview of Word Logic Operations

Word logic operation	With a word variable	doubleword variable
AND	WAND_W	WAND_DW
OR	WOR_W	WOR_DW
Exclusive OR	WXOR_W	WXOR_DW

Function

The word logic operation is executed when "1" is present at the enable input (when power flows through the input EN). If execution of the operation is not enabled (EN = "0"), the operation does not take place and ENO is also "0".

IF EN == "1" or not wired	
THEN	ELSE
OUT := IN1 Wlog IN2	
ENO := "1"	ENO := "0"

with Wlog as word logic

If the Master Control Relay (MCR) is active, output OUT is set to zero when the word logic operation is executed (EN = "1"). The MCR does not affect the ENO output.

Word logic operations generate a result bit by bit. Bit 0 of input IN1 is combined with bit 0 of input IN2, and the result is stored in bit 0 of output OUT. The same is done with bit 1, bit 2, and so on, up to bit 15 resp. 31. Table 14.2 shows the result formation for a single bit.

Chapter 15, "Status Bits", explains how the word logic operations set the status bits.

Examples

Figure 14.1 shows one example for each word logic operation.

In the case of incremental programming, you will find the word logic operations in the Pro-

14 Word Logic

Table 14.2 Result Formation in Word Logic Operations

Contents of input IN1	0	0	1	1
Contents of input IN2	0	1	0	1
Result with AND	0	0	0	1
Result with OR	0	1	1	1
Result with Exclusive OR	0	1	1	0

gram Elements Catalog (with VIEW → CATALOG [Ctrl – K] or INSERT → PROGRAM ELEMENTS) under "Word Logic".

Word logic in a rung (LAD)

You can arrange contacts in series and in parallel before input EN and after output ENO.

The word logic box itself may be placed after a T-branch or in a branch that leads directly to the left power rail. This branch can also have contacts before input EN and it need not be the uppermost branch.

The direct connection to the left power rail allows you to connect word logic boxes in parallel. When connecting boxes in parallel, you require a coil to terminate the rung. If you have not provided error evaluation, assign a "dummy" operand to the coil, for example a temporary local data bit.

You can connect word logic boxes in series. If the ENO output from the preceding box leads to the EN input of the subsequent box, the subsequent box is always processed. If you want to use the result of the preceding box as input value for a subsequent box, variables from the temporary local data area make convenient intermediate buffers.

If you arrange several word logic boxes in one rung (parallel to the left power rail and then further in series), the boxes in the uppermost branch are processed first from left to right, followed by the boxes in the second branch from left to right, and so on.

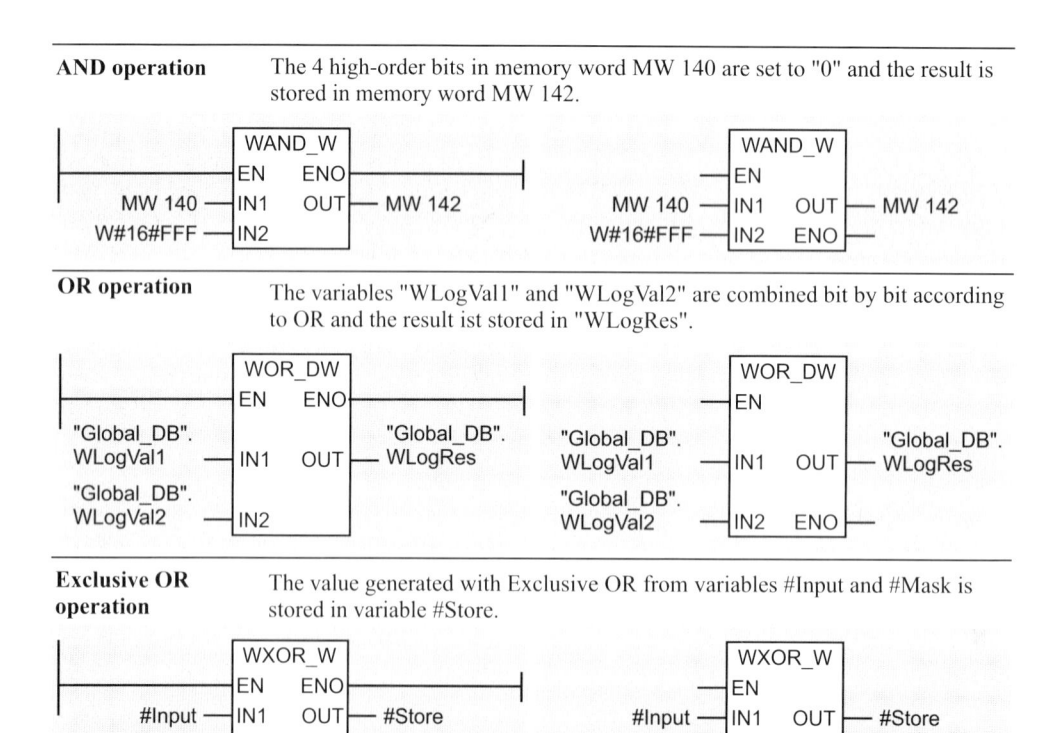

AND operation The 4 high-order bits in memory word MW 140 are set to "0" and the result is stored in memory word MW 142.

OR operation The variables "WLogVal1" and "WLogVal2" are combined bit by bit according to OR and the result ist stored in "WLogRes".

Exclusive OR operation The value generated with Exclusive OR from variables #Input and #Mask is stored in variable #Store.

Figure 14.1 Examples of Word Logic Operations

The "LAD_Book" library on the diskette supplied with the book contains further examples for the representation and arrangement of word logic operations (FB 114 in the "Digital Functions" program).

Word logic in a logic circuit (FBD)

EN and ENO need not be wired. If you want to have the word logic box processed in dependence on specific conditions, you can arrange binary logic operations before the EN input. You can connect the ENO output with binary inputs from other functions; for example, you can arrange word logic boxes in series, whereby the ENO output of the preceding box leads to the EN input of the subsequent box. If you want to use the result from the preceding box as input value for a subsequent box, variables in the temporary local data area make convenient intermediate buffers.

The "FBD_Book" library on the diskette supplied with the book contains further examples for representation and arrangement of word logic operations (FB 114 in the "Digital Functions" program).

14.2 Description of the Word Logic Operations

AND operation

AND combines the individual bits of the value at input IN1 with the corresponding bits of the value at input IN2 according to AND. A bit in result word OUT will be "1" only if the corresponding bit in both values to be combined is "1".

Since the bits that are "0" at input IN2 also set the corresponding result bits to "0" regardless of what value these bits have at input IN1, we also refer to these bits as being "masked". Masking is the primary use for the (digital) AND operation.

OR operation

OR combines the individual bits of the value at input IN1 with the corresponding bits of the value at input IN2 according to OR. A bit result word OUT will be "0" only if the corresponding bit in both values to be combined is "0".

Since the bits that are "1" at input IN2 also set the corresponding result bits to "1" regardless of what value these bits have at input IN1, we also refer to these bits as being "masked". Masking is the primary use for the (digital) OR operation.

Exclusive OR operation

Exclusive OR combines the individual bits of the value at input IN1 with the corresponding bits of the value at input IN2 according to Exclusive OR. A bit in result word OUT will be "1" only if the corresponding bit in only one of the two values to be combined is "1". If a bit at input IN2 is "1", the corresponding bit in the result is the reverse of the bit at the same position in IN1.

In the result, only those bits with opposing signal states in IN1 and IN2 prior to execution of the digital Exclusive OR operation will be "1". Locating bits with opposing signal states or "negating" the signal states of individual bits is the primary use for the (digital) Exclusive OR operation.

Program Flow Control

LAD and FBD provide you with a variety of options for controlling the flow of the program. You can exit linear program execution within a block or you can structure the program with programmable block calls. You can affect program execution in dependence on values calculated at runtime, or in dependence on process parameters, or in accordance with your plant status.

The **status bits** provide information on the result of an arithmetic or mathematical function and on errors (such as a range violation during a calculation). You can incorporate the signal states of the status bits directly in your program using contacts.

You can use **jump functions** to branch unconditionally or in dependence on the RLO.

A further method of affecting program execution is provided by the **Master Control Relay** (MCR). Originally developed for relay contactor controls, LAD and FBD offer a software version of this program control method.

You can use **block functions** to structure your program. You can use functions and function blocks again and again by defining **block parameters**.

Chapter 19, "Block Parameters", contains the examples shown in Chapters 5, "Memory Functions", and Chapter 8, "Counters", this time programmed as function blocks with block parameters. These function blocks are then also called in the "Feed" example as local instances.

15 Status Bits

Status bits RLO, BR, CC0, CC1 and overflow; setting and evaluating the status bits; using the binary result; EN/ENO

16 Jump Functions

Jump unconditionally; jump in dependence on the RLO

17 Master Control Relay

MCR-dependence; MCR range; MCR zone

18 Block Functions

Block types, block call, block end; status local data; data block register, using data operands; handling data blocks

19 Block Parameters

Parameter declaration; formal parameters, actual parameters; passing parameters to called blocks; Examples: Conveyor belt, parts counter and supply

15 Status Bits

The status bits are binary "flags" (condition code bits). The CPU uses them for controlling the binary logic operations and sets them during digital processing. You can check these status bits or act upon specific bits. The status bits are combined into a word, the status word. However, you cannot access this status word with LAD or FBD.

15.1 Description of the Status Bits

Table 15.1 shows the available status bits. The CPU uses the binary flags for controlling the binary functions; the digital flags indicate mainly results of arithmetic and mathematical functions.

First check

The /FC status bit steers the binary logic within a logic control system. A bit logic step always starts with /FC = "0" and a binary check instruction, the first check. The first check sets /FC = "1".

Table 15.1 Status Bits

Binary Flags	
/FC	First Check
RLO	Result of logic operation
STA	Status
OR	Status bit OR
BR	Binary result
Digital Flags	
OS	Stored overflow
OV	Overflow
CC0	Condition code (status) bit 0
CC1	Condition code (status) bit 1

The first check corresponds in LAD to the first contact in a network, and in FBD to the first binary function input.

A bit logic step ends with a binary value assignment (e.g. of a single coil or an assignment) or with a conditional jump or a block change. These set /FC = "0".

Result of the logic operation (RLO)

The RLO status bit is the intermediate buffer in binary logic operations. In the first check, the CPU transfers the check result to the RLO, combines the check result with the stored RLO on each subsequent check, and stores the result, in turn, in the RLO.

You can store the RLO with the SAVE coil/box in the binary result BR. Memory functions, timers and counters are controlled using the RLO and certain jump functions are executed. The RLO corresponds in LAD to the power flowing in the rung (RLO = "1" is the same as "power flowing").

Status

Status bit STA corresponds to the signal state of the checked binary operand. In the case of memory functions, the value of STA is the same as the written value or (if no write operation takes place, for example if the RLO = "0" or the MCR is active), STA corresponds to the value of the addressed (and unmodified) binary operand.

In the case if edge evaluations FP or FN, the value of the RLO prior to the edge evaluation is stored in STA. All other binary functions set STA = "1".

The STA status bit has no effect on the processing of the LAD or FBD functions.

Status bit OR

Status bit OR stores the result of a fulfilled series circuit or a fulfilled AND condition and indicates to a subsequently processed parallel circuit or OR function that the result has already been determined. All other binary functions reset the OR status bit.

Overflow

Status bit OV indicates a range violation or the use of invalid REAL numbers. The following functions affect the OV status bit: Arithmetic functions, mathematical functions, some conversion functions, REAL comparison functions.

You can check the OV status bit directly.

Stored overflow

The OS status bit stores a set OV status bit: Whenever the CPU sets status bit OV, it also sets status bit OS. However, while the next properly executed operation resets OV, OS remains set. This provides you with the opportunity of evaluating a range violation or an operation with an invalid REAL number, even at a later point in your program.

You can check status bit OS directly. A block change resets the OS status bit.

Status bits CC0 and CC1 (condition code bits)

Status bits CC0 and CC1 provide information on the result of a comparison function, an arithmetic or mathematical function, a word logic operation, or on the bit shifted out by a shift function.

You can check all combinations of CC0 and CC1 directly (see below).

Binary result

LAD and FBD use status bit BR to implement the EN/ENO mechanism for boxes. You can also set, reset or check status bit BR yourself.

15.2 Setting the Status Bits

The digital functions affect status bits CC0, CC1, OV and OS as shown in Table 15.2. You can check these status bits immediately following the function box.

Status bits in INT and DINT calculations

The arithmetic functions with data formats INT and DINT set all digital status bits. A result of zero sets CC0 and CC1 to "0". CC0 = "0" and CC1 = "1" indicates a positive result, CC0 = "1" and CC1 = "0" indicates a negative result. A range violation sets OV and OS (please note the other meaning of CC0 and CC1 in the case of overflow). Division by zero is indicated by all digital status bits being set to "1".

Status bits in REAL calculations

The arithmetic functions with data format REAL and the mathematical functions set all digital status bits. A result of zero sets CC0 and CC1 to "0". CC0 = "0" and CC1 = "1" indicates a positive result, CC0 = "1" and CC1 = "0" indicates a negative result. A range violation sets OV and OS (please note the other meaning of CC0 and CC1 in the case of overflow). An invalid REAL number is indicated when all digital status bits are set to "1".

A REAL number is referred to as "denormalized" if it is represented with reduced accuracy. The exponent is then zero; the absolute value of a denormalized REAL number is less than $1.175\ 494 \times 10^{-38}$. S7-300 CPUs treat denormalized REAL numbers as through they were zero (also see Section 3.4, "Data Types").

Status bits for conversion functions

Of the conversion functions, the two's complements affect all digital status bits. In addition, the following conversion functions set status bits OV and OS in the event of an error (range violation or invalid REAL number):

▷ I_BCD and DI_BCD:
  Conversion of INT and DINT to BCD

▷ CEIL, FLOOR, ROUND, TRUNC:
  Conversion of REAL to DINT

15.2 Setting the Status Bits

Table 15.2 Setting the Status Bits

INT calculation

The result is:	CC0	CC1	OV	OS
< −32 768 (ADD_I, SUB_I)	0	1	1	1
< −32 768 (MUL_I)	1	0	1	1
−32 768 to −1	1	0	0	-
0	0	0	0	-
+1 to +32 767	0	1	0	-
> +32 767 (ADD_I, SUB_I)	1	0	1	1
> +32 767 (MUL_I)	0	1	1	1
32 768 (DIV_I)	0	1	1	1
(−) 65 536	0	0	1	1
Division by zero	1	1	1	1

DINT calculation

The result is:	CC0	CC1	OV	OS
< −2 147 483 648 (ADD_DI, SUB_DI)	0	1	1	1
< −2 147 483 648 (MUL_DI)	1	0	1	1
−2 147 483 648 to −1	1	0	0	-
0	0	0	0	-
+1 to +2 147 483 647	0	1	0	-
> +2 147 483 647 (ADD_DI, SUB_DI)	1	0	1	1
> +2 147 483 647 (MUL_DI)	0	1	1	1
2 147 483 648(DIV_DI)	0	1	1	1
(−) 4 294 967 296	0	0	1	1
Division by zero (DIV_DI, MOD_DI)	1	1	1	1

REAL calculation

The result is:	CC0	CC1	OV	OS
+ normalized	0	1	0	-
± denormalized	0	0	1	1
± zero	0	0	0	-
− normalized	1	0	0	-
+ infinite (division by zero)	0	1	1	1
− infinite (division by zero)	1	0	1	1
± invalid REAL number	1	1	1	1

Comparison

The result is:	CC0	CC1	OV	OS
equal to	0	0	0	-
greater than	0	1	0	-
less than	1	0	0	-
invalid REAL number	1	1	1	1

Conversion NEG_I

The result is:	CC0	CC1	OV	OS
+1 to +32 767	0	1	0	-
0	0	0	0	-
−1 to −32 767	1	0	0	-
(−) 32 768	1	0	1	1

Conversion NEG_D

The result is:	CC0	CC1	OV	OS
+1 to +2 147 483 647	0	1	0	-
0	0	0	0	-
−1 to −2 147 483 647	1	0	0	-
(−) 2 147 483 648	1	0	1	1

Shift function

The shifted out bit is:	CC0	CC1	OV	OS
"0"	0	0	0	-
"1"	0	1	0	-
with shift number 0	-	-	-	-

Word logic

The result is:	CC0	CC1	OV	OS
zero	0	0	0	-
not zero	0	1	0	-

Status bits for comparison functions

The comparison functions set the CC0 and CC1 status bits. The flags are set independently of the executed comparison function.

Status bits for shift functions

In the case of the shift functions, the signal state of the last bit to be shifted out is transferred to status bit CC1. CC0 and OV are reset.

Status bits for word logic

If the result of the word logic operation is zero (all bits are "0"), CC1 is reset; if at least one bit in the result is "1", CC1 is set. CC0 and OV are reset.

15.3 Evaluating the Status Bits

LAD: You can use the normally-open (NO) and the normally-closed (NC) contact to check the digital status bits and the binary result. Figure 15.1 shows the check with a normally-open contact. The check with a normally-closed contact returns the negated check result. You can handle the NO and NC contacts for evaluating the status bits in exactly the same way as the "normal" contacts. The "LAD_Book" library on the diskette accompanying the book contains examples for evaluating the status bits (FB 115 in the program "Program Flow Control").

LAD representation	FBD representation	Power flows or the check is successful when
>0	>0	Result greater than zero [(A0=0) & (A1=1)]
>=0	>=0	Result greater than or equal to zero [(A0=0)]
<0	<0	Result less than zero [(A0=1) & (A1=0)]
<=0	<=0	Result less than or equal to zero [(A1=0)]
<>0	<>0	Result not equal to zero [(A0=0) & (A1=1) v (A0=1) & (A1=0)]
==0	==0	Result equal to zero [(A0=0) & (A1=0)]
UO	UO	Result invalid (unordered) [(A0=1) & (A1=1)]
OV	OV	Number range overflow [OV=1]
OS	OS	Stored overflow [OS=1]
BR	BR	Binary result [BR=1]

Figure 15.1 Evaluating the Status Bits

FBD: Direct or negated checking of the digital status bits and the binary result is possible. This is shown by direct checking for signal state "1". The check for signal state "0" returns the negated check result. The checks for evaluation of the status bits can be treated exactly the same as the "normal" checks for binary operands. The "FBD_Book" library on the diskette supplied with the book contains examples of status bit evaluation (FB 115 in the "Program Flow Control" program).

In the case of incremental programming, you will find these checks in the Program Elements Catalog (with VIEW → CATALOG [Ctrl – K] or INSERT → PROGRAM ELEMENTS) under "Status Bits".

15.4 Using the Binary Result

15.4.1 Setting the Binary Result BR

Assignment of binary result

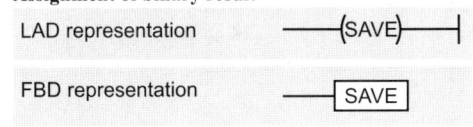

| LAD representation | ——(SAVE)——| |
| FBD representation | ——[SAVE] |

SAVE coil in a rung (LAD)

You store the RLO in the binary result using the SAVE coil. If power flows into the SAVE coil, BR is set, otherwise it is reset. You program the SAVE coil in the same way as a "single" coil without binary operands.

Please note that the SAVE coil does not terminate the logic operation (the /FC status bit is not set to "0"). This means that the logic operation preceding the SAVE coil is also the preceding logic operation for the next network.

The SAVE coil cannot be programmed in conjunction with a T-branch.

SAVE box in a logic circuit (FBD)

With the SAVE box, you save the RLO in the binary result. If the RLO is "1" before the SAVE box, BR is set; otherwise, BR is reset. You program the SAVE box the same way you program an assign box without binary operands.

Please note that the SAVE box does not terminate a logic operation (status bit /FC is not set to "0"). The next network thus begins with an "open" logic operation.

The SAVE box is not permitted after a T-branch.

Controlling the binary result

LAD and FBD affect even the binary result in order to control the ENO output (Figure 15.2). If enable output ENO is wired, its signal state is the same as that of the BR. In certain cases ("BR corresponds to function"), the LAD or FBD function executed sets the binary result as follows:

▷ BR := "1"
   for MOVE, for the shift functions and for the word logic operations

▷ BR := OV
   for the arithmetic and mathematical functions

▷ BR := OV or "1"
   for the conversion functions

▷ BR := BR of the called block in the case of block calls

15.4.2 Main Rung, EN/ENO Mechanism

In programming languages LAD and FBD, many boxes have an enable input EN and an enable output ENO. If the enable input is "1", the function in the box is processed. When the box is processed correctly, the enable output also has signal state "1". If an error occurs during processing of a box (for example, overflow during execution of an arithmetic function), ENO is set to "0". If EN has signal state "0", ENO is also set to "0".

These characteristics of EN and ENO can be used in order to connect several boxes together in a chain, with the enable output leading to the enable input of the next box (Figure 15.3). This means, for example, that the entire chain can be "disabled" (no boxes are processed if input I 1.0 in the example has signal state "0") or the rest of the chain is no longer processed if one box signals an error.

The input EN and the output ENO are not block parameters but statement sequences that the Program Editor itself generates prior to and fol-

15 Status Bits

Is ENO switched ?						
YES			NO			
Is EN switched ?			Is EN switched ?			
YES		NO	YES		NO	
Is EN == "1" ?			Is EN == "1" ?			
YES	NO		YES	NO		
BR corresponds to function	BR := "0"	BR corresponds to function	BR := "1"	BR := "0"		BR not affected

Figure 15.2 General Schematic for Setting the Binary Result

lowing all boxes (also in the case of functions and function blocks). The Program Editor uses the binary result to store the signal state at EN while the block is being processed or to check the error flag from the box.

15.4.3 ENO in the Case of User-written Blocks

The Program Editor provides the call of your own blocks with the enable input EN and the enable output ENO. You can use the enable input EN to call the block conditionally. You can use the ENO output, for example, to signal a group error (signal state "1", if the block has been properly processed; "0" if an error has occurred during processing of the block). All system blocks also signal group errors via BR.

You control the ENO output with the binary result BR. The ENO output has the same signal state that BR has when the block is exited.

For example, BR could be set to "1" at the start of a block. If an error then occurs during processing of the block, for example if a result exceeds the fixed range so that further processing must be prevented, set the binary result to "0" with the SAVE coil/box and jump to the end of the block where the block will be exited (in the event of an error, the condition must supply signal state "0"). Please note that the RET coil/box sets the BR to "1" if you exit the block via this coil/box.

Example of a main rung (LAD)

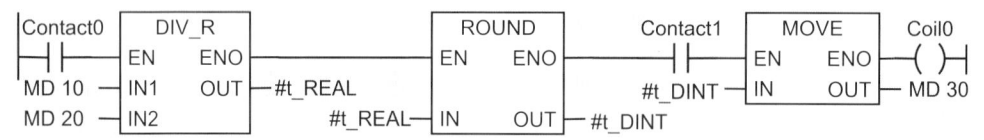

```
Contact0      DIV_R                    ROUND              Contact1      MOVE           Coil0
──┤├──────┤EN    ENO├──────────────┤EN    ENO├──────────┤├───────┤EN    ENO├──────( )──
   MD 10 ──┤IN1   OUT├── #t_REAL        EN    ENO           #t_DINT ──┤IN    OUT├── MD 30
   MD 20 ──┤IN2                   #t_REAL──┤IN   OUT├── #t_DINT
```

Example for the series connection of boxes (FBD)

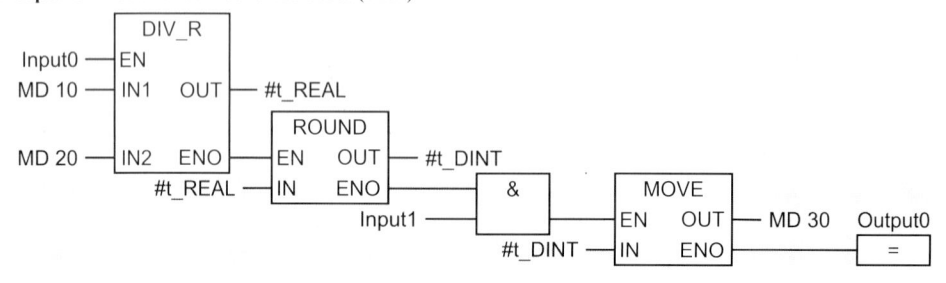

```
                 DIV_R
   Input0 ──────┤EN
   MD 10 ──────┤IN1   OUT├── #t_REAL
                              ROUND
   MD 20 ──────┤IN2   ENO├──┤EN    OUT├── #t_DINT
        #t_REAL ──────────┤IN    ENO├──
                                        &                    MOVE
                               Input1 ──┤                 ──┤EN    OUT├── MD 30  Output0
                              #t_DINT ──┤IN    ENO├──        ──────────────────────┤=├──
```

Figure 15.3 Example for the Series Connection of EN and ENO

16 Jump Functions

You can use jump functions to interrupt the linear flow of the program and continue at another point in the block. This program branching can be executed unconditionally or dependent on the RLO.

16.1 Processing a Jump Function

Representation

LAD representation

Jump if RLO = "1" ——(JMP)—— Dest

Jump if RLO = "0" ——(JMPN)—— Dest

Destination, Jump label | Dest |

FBD representation

Jump if RLO = "1" — | JMP | Dest

Jump if RLO = "0" — | JMPN | Dest

Destination, Jump label | Dest |

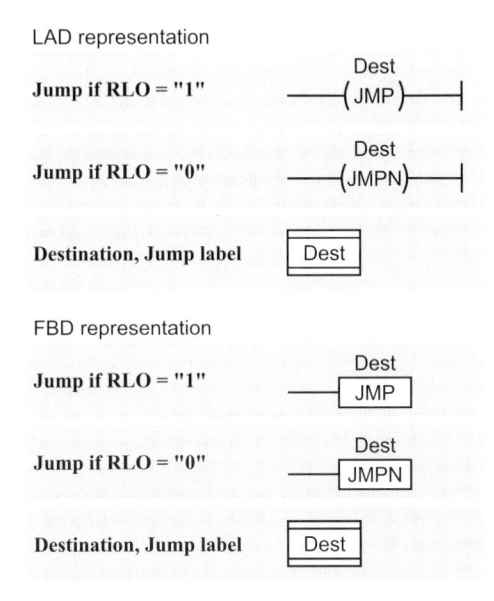

A jump function consists of the jump operation in the form of a coil (LAD) or box (FBD) and a jump label designating the program location at which processing is to continue after the jump. The jump label is above the jump operation.

Jumps cannot be programmed in conjunction with a T-branch.

A jump label consists of up to 4 characters that can include letters, digits, and the underscore. It begins with letter. A jump label in a box designates the network that is to be processed following completion of the jump operation. The box with the jump label must be at the start of a network ("LABEL" in the Program Elements Catalog).

In the case of incremental programming, you will find the jump functions in the Program Elements Catalog (with VIEW → CATALOG [Ctrl - K] or INSERT → PROGRAM ELEMENTS) under "Logic Control / Jump".

Function

A jump is either always executed (absolute or unconditional jump) or it is executed depending on the result of the logic operation (RLO) (conditional jump). In the case of a jump dependent on the RLO, you can decide whether the jump is to be executed if the RLO is "1" or if the RLO is "0".

You can execute both forward (in the direction of program processing; in the direction of ascending network numbers) and backward jumps. The jump can take place only within a block, that is, the jump destination must be in the same block as the jump function. If you use the Master Control Relay (MCR), the jump label must be in the same MCR zone or in the same MCR area as the jump function.

The jump destination must be designated unambiguously, that is, you must only assign a jump label once in a block. The jump destination can be jumped to from several locations.

The Program Editor stores the names of the jump labels in the non-executable section of the relevant blocks on the programming device's data medium. Only the widths of the jumps (jump displacement) are stored in the CPU's ser memory (in the compiled block). When program modifications are made online to blocks in the CPU, these modifications must therefore always be updated on the programming device's data medium in order to retain the original label names. If this update is not

195

made, or if blocks are transferred from the CPU to the programming device, the non-executable block sections will be overwritten or deleted. On the display or the printout, the Editor then generates replacement symbols for the jump labels (M001, M002 etc.).

16.2 Unconditional Jump

The absolute jump, which is the jump that is always executed, is jump function JMP, whose coil is connected to the left power rail (LAD) or whose box has no preceding logic operation (FBD). This jump function is always executed when encountered in the program. The CPU interrupts the linear flow of the program and continues in the network designated by the jump label.

Example (Figure 16.1 and Figure 16.2): In Network 3, there is an unconditional jump to jump label M2. After this network has been processed, the CPU continues program execution at jump label M2 in Network 5. A jump from another program location is required in order to process Network 4.

Network 1 : Conditional jump if RLO = "1"

```
         CMP ==I                              M1
                                            (JMP)
#CompVal1 —IN1
#CompVal2 —IN2
```

Network 2 : Jump to M1 is not executed

```
Contact0                                    Coil0
  ——| |——————————————————————————————————————( )——
```

Network 3 : Unconditional Jump

```
                                              M2
                                            (JMP)
```

Network 4 : Conditional jump if RLO = "0"

```
  M1
          ADD_I                               M3
         EN   ENO                          (JMPN)
#CalcVal1 —IN1  OUT— #CalcRes
#CalcVal2 —IN2
```

Network 5 : Jump from M2

```
  M2
Contact1                                    Coil1
  ——| |——————————————————————————————————————( )——
```

Network 6 : Jump from M3

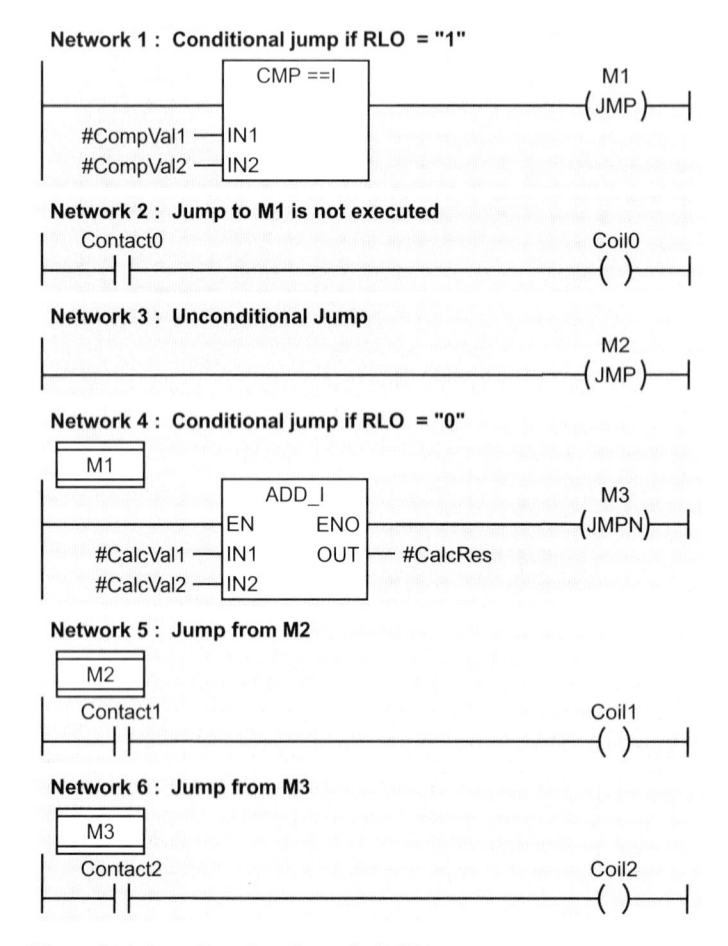

```
  M3
Contact2                                    Coil2
  ——| |——————————————————————————————————————( )——
```

Figure 16.1 Jump Functions Example (LAD)

16.3 Jump if RLO = "1"

The conditional jump if RLO = "1" is jump function JMP, whose coil is not directly connected to the left power rail (LAD) or whose box has a preceding logic operation (FBD). The logic operation preceding this coil can be implemented in any way. If the RLO = "1" (if the preceding logic operation is fulfilled), the CPU interrupts the linear flow of the program and continues in the network designated by the jump label. If the preceding logic operation is not fulfilled, the CPU continues program execution in the next network.

Example (Figure 16.1 and Figure 16.2): If the compare condition in Network 1 is fulfilled, program execution continues in Network 4. If the compare condition is not fulfilled, the next network, which is Network 2, is processed.

16.4 Jump if RLO = "0"

The conditional jump if RLO = "0" is jump function JMPN, whose coil is not directly connected to the left power rail (LAD) or whose box has a preceding logic operation (FBD). The logic operation preceding this coil can be implemented in any way. If the RLO = "0" (if the preceding logic operation is not fulfilled), the CPU interrupts the linear flow of the program and continues in the network designated by the jump label. If the preceding logic operation is fulfilled, the CPU continues program execution in the next network.

Example (Figure 16.1 and Figure 16.2): If the adder in Network 4 signals an error, the jump to Network 6 (jump label M3) is executed. If no error is signaled, Network 5 is the next to be processed.

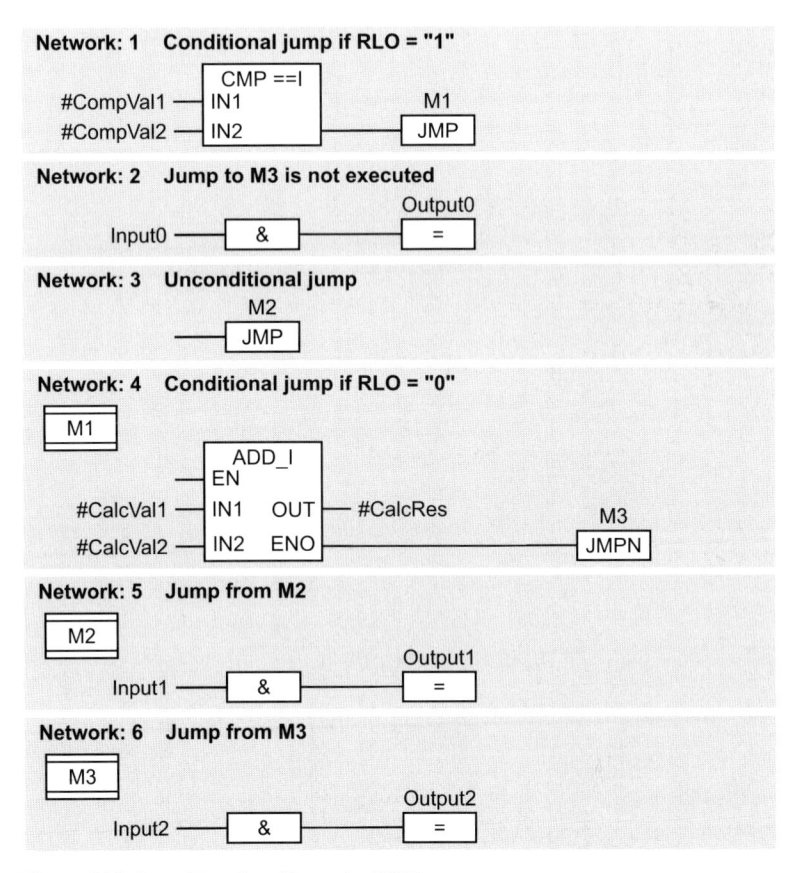

Figure 16.2 Jump Functions Example (FBD)

17 Master Control Relay

In contact control systems, a Master Control Relay activates or deactivates a section of the control system that can consist of one or more rungs.

A deactivated rung

▷ deenergizes all non-retentive contactors and
▷ retains the states of retentive contactors.

You cannot change the state of the contactors again until the Master Control Relay (MCR) is active.

Because of this characteristic feature, the MCR is used most frequently in the LAD programming language, although the FBD programming language also provides the functions required for the MCR.

LAD representation

Activate MCR area	─┤(MCRA)├─
Open MCR zone	──(MCR<)──
Close MCR zone	─┤(MCR>)├─
De-activate MCR area	─┤(MCRD)├─

FBD representation

Activate MCR area	[MCRA]
Open MCR zone	── [MCR<]
Close MCR zone	[MCR>]
De-activate MCR area	[MCRD]

Please note that deenergizing with the "software" Master Control Relay is no substitute for an EMERGENCY OFF or safety facility! Treat Master Control Relay switching in exactly the same way as switching with a memory function!

With MCRA and MCRD, you designate an area in your program in which MCR dependency is to take effect. Within this area, you use MCR< and MCR> to define one or more zones in which the MCR dependency can be enabled and disabled. You can also nest the MCR zones. The result of the logic operation (RLO) immediately preceding an MCR zone enables or disables MCR dependency within that zone.

17.1 MCR Dependency

MCR dependency affects coils and boxes. If MCR dependency is enabled (corresponds to Master Control Relay disabled),

▷ a single coil or assign box and a midline output set the binary operand to signal state "0" (following the midline output, the RLO is then = "0", that is, power no longer flows)

▷ a set and reset coil or box no longer affect the signal state of the binary operand ("freeze it")

▷ an SR and RS box no longer affect the signal state of the binary operand ("freeze it")

▷ a transfer operation writes zero to the digital operand (every function output of a box of digital data type then writes zero to the operand or to the variable)

Additional to the direct influencing of an operand the RLO is then "0" (power no longer flows) behind a midline output and a T-branch.

Some LAD and FBD functions use transfer statements (invisible to the user). Since a transfer statement writes the value zero if MCR dependency is switched on, the corresponding function can no longer be guaranteed.

You must exclude the following program sections from MCR dependency otherwise the CPU will go to STOP or undefined runtime behavior can occur:

▷ Block calls with block parameters

▷ Accesses to block parameters that are parameter types (e.g. BLOCK_DB).

▷ Accesses to block parameters that are components or elements of complex data types or UDTs.

You enable MCR dependency in a zone if the RLO is "0" immediately prior to opening the zone (analogous to disabling the Master Control Relay). If you open an MCR zone with RLO "1" (Master Control Relay enabled), processing within this MCR zone takes place without MCR dependency. MCR dependency is effective only within an MCR zone.

In the case of incremental programming, you will find the MCR functions in the Program Elements Catalog (with VIEW → CATALOG [Ctrl - K] or INSERT → PROGRAM ELEMENTS) under "Program Control".

17.2 MCR Area

To be able to use the characteristic features of the Master Control Relay, define an MCR area with MCRA (start) and MCRD (end of the MCR area). MCR dependency is active within an MCR area (but not yet enabled).

The MCRA coil/box and the MCRD coil/box always stand alone in separate networks.

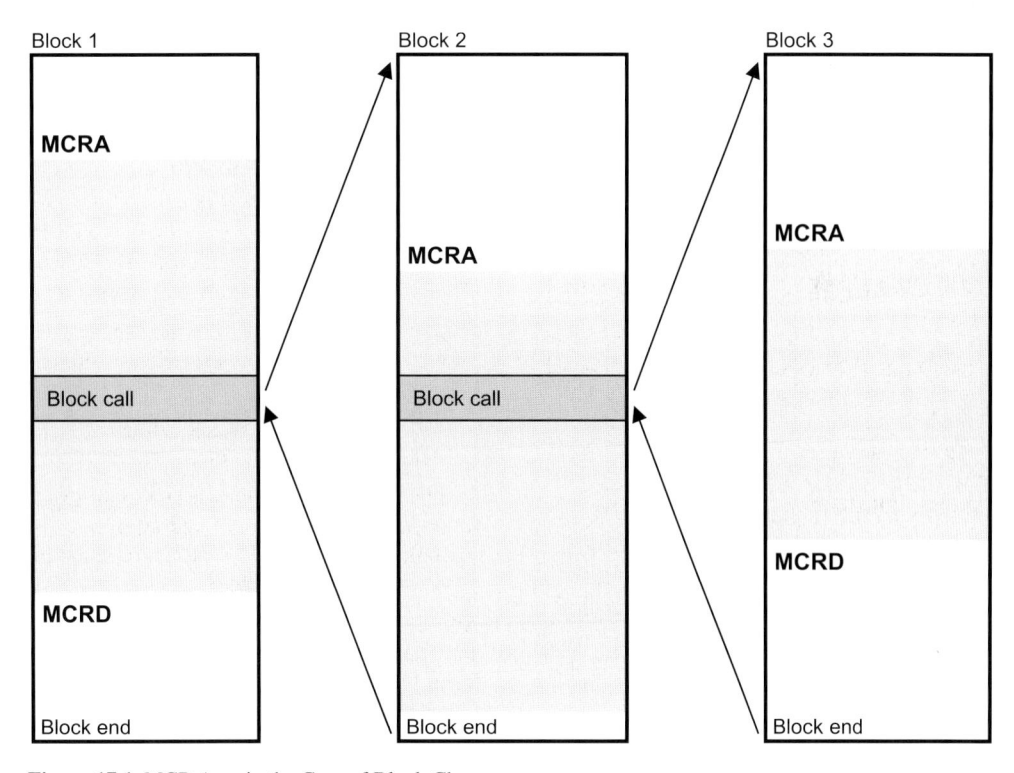

Figure 17.1 MCR Area in the Case of Block Change

If you call a block within an MCR area, MCR dependency is deactivated in the block called (Figure 17.1). An MCR area only starts again with the MCRA coil/box. When a block is exited, MCR dependency is set as it was before the block was called, regardless of the MCR dependency with which the called block was exited.

17.3 MCR Zone

LAD: You define an MCR zone with the MCR< coil (start) and the MCR> coil (end of the MCR zone). The MCR< coil requires a preceding logic operation; the MCR> coil is connected directly to the left power rail. Both coils terminate a rung. Within this zone, you control MCR dependency with the RLO prior to the MCR< coil: If power flows into the coil, MCR dependency is disabled ("normal" processing); if power does not flow into the coil, MCR dependency is enabled.

FBD: You define an MCR zone with the MCR< box at the beginning and the MCR> box at the end of the MCR zone. The MCR< box requires a preceding logic operation; the MCR> box stands alone in a network. Within this zone, you control MCR dependency with the RLO preceding the MCR< box: If it is "1", MCR dependency is disabled ("normal" processing); if it is "0", MCR dependency is enabled.

You can open an MCR zone within another MCR zone. The nesting depth for MCR zones is 8; that is, you can open a zone up to eight times before you close a zone.

You control the MCR dependency of a nested MCR zone with the RLO on opening the zone. However, if MCR dependency is enabled in a "supraordinate" zone, you cannot disable MCR dependency in a "subordinate" MCR zone. The Master Control Relay of the first MCR zone controls the MCR dependency in all nested (Figure 17.2).

A block call within an MCR zone does not change the nesting depth of an MCR zone. The program in the called block is still in the MCR zone that was open when the block was called

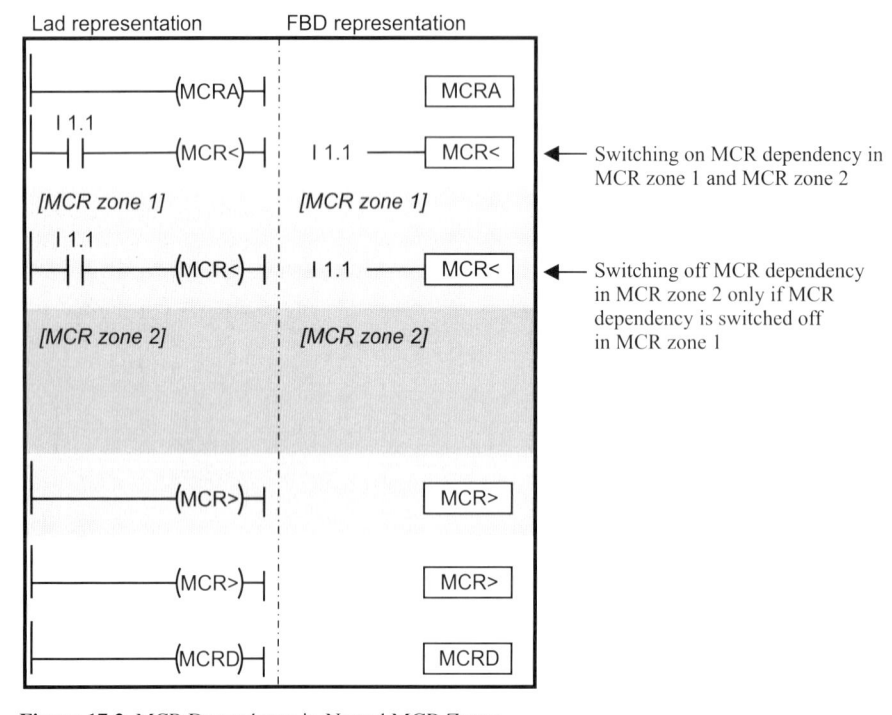

Figure 17.2 MCR Dependency in Nested MCR Zones

17.3 MCR Zone

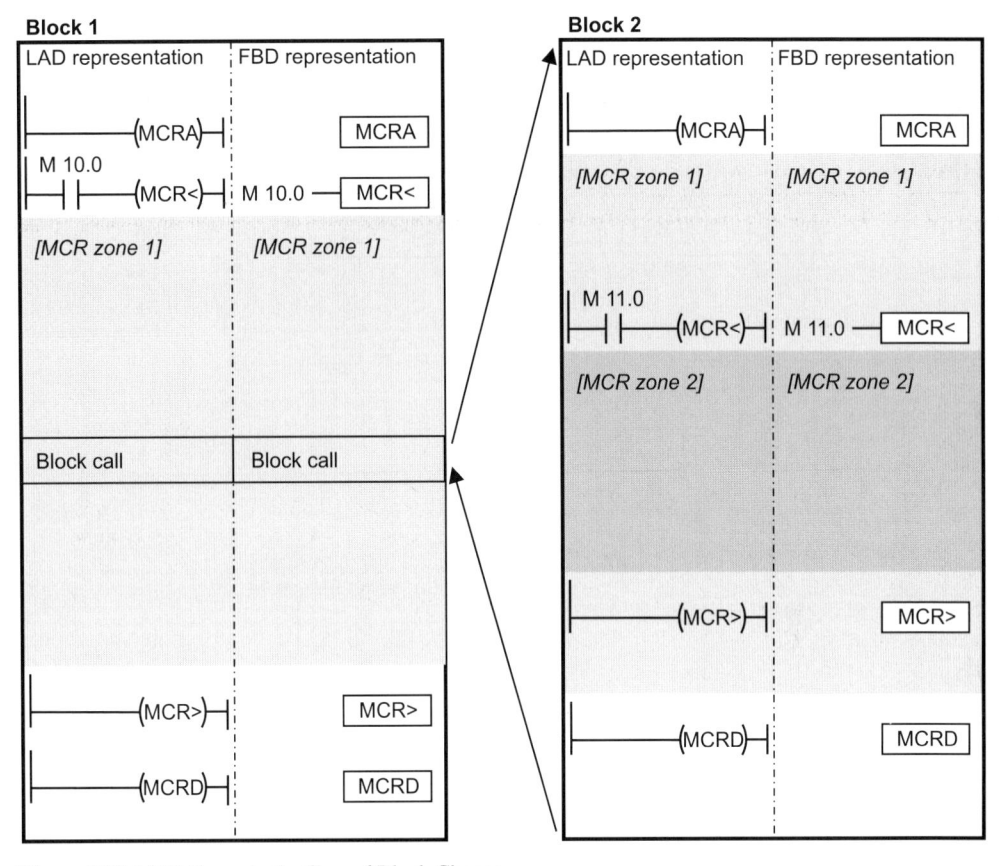

Figure 17.3 MCR Zones in the Case of Block Change

(and is controlled form here). However, you must reactivate MCR dependency in a called block by opening the MCR area.

In Figure 17.3, memory bits M 10.0 and M 11.0 control the MCR dependencies. With memory bit M 10.0, you can enable MCR dependency in both zones (with "0") regardless of the signal state of memory bit M 11.0. If the MCR dependency for zone 1 is disabled with M 10.0 = "1", you can control the MCR dependency of zone 2 with memory bit M 11.0 (Table 17.1).

Table 17.1 MCR Dependency in the Case of Nested MCR Zones (Example)

M10.0	M11.0	Zone 1	Zone 2
"1"	"1"	No MCR dependency	
"1"	"0"	No MCR dependency	MCR dependency enabled
"0"	"1" or "0"	MCR dependency enabled	

17.4 Setting and Resetting I/O Bits

Despite enabled MCR dependency, you can set or reset the bits of an I/O area with the system functions. A requirement for this is that the bits to be controlled are in the process-image output table or a process-image output table has been defined for the I/O area to be controlled.

The system function **SFC 79 SET** is available for setting and **SFC 80 RSET** for resetting the I/O bits (Table 17.2). You call these system functions in an MCR zone. The system functions are only effective if MCR dependency is enabled; if MCR dependency is disabled, calling these SFCs has no effect.

Setting and resetting the I/O bits also simultaneously updates the process-image output table. The I/O are affected byte by byte. The bits not selected with the SFCs (in the first and in the last byte) retain the signal states as they are currently available in the process image.

The "LAD_Book" and "FBD_Book" libraries on the diskette accompanying the book contain examples of the Master Control Relay and of the system functions SFC 79 and SFC 80 (FB 117 in the program "Program Flow Control").

Table 17.2 Parameters of the SFCs for Controlling the I/O Bits

SFC	Parameter	Declaration	Data Type	Assignment, Description
79	N	INPUT	INT	Number of bits to be set
	RET_VAL	OUTPUT	INT	Error information
	SA	OUTPUT	POINTER	Pointer to the first bit to be set
80	N	INPUT	INT	Number of bits to be reset
	RET_VAL	OUTPUT	INT	Error information
	SA	OUTPUT	POINTER	Pointer to the first bit to be reset

18 Block Functions

In this chapter, you will learn how to call and terminate code blocks and how to work with operands from data blocks. The next chapter then deals with using block parameters.

18.1 Block Functions for Code Blocks

Block functions for code blocks include instructions for calling and terminating blocks (Figure 18.1). Code blocks are called with the call box. If functions or system functions have no block parameters, they can also be called with the CALL coil or with the CALL box. In both cases, a preceding logic operation enabling a conditional call (call dependent on conditions) is permissible. The block end function RET always requires a preceding logic operation.

In addition to the block change, the call box also contains the transfer of block parameters. When function blocks are called, it also opens the instance data block. The CALL coil/box is no more than a change to another block and is only meaningful (and permissible) in the case of functions and system functions.

After a block has been terminated, and following the call function, the CPU continues program execution in the block that made the call (the calling block). If an organization block is terminated, the CPU returns to the operating system.

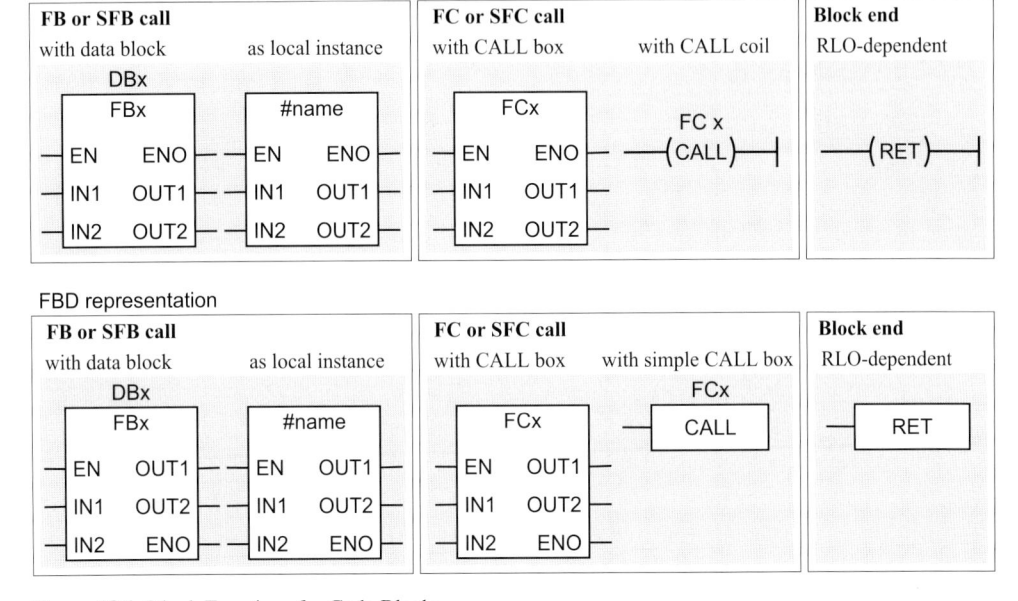

Figure 18.1 Block Functions for Code Blocks

18 Block Functions

In the case of incremental programming, you will find the CALL coil/box and the RET coil/box in the Program Elements Catalog (with VIEW → CATALOG [Ctrl - K] or INSERT → PROGRAM ELEMENTS) under "Program Control"; you insert block calls with call boxes into your program when you select blocks from "FC/FB/SFC/SFB blocks", "Multiple Instances" or "Libraries".

18.1.1 Block Calls: General

If a code block is to be processed, it must be "called". Figure 18.2 gives an example for calling function FC 10 in organization block OB 1.

A block call consists of the call box that contains the address of the called block (here: FC 10), the enable input EN, the enable output ENO, and any block parameters. Following processing of the call function, the CPU continues program execution in the called block. The block is processed to the end or until a block end function is encountered. The CPU then returns to the calling block (here: OB 1) and continues processing this block after the call box.

The information the CPU requires to find its way back to the calling block is stored in the block stack (B stack). With every new block call, a new stack element is created that includes the return address, the contents of the data block register and the address of the local da-

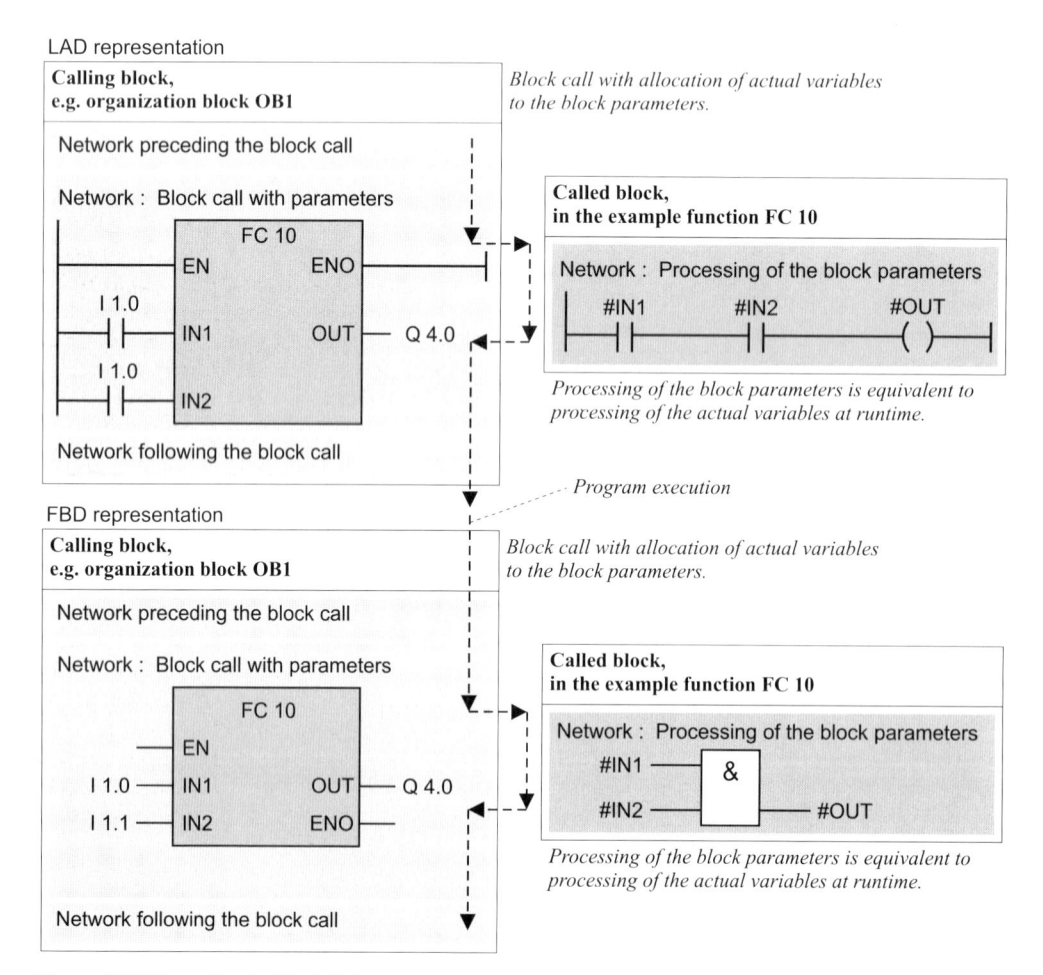

Figure 18.2 Example of a Block Call

ta stack of the calling block. If the CPU goes to the Stop state as a result of an error, you can use the programming device to see from the contents of the B stack which blocks were processed up to the error.

You can transfer data to and from the called block for processing. These data are transferred via block parameters. With the call box, you can also call blocks without block parameters.

18.1.2 Call Box

You use the call box to call FBs, FCs, SFBs and SFCs. (You cannot call organization blocks since they are event-driven and are started by the operating system.)

You can use the EN input to make the block call subject to conditions. If the EN input is connected direct to the left power rail, the call is absolute; it is always executed. If there is a logic operation preceding EN, the block call is only executed if the preceding logic operation is fulfilled. The ENO output has the same signal state as the binary result BR on exiting the called block.

IF EN == "1" or not connected		
THEN		ELSE
Called block is processed		Called block is not processed
IF called block returns BR = "1"		
THEN	ELSE	
ENO := "1"	ENO := "0"	ENO := "0"

You label the parameters of the called block with the absolute or symbolic operand current for the call. If a parameter is of data type BOOL, precede this parameter with

▷ a contact or a rung (LAD) or

▷ a binary variable or a binary logic operation (FBD).

A Boolean output parameter cannot be combined further.

LAD: You can arrange several call boxes in series, connecting them with each other via EN or ENO. You can only insert a call box in a parallel rung if it is connected directly to the left power rail.

FBD: You can connect call boxes in series by connecting the ENO output of one box with the EN input of the next. The ENO outputs of several boxes can be combined with AND or OR.

MCR dependency is de-activated when a block is called. The MCR is disabled in the called block regardless of whether the MCR was enabled or disabled prior to the block call. When exiting a block, MCR dependency assumes the same setting that it had prior to the block call.

Depending on the block parameters, you can modify the contents of the data block registers when the block change is made. If the *called* block is a function block, the instance block is always opened in this block via the DI register. If the *calling* block is a function block, the contents of the DI register (the instance data block) are retained after the block call. The contents of the DB register depend, among other things, on the block parameters that were passed.

Calling function blocks

You call a function block by selecting the relevant function block from the Program Elements Catalog under "FB Blocks". Prerequisite is that the function block to be called must already be in the user program. You write the instance data block belonging to the call above the box. Both blocks (function block and instance data block) can have absolute or symbolic addresses.

In the case of function blocks, you do not need to initialize all block parameters at the call. The uninitialized block parameters retain their current value.

You can also call "function blocks with multiple instance capability" within other "function blocks with multiple instance capability" as local instance. In doing so, the called function block uses the instance data block of the calling function block as the store for its local data. Prior to the call, you declare the local instance in the static local data of the calling function block (the block you are currently programming). The local instance is called by selecting one of the available local instances under 'Multiple Instances' in the Program Elements Catalog; it is not necessary to specify an instance data block (see also Section 18.1.6, "Static Local Data").

18 Block Functions

Calling functions

You call a function by selecting the relevant function under "FC Blocks" in the Program Elements Catalog. The function can have an absolute or a symbolic address.

When you call functions, you must initialize all available parameters.

Calling functions with a function value takes exactly the same form as calling functions with no function value. Only the first output parameter, corresponding to the function value, has the name RET_VAL.

Calling system blocks

The CPU operating system contains system functions (SFCs) and system function blocks (SFBs) that you can use. The number and type of system blocks depends on the CPU. You can call all system blocks with the call box.

You call a system function block in the same way as one you have written yourself; set up the associated instance data block in the user memory with the same data type as the SFB.

You call a system function in the same way as a function you have written yourself.

System blocks exist only in the CPU operating system. If you want to call system blocks during offline programming, the Program Editor needs a description of the call interface so it can initialize the parameters. You will find this interface description under "System Function Blocks" on the library named "Standard Library". From here, the Program Editor copies the interface description to the offline block container when you call a system block. The interface description thus copied then appears as "normal" block object.

The Program Elements Catalog provides the system blocks currently available in the user program under "SFC Blocks" or "SFB Blocks". You can, for example, select a system block from the Program Elements Catalog with the mouse and drag it to the block currently being processed block, where it is then called. At the same time, this block (or, more precisely: its interface description) is copied into the block container.

18.1.3 CALL Coil/Box

You can call functions and system functions using the CALL coil/box. It is a requirement that the called blocks have no block parameters. You can use the CALL coil/box if a block is too long or not clear enough for you by simply "breaking down" the block into sections and calling the sections one after the other. One single CALL coil or box is permitted per network.

LAD: If the CALL coil is connected directly to the left power rail, the call is always executed (unconditional call).

If there is a logic operation preceding the CALL coil, the call is only executed if the preceding logic operation is fulfilled, that is, if power flows into the CALL coil. If the preceding logic operation is not fulfilled, the call is not executed and the next network is processed.

FBD: If there is a logic operation preceding the CALL box, the call is executed only when the preceding logic operation is fulfilled, that is, when RLO = "1" is present at the CALL box. If the preceding logic operation is not fulfilled, the call is not executed and the next network is immediately processed.

When a block change is made, status bit OS is reset; status bits CC0, CC1 and OV are not affected.

MCR dependency is deactivated when a block is called. The MCR is disabled in the called block regardless of whether the MCR was enabled or disabled prior to the block call. When a block is exited, MCR dependency assumes the same setting it had prior to the block call.

Calling a block with the CALL coil/box saves the data block registers in the B stack; the block end restores their contents when the called block is exited. The global data block current prior to the block call and the instance data block are also open following the block call. If no data block was open prior to the block call (for example, no instance data block in OB 1), no data block is open following the block call either, regardless of which data blocks may be open in the called block.

18.1.4 Block End Function

You can terminate processing in a block prematurely with the block end function RET.

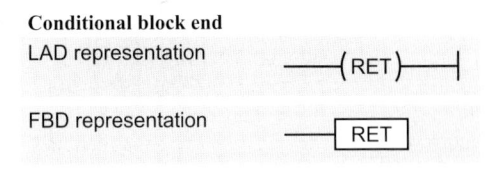

Conditional block end
LAD representation —(RET)—|

FBD representation —[RET]

The block end function is represented as a coil or box requiring a preceding logic operation. The RET coil/box must be alone in a network.

If the preceding logic operation is fulfilled, the block is exited. A return jump is made to the previously processed block in which the block call took place. If an organization block is terminated, the CPU returns to the system program.

If the preceding logic operation is not fulfilled, the next network in the block is processed.

IF preceding logic operation == "1"	
THEN	ELSE
The block is exited	The next network is processed
BR := "1"	BR := "0"

The RET coil/box simultaneously stores the RLO (whether power flows or not) in binary result BR, regardless of whether or not the logic operation was fulfilled. The binary result is decisive for controlling the ENO output at the call box (see also Chapter 15, "Status Bits").

18.1.5 Temporary Local Data

You use the temporary local data to buffer results arising during processing of a block. Temporary local data are available only during block processing; once a block has been processed, your buffered data are lost.

Temporary local data are operands which lie in the local data stack (L stack) in system memory. The CPU's operating system makes the temporary local data for a code block available when that code block is called. When a block is called, the values in the L stack are virtually coincidental. In order to be able to make sensible use of the local data, you must first write them prior to reading. When the block is terminated, the L stack is assigned to the next block called.

The number of temporary local data bytes a block requires is in the block header. Reading the header tells the operating system how many bytes have to be reserved in the L stack when the block is called. You, too, can tell from the entry in the block header just how many local data bytes the block requires (using the Editor, with the block open, by invoking FILE → PROPERTIES or in the SIMATIC Manager with EDIT → OBJECT PROPERTIES, in each case on tab "General – Part 2").

Declaring temporary local data

You declare the temporary local data in the declaration section of the code block:

▷ under "temp" in the case of incremental programming, or
▷ between VAR_TEMP and END_VAR in the case of source-oriented programming.

Address	Declaration	Name	Type	Initial Value	Comment
0.0	in	Man_on	BOOL	FALSE	Input parameter
2.0	out	Switch_on	BOOL	FALSE	Output parameter
4.0	in_out	Length	INT	0	I/O parameter
6.0	stat	Total	INT	0	Static lokal data
8.0	stat	Setpoint	DINT	L#0	
0.0	temp	Deviation	INT		Temporary lokal data
2.0	temp	Intermediate	REAL		

Figure 18.3 Example for the Declaration of Local Data in a Function Block

18 Block Functions

Figure 18.3 shows an example for the declaration of temporary local data. The variable *temp1* lies in the temporary local data and is of data type INT; the variable *temp2* is of data type REAL.

The temporary local data are stored in the L stack order of their declaration in accordance with their data type. You will find more detailed information on the storing of these data in the L stack in Section 26.2, "Storing Variables".

Symbolic addressing of temporary local data

You reference temporary local data using their symbolic names. You assign these names in accordance with the rules for block-local symbols.

All operations allowed for bit memory are also allowed for temporary local data. Please note, however, that a temporary local data bit is not suitable for use as an edge memory bit because it does not retain its signal state outside the relevant block.

You can address the temporary local data for a block only within that block (exception: the temporary local data for the calling block can be accessed via block parameters).

Size of the L stack

The total size of the L stack is CPU-specific. The available number of temporary local data bytes in a priority class, that is, in the program of an organization block, is also predetermined. On an S7-300, the number of temporary local data bytes is fixed, for example there are 256 bytes per priority class on the CPU 341. On an S7-400, you can specify the number of temporary local data bytes you will need when you initialize the CPU. These bytes must be shared by the blocks called in the relevant organization block as well as by the blocks which they call.

Please note that the Editor also uses temporary local data, for instance for passing block parameters, a fact which goes unnoticed on the programming interface.

Start information

When an organization block is called, the CPU operating system passes start information in the temporary local data. This start information comprises 20 bytes for every organization block, and is nearly identical for all OBs. The start information for the various organization blocks is described in detail in Chapters 20, "Main Program", 21, "Interrupt Handling", 22, "Start-up Characteristics", and 23, "Error Handling".

These 20 bytes of information must always be available in each priority class used. If you program a routine for the evaluation of synchronization errors (programming and access errors), you must set aside an additional 20 bytes at least for the start information of these error organization blocks, as these error OBs belong to the same priority class.

You declare the start information for an organization block when you program that block. The information is mandatory. Sample declarations in English can be found on the *Standard Library* under *Organization Blocks*. If you do not need the start information, simply declare the first 20 bytes as something else, for example as an array (as shown in Figure 18.4).

Absolute addressing of temporary local data

Normally, you reference temporary local data by their symbolic names. The use of absolute addresses is the exception. Once you are familiar with the way data are stored in the L stack, you can compute the addresses of the static local data yourself. You will also see the addresses listed in the variable declaration table of the compiled block.

The operand identifier for temporary local data is L; a bit is addressed with L, a byte with LB, a word with LW, and a doubleword with LD.

Address	Declaration	Name	Type
0.0	temp	SINFO	ARRAY [1..20]
*1.0	temp		BYTE
20.0	temp	LByte	ARRAY [1..16]
*1.0	temp		BYTE

Figure 18.4
Example for the Declaration of Temporary Local Data in an Organization Block

Example: You want to reserve 16 bytes of temporary local data for absolute addressing, and you want to reference the values in these bytes by both byte and bit. To do this, create an array at the beginning of the local data area so that addressing begins at 0. In an organization block, you would place this array declaration immediately behind the declaration for the start information, in which case addressing would begin at 20.

Note: Temporary local data can be referenced using absolute addresses only in the basic languages STL, LAD, and FBD. You must use symbolic addresses to reference temporary local data in SCL.

Chapter 26, "Direct Variable Access", describes how you can ascertain the address of a variable in the temporary local data area at runtime.

Data type ANY

A variable in the temporary local data can – although this is an exception – be declared as data type ANY. You can use this feature to modify the ANY pointer at runtime (see Section 24.2.5, "Variable ANY Pointer").

18.1.6 Static Local Data

Static local data are operands which a function block stores in its instance data block.

Static local data are a function block's "memory". They retain their values until those values are changed by the program, just like data operands in global data blocks.

The number of static local data bytes is limited by the data types of the variables and by the CPU-specific length of a data block.

Declaring static local data

You declare static local data in the declaration section of the function block:

▷ under "stat" in the case of incremental programming or

▷ between VAR and END_VAR in the case of source-oriented programming.

Figure 18.3 in Section 18.1.5 shows an example of a variable declaration in a function block. The block parameters are declared first, then the static local data, and finally the temporary local data.

The static local data are stored in the instance data block behind the block parameters in the order of their declarations and in accordance with their data types. You will find more detailed information on how data is stored in data blocks in Section 26.2, "Storing Variables".

Symbolic addressing of static local data

You reference static local data with symbolic names. You assign these names in accordance with the rules for block-local symbols.

All the same operations that can address data operands in global data blocks can also address static local data.

Example: The function block "Totalizer" adds an input value to a value stored in the static local data and then stores the total in the static local data. At the next call, the input value is again added to this total, and so on (Figure 18.5 above).

Total is a variable in the data block "TotalizerData", the instance data block for the "Totalizer" function block (you can define the names of all blocks yourself in the Symbol Table, but you must stick to the applicable rules when doing so). The instance data block has the data structure of the function block; in the example, it contains two INT variables with the names *In* and *Total*.

Accessing static data outside the function block

As a rule, the static local data are processed only in the function block itself. Because they are stored in a data block, however, you can access the static local data at any time with "*Data Block Name*".*Operand Name* just as you would a variable in a global data block.

209

18 Block Functions

FB "Totalizer"

Address	Declaration	Name	Type
+ 0.0	in	In	INT
+ 2.0	stat	Total	INT

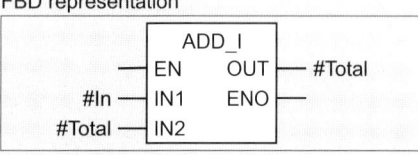

DB "TotalizerData"

Address	Declaration	Name	Type
+ 0.0	in	In	INT
+ 2.0	stat	Total	INT

LAD representation

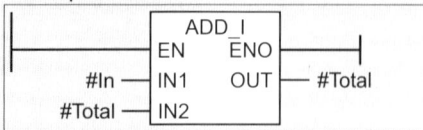

```
         ADD_I
        EN  ENO
  #In —IN1  OUT— #Total
#Total —IN2
```

FBD representation

```
         ADD_I
        EN  OUT— #Total
  #In —IN1  ENO
#Total —IN2
```

*In the **data view**, the data block shows each individual variable, so that the variables in a local instance appear with their full names.*

At the same time, you can see the absolute address of each variable.

FB "Evaluation"

Address	Declaration	Name	Type
0.0	in	Add	BOOL
0.1	in	Delete	BOOL
2.0	stat	EM_Add	BOOL
2.1	stat	EM_Del	BOOL
4.0	stat	Memory	Totalizer

DB "EvaluationData"

Address	Declaration	Name	Type
0.0	in	Add	BOOL
0.1	in	Delete	BOOL
2.0	stat	EM_Add	BOOL
2.1	stat	EM_Del	BOOL
4.0	stat:in	Memory.In	INT
6.0	stat	Memory.Total	INT

LAD representation

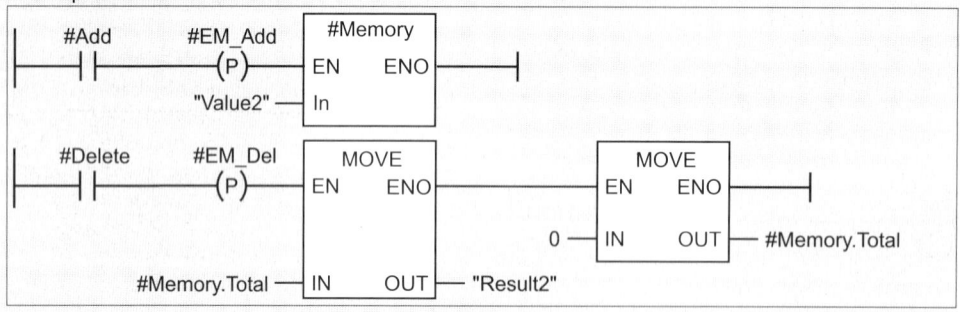

FBD representation

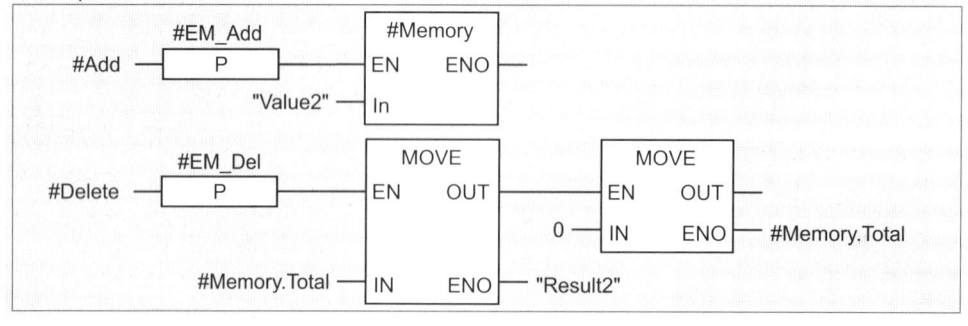

Figure 18.5 Example for Static Local Data and Local Instances

In our little example, the data block is named *TotalizerData* and the data operand is named *Total*. The applicable access instructions might be as follows:

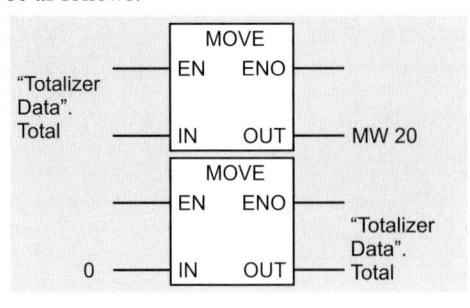

```
              MOVE
              EN  ENO
"Totalizer
Data".
Total  ─────  IN  OUT ─── MW 20

              MOVE
              EN  ENO
                              "Totalizer
                              Data".
        0 ──  IN  OUT ─────── Total
```

Local instances

When you call a function block, you normally specify the instance data block for that call. The function block then stores its block parameters and its static local data in this instance data block.

Beginning STEP 7 V2, you can generate "multiple instances", that is, you can call a function block as a local instance in another function block. The static local data (and the block parameters) of the function block called are then a subset of the calling block's static local data. Prerequisite is that both the calling function block and the called function block have block version 2, that is, that both have "multiple instance capability". This allows you to "nest" function block calls up to a depth of eight.

Example (Figure 18.5 below): In the static local data of the function block "Evaluation", you declare a variable *Memory* that corresponds to the function block "Totalizer" and has its data structure. Now you can call the function block "Totalizer" via the variable *Memory*, but without specifying a data block because the data for *Memory* are stored "block-locally" in the static local data (*Memory* is the local instance of the function block "Totalizer").

You access *Memory*'s static local data in the program in function block "Evaluation" the same way you would access structure components, which is by specifying the structure name (*Memory*) and the component name (*Total*).

The instance data block "EvaluationData" thus contains the variables *Memory.In* and *Memo-*

ry.Total, which you can also address as global variables, for example as *"EvaluationData".Memory.Total*.

You will find this example for the use of a local instance in function blocks FB 6, FB 7 and FB 8 in program "Program Flow Control" on the diskette supplied with the book. The "Feed" example in Section 19.5.3 contains additional usage of local instances.

Absolute addressing of static local data

As a rule, the static local data are addressed symbolically. Absolute addressing is the exception. Within a function block, the instance data block is opened via the DI register. Operands in this data block, and this includes the static local data as well as the block parameters, therefore have DI as operand identifier. A bit is addressed with DIX, a byte with DIB, a word with DIW, and a doubleword with DID.

Once you know just how data are stored in a data block, you can compute the absolute addresses of the static local data yourself. You can also find the addresses in the compiled block in the variable declaration table. But be very careful ! *These addresses are addresses that are relative to the start of the instance.* They apply only when you call the function block with a data block. If you call the function block as local instance, the local instance's local data are located right in the middle of the calling function block's instance data block. You can view the absolute addresses, for example, in the compiled instance data block which contains all local instances. Select VIEW → DATA VIEW when you want to read up on the addresses of individual local data operands.

Using our example as a basis, the variable Total in the function block "Totalizer" could be addressed with DIW 2 if function block "Totalizer" were called with a data block (also see the operands in the data block "TotalizerData"), and with DIW 6 if function block "Totalizer" were called as local instance in function block "Evaluation" (also see the operands in data block "EvaluationData").

However, when we program a function block without knowing whether it will be called with a data block or as local instance, that is, a function block that is to have "multiple instance capability", how can we use absolute addresses to access the static local data? Put simply, we add the offset of the local instance from address register AR2 to the address of the variable. For further details, refer to Chapter 25, "Indirect Addressing", and Chapter 26, "Direct Variable Access".

Note: The absolute addressing of static local data is possible only in the basic languages STL, LAD, and FBD. Static local data must be addressed symbolically in SCL.

18.2 Block Functions for Data Blocks

You store your program data in the data blocks. In principle, you can also use the bit memory area for storing data; however, with the data blocks, you have significantly more possibilities with regard to data volume, data structuring and data types. This chapter shows you

▷ how to work with data operands,

▷ how to call data blocks and

▷ how to create, delete and test data blocks at runtime.

You can use data blocks in two versions: as *global data blocks* that are not assigned to any code block, and as *instance data blocks* that are assigned to a function block. The data in the global data blocks are "free" data that every code block can make use of. You yourself determine their volume and structure directly through programming the global data block. An instance data block contains only the data with which the associated function block works; this function block then also determines the structure and storage location of the data in "its" instance data block.

The number and length of data blocks are CPU-specific. The numbering of the data block begins at 1; there is no data block DB 0. You can use each data block either as a global data block or as an instance data block.

You must first create ("set up") the data blocks you use in your program, either by programming, such as code blocks, or at runtime using the system function SFC 22 CREAT_DB.

18.2.1 Two Data Block Registers

Each S7-CPU has two data block registers. These registers contain the numbers of the current data blocks; these are the data blocks with whose operands processing is currently taking place. Before accessing a data block operand, you must open the data block containing the operand. If you use fully-addressed access to data operands (with specification of the data block, see below), you need not be concerned with opening the data blocks or with the contents of the data block registers. The Editor generates the necessary instructions from your specifications.

The Program Editor uses the first data block register preferably for accessing global data blocks and the second data block register for accessing instance data blocks. For this reason, these registers are given the names "Global data block register" (DB register for short) and "Instance Data Block Register" (DI register for short). The handling of the registers by the CPU is absolutely identical. Each data block can be opened via one of the two registers (or also via both simultaneously).

To come to the essential point first: You can affect only the DB register with LAD or FBD. If you open a data block with the OPN coil/box (see below), you always open it via the DB register. In LAD and FBD, the instance data block is always opened via the DI register; this happens with the call box when the block is called.

When you load a data word, you must specify which of the two possible open data blocks contains the data word. If the data block has been opened via the DB register, the data word is called DBW; if the data word is in the data block opened via the DI register, it is called DIW. The other data operands are named accordingly (Table 18.1).

18.2.2 Accessing Data Operands

You can use the following methods for accessing data operands:

▷ Symbolic addressing with full addressing,

▷ Absolute addressing with full addressing and

▷ Absolute addressing with partial addressing

Table 18.1 Data Operands

Data operand	Located in a data block opened via the	
	DB register	DI register
Data bit	DBX y.x	DIX y.x
Data byte	DBB y	DIB y
Data word	DBW y	DIW y
Data doubleword	DBD y	DID y

x = Bit address, y = Byte address

Symbolic access to the data operands in global data blocks requires the minimum system knowledge. For absolute access or for using both data block registers, you must observe the notes below.

Symbolic addressing of data operands

It is recommended that you address data operands symbolically whenever possible. Symbolic addressing

▷ makes it easier to read and understand the program (if meaningful terms are used as symbols),

▷ reduces write errors made during programming (the Program Editor compares the terms used in the Symbol Table and in the program; "number-switching errors" such as DBB 156 and DBB 165 that can occur when using absolute addresses cannot occur here) and

▷ does not require programming knowledge at the machine code level (which data block has the CPU opened currently?).

Symbolic addressing uses fully-addressed access (data block together with data operand), so that the data operand always has a unique address.

You determine the symbolic address of a data operand in two steps:

1) Assignment of the data block in the Symbol Table
Data blocks are global data that have unique addresses within a program. In the Symbol Table, you assign a symbol (e.g. Motor1) to the absolute address of the data block (e.g. DB 51).

2) Assignment of the data operands in the data block
You define the names of the data operands (and the data type) during programming of the data block. The name applies only in the associated block (it is "block-local"). You can also use the same name in another block for another variable.

Fully-addressed access to data operands

In the case of fully-addressed access, you specify the data block together with the data operand. This method of addressing can be symbolic or absolute:

```
"MOTOR1".ACTVAL
DB 51.DBW 20
```

MOTOR1 is the symbolic address that you have assigned to a data block in the Symbol Table. ACTVAL is the data operand you defined when programming the data block. The symbolic name "MOTOR1".ACTVAL is just as unique a specification of the data operand as the specification DB 51.DBW 20.

Fully-addressed data access is only possible in conjunction with the global data block register (DB register). In the case of fully-addressed data operands, the Program Editor first opens the data block via the DB register and then accesses the data operand.

You can use fully-addressed access with all functions permissible for the data type of the addressed data operand. So with block parameters, for example, you can also specify a fully-addressed data operand as actual parameter.

Absolute addressing of data operands

For absolute addressing of data operands, you must know the addresses at which the Program Editor places the data operands when setting up. You can find out the addresses by outputting them after programming and compiling the data block. You will then see from the address column the absolute address at which the relevant variable begins. This procedure is suitable for all data blocks, both those you use as global data blocks as well as those you use as instance data blocks (for local instances see Section 18.2.4 below). In this way, you can also see where the Program Editor stores the block parameters and the static local data for function blocks.

Data operands are addressed byte by byte like, for example, bit memory; the functions used on the data operands are also the same as those used with bit memory.

18.2.3 Opening a Data Block

You use the OPN coil (LAD) or OPN box (FBD) to open a data block via the DB register.

Open data block

LAD representation	DB x ┤—(OPN)—├
FBD representation	DB x [OPN]

The OPN coil/box is connected directly to the left power rail and stands alone in a rung. The specified data block is always opened via the DB register. In LAD or FBD, it is not possible to open a data block via the DI register with OPN coil/box. (The DI register uses the call box to open the current instance data block.)

In the case of incremental programming, you will find the OPN coil/box in the Program Elements Catalog under "DB Call".

The open data block must be in the user memory at runtime.

In the networks following the OPN coil/box, you can use partial addressing to access only those data operands that are located in the open data block. If you want to copy from one data block to another, you can, for example go into the temporary local data via an intermediate buffer or (better) use fully-addressed data operands. Please note: A fully-addressed data operand overwrites the DB register with "its" data block.

Example: The value of data word DBW 10 of data block DB 13 is to be transferred to data word DBW 16 of data block DB 14. With the MOVE box, you can only copy partially-addressed data operands within the currently open data block. For copying between data blocks, you require an intermediate buffer, such as a local data word (see Figure 18.6). It is better to use full addressing.

When you open a data block, it remains "valid" until another data block is opened. Under certain circumstances, you may not be able to see this via the Program Editor, for example if a block is changed with the call box (see Section 18.2.4).

In the case of a block change with the CALL coil/box, the contents of the data block registers are retained. On returning to the calling block, the block change restores the old contents of the registers.

18.2.4 Special Points in Data Addressing

Changes in the contents of the DB registers

With the following functions, the Program Editor generates additional instructions which can alter the contents of the two DB registers:

Full addressing of data operands

Each time data operands are fully addressed, the Program Editor first opens the data block and then accesses the data operand. The DB register is overwritten each time. This also applies to supplying block parameters with fully-addressed data operands.

Accessing block parameters

Access to the following block parameters changes the contents of the DB register: All block parameters of complex data type in the case of functions, and in/out parameters of complex data type in the case of function blocks.

Calling a function block with the call box

Before the actual function block call, the call box stores the number of the current instance data block in the DB register (by swapping the data block registers) and opens the instance data block for the called function block. In this way, the instance data block associated with a called function block is always open. After the actual block call, the call box swaps data block registers again, so that the current instance data block is available again in the calling function block. In this way, the call box changes the contents of the DB register.

18.2 Block Functions for Data Blocks

Example: Partially- and fully-addressed data operands

Network: 5 Open data block DB 13

```
      DB 13                          DB 13
──────(OPN)─────┤                   ┌─────┐
                                    │ OPN │
                                    └─────┘
```

Network: 6 Load from DB 13 and copy to an intermediate buffer

```
           ┌──MOVE──┐                          ┌──MOVE──┐
          ─┤EN   ENO├─                        ─┤EN   OUT├─ #t_WORD
  DBW 10 ──┤IN   OUT├── #t_WORD       DBW 10 ──┤IN   ENO├─
           └────────┘                          └────────┘
```

Network: 7 Open data block DB 14

```
      DB 14                          DB 14
──────(OPN)─────┤                   ┌─────┐
                                    │ OPN │
                                    └─────┘
```

Network: 8 Copy from the intermediate buffer and transfer to DB 14

```
            ┌──MOVE──┐                          ┌──MOVE──┐
           ─┤EN   ENO├─                        ─┤EN   OUT├── DBW 16
  #t_WORD ──┤IN   OUT├── DBW 16      #t_WORD ──┤IN   ENO├─
            └────────┘                          └────────┘
```

Network: 9 Copy with fully-addressed data operands

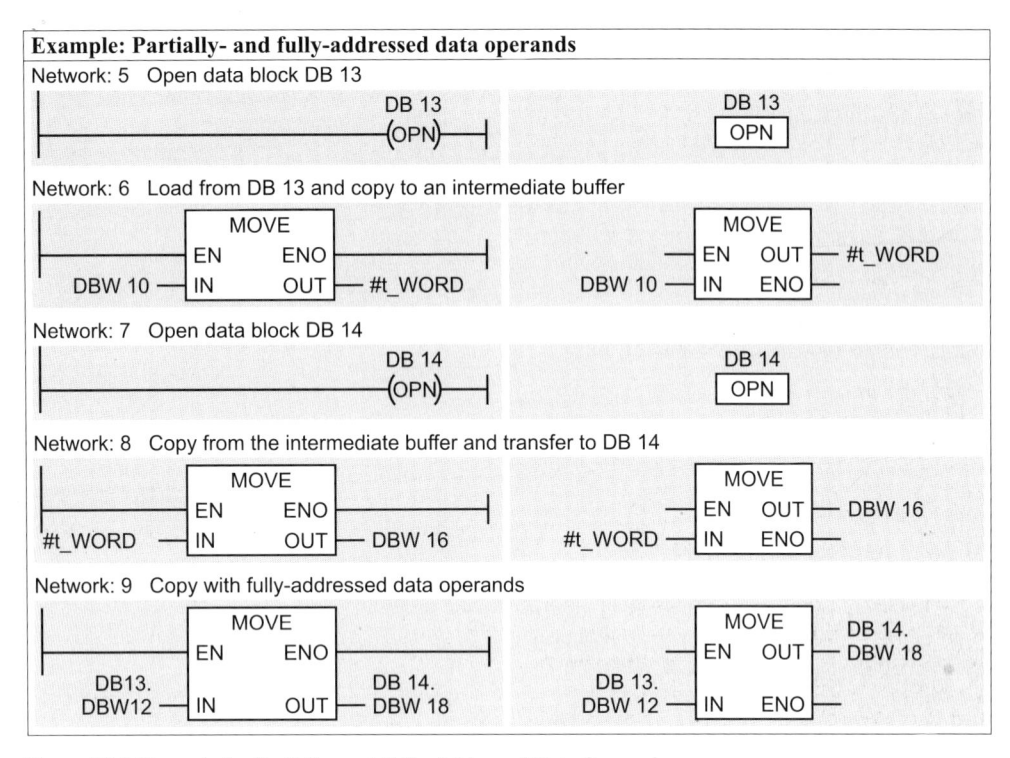

```
           ┌──MOVE──┐                          ┌──MOVE──┐  DB 14.
          ─┤EN   ENO├─                        ─┤EN   OUT├── DBW 18
  DB13.    │        │  DB 14.        DB 13.    │        │
  DBW12 ───┤IN   OUT├── DBW 18       DBW 12 ──┤IN   ENO├─
           └────────┘                          └────────┘
```

Figure 18.6 Example for Partially- and Fully-Addressed Data Operands

DI register in function blocks

In function blocks, the DI register always contains the number of the current instance data block. All access to block parameters or static local data is over the DI register. The address of the local data given in the Declaration Table of the function block applies only when you open the function block with an instance data block, in which case the data operands begin at byte zero.

When you call a function block as local instance, its data are "in the middle" of the calling data block's instance data block. They contain the absolute address of the local data when you output the instance data block in data view format. Each individual variable is then displayed with name and address, including the variables of the local instance called.

When you program a function block that you also want to have available later as local instance, you do not yet know the absolute address of the variable at the time of programming. In this case – just as the Program Editor does for symbolic programming – the contents of address register AR2 are added to the variable address. However, this is possible only in the STL programming language.

Changing the contents of data blocks at a later time

In the Properties window of the offline object container *Blocks* in the "Blocks" register, you can specify whether the absolute address or the symbolic address of the data operand is to have priority for subsequent displays and saves when there is a change in the contents of the data blocks for code blocks that have already been stored.

The default is "Absolute address has priority" (the same as in previous STEP 7 versions). This default means that when there is a change in the declaration, the absolute address is retained in the program while the symbolic address changes accordingly. If the "Symbolic address has priority" default is chosen, the absolute address changes while the symbolic address is retained.

Example: Assuming that data word DBW 10 in data block DB 1 contains the symbolic address of *ActValue*. In the program, you might address this data word with

`"Data".ActValue    DB1.DBW 10`

where "Data" is the symbolic address for data block DB 1. If you would now add an additional data word with the symbolic address *MaxCurrent* immediately in front of data word DBW 10, the result upon the next opening (or storing) of the code block will be as follows:

In the case of "Absolute address has priority":

`"Data".MaxCurrent DB1.`**`DBW 10`**

In the case of "Symbolic address has priority":

`"Data".`**`ActValue`**`   DB1.DBW 12`

As for access to data operands in global data blocks, the same applies as for access to global operands (for instance inputs) for which a symbolic address has been entered in the Symbol Table. You will find detailed information on this subject in Section 2.5.6, "Operand Priority".

18.3 System Functions for Data Blocks

There are three system functions for handling data blocks. Their parameters are described in Table 18.2.

▷ SFC 22 CREAT_DB
  Create data block

▷ SFC 23 DEL_DB
  Delete data block

▷ SFC 24 TEST_DB
  Test data block

18.3.1 Creating a Data Block

System function SFC 22 creates a data block in the user memory. As the data block number, the system function takes the lowest free number in the number band given by the input parameters LOW_LIMIT and UP_LIMIT. The numbers specified at these parameters are included in the number band. If both values are the same, the data block is generated with this number. The output parameter DB_NUMBER supplies the number of the data block actually created. With the input parameter COUNT, you specify the length of the data block to be created. The length corresponds to the number of data bytes and must be an even number.

Creating the data block is not the same as calling it. The current data block is still valid. A data block created with this system function contains random data. For meaningful use, data must first be written to a data block created in this way before the data can be read.

In the event of an error, the data block is not created, the contents of the output parameter are undefined, and an error number is returned as function value.

18.3.2 Deleting a Data Block

System function SFC 23 deletes the data block in RAM (work and load memory) whose number is specified in the input parameter DB_NUMBER. The data block must not be open at the time, as otherwise the CPU will go to STOP.

Data blocks created with the keyword UNLINKED and data blocks on an FEPROM memory card cannot be deleted with system function SFC 23.

In the event of an error, the data block is not deleted and an error number is returned as function value.

18.3.3 Testing a Data Block

System function SFC 24 returns the number of available data bytes for a data block in the user memory (in output parameter DB_LENGTH) and the write-protected status (in output parameter WRITE_PROT). You specify the number of the selected data block in input parameter DB_NUMBER.

In the event of an error, the contents of the output parameters are undefined and an error number is returned as function value.

18.3 System Functions for Data Blocks

Table 18.2 SFCs for Handling Data Blocks

SFC	Name	Decl.	Data Type	Assignment, Description
22	LOW_LIMIT	INPUT	WORD	Lowest number of the data block to be created
	UP_LIMIT	INPUT	WORD	Highest number of the data block to be created
	COUNT	INPUT	WORD	Length of the data block in bytes (even number)
	RET_VAL	OUTPUT	INT	Error information
	DB_NUMBER	OUTPUT	WORD	Number of the created data block
23	DB_NUMBER	INPUT	WORD	Number of the data block to be deleted
	RET_VAL	OUTPUT	INT	Error information
24	DB_NUMBER	INPUT	WORD	Number of the data block to be tested
	RET_VAL	OUTPUT	INT	Error information
	DB_LENGTH	OUTPUT	WORD	Length of the data block (in bytes)
	WRITE_PROT	OUTPUT	BOOL	"1" = write-protected

19 Block Parameters

This chapter describes how to use block parameters. You will learn

▷ how to declare block parameters,
▷ how to work with block parameters,
▷ how to initialize block parameters and
▷ how to 'forward' block parameters.

Block parameters represent the transfer interface between the calling and the called block. All functions and function blocks can be provided with block parameters.

19.1 Block Parameters in General

19.1.1 Defining the Block Parameters

Block parameters make it possible to parameterize the processing instruction in a block, the block function. Example: You want to write a block as an adder that you can use in your program several times with different variables. You transfer the variables as block parameters; in our example, three input parameters and one output parameter (Figure 19.1).

Since the adder need not store any values internally, a function is suitable as the block type.

You define a block parameter as an input parameter if you only check or load its value in the block program. If you only describe a block parameter (assign, set, reset, transfer), you use an output parameter.

You must always use an in/out parameter if a block parameter is to be both checked and overwritten. The Program Editor does not check the use of the block parameters.

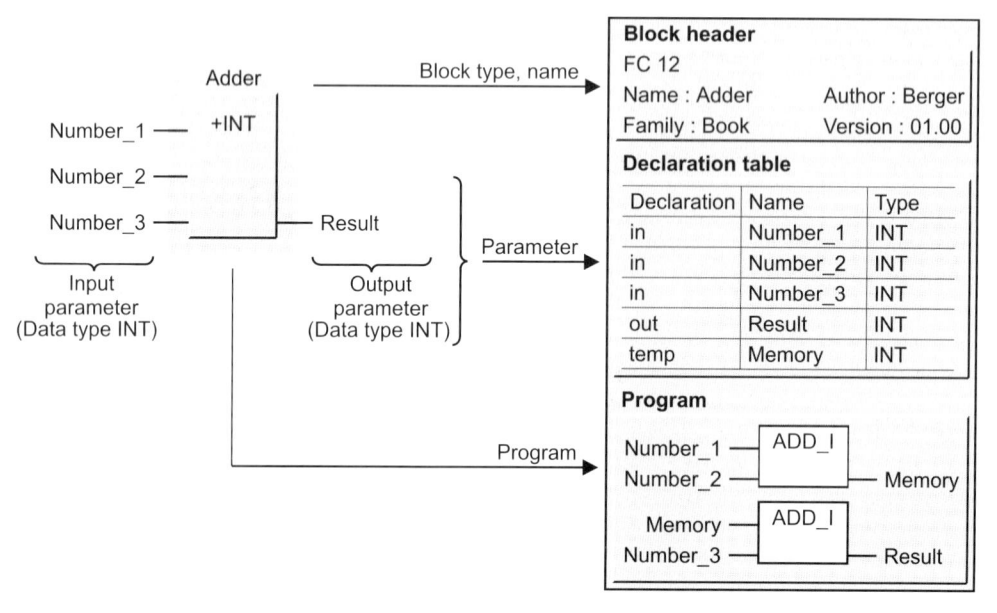

Figure 19.1 Example of Block Parameters

19.1.2 Processing the Block Parameters

In the adder program, the names of the block parameters stand as place holders for the latest actual variables. You use the block parameters in the same way as symbolically addressed variables; in the program, they are called formal parameters.

You can call the Adder function several times in your program. With each call, you transfer other values to the adder in the block parameters (Figure 19.2). The values can be constants, operands or variables; they are called actual parameters.

At runtime, the CPU replaces the formal parameters with the actual parameters. The first call in the example adds the contents of memory words MW 30, MW 32 and MW 34 and stores the result in memory word MW 40. The same block with the actual parameters of the second call adds data words DBW 30, DBW 32 and DBW 34 of data block DB 13 and stores the result in data word DBW 40 of data block DB 14.

19.1.3 Declaration of the Block Parameters

You define the block parameters in the declaration section of the block when you program the block. An empty declaration table is shown. In addition to the block parameters (in, out, in_out), you also declare the static local data (stat) and the temporary local data (temp) in this table.

Default values are optional and serve a practical purpose only for function blocks if a block parameter is stored as a value. This applies to all block parameters of elementary data type and to input and output parameters of complex data type. A parameter comment can also be given.

The *block parameter name* can be up to 24 characters in length. It must consist only of letters (without national characters such as the German Umlaut), digits, and the underscore. No distinction is made between upper and lower case. The name must not be a keyword.

For the *data type* of a block parameter, all elementary, complex and user-defined data types are permissible as well as the parameter types (see Chapter 3.4, "Data Types").

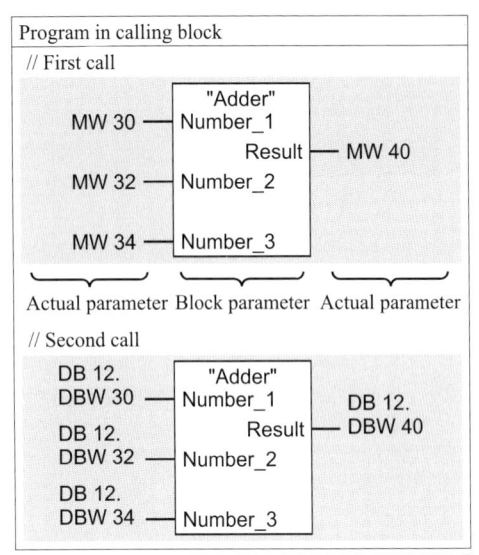

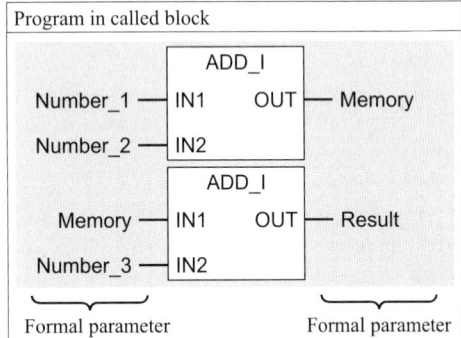

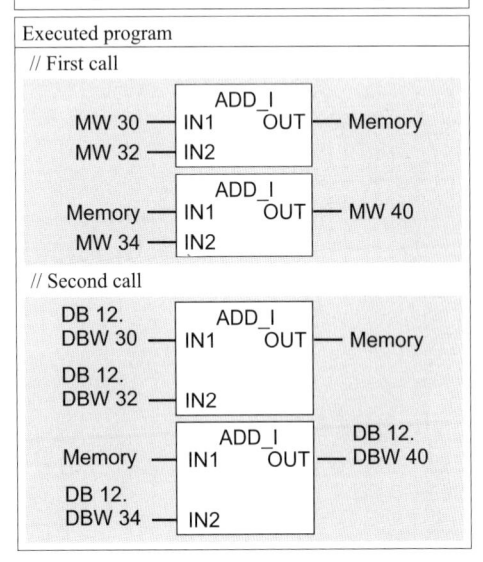

Figure 19.2 Block Call with Block Parameters

Table 19.1 Empty Declaration Table

Address	Declaration	Name	Type	Initial Value	Comment
	in				Input parameter (for FCs and FBs)
	out				Output parameter (for FCs and FBs)
	in_out				In/out parameter (for FCs and FBs)
	stat				Static local data (for FBs)
	temp				Temporary local data (for OBs, FCs and FBs)

STEP 7 stores the names of the block parameters in the non-executable section of the blocks on the programming device's storage medium. The user memory of the CPU (in the compiled block) contains only the declaration types and the data types. For this reason, program changes made to blocks online in the CPU must always be updated on the programming device's data medium in order to retain the original names. If the update is not made, or if blocks are transferred from the CPU to the programming device, the non-executable-block sections are overwritten or deleted. The Program Editor then generates replacement symbols for display or printout.

19.1.4 Declaration of the Function Value

In functions, the function value is a specially treated output parameter. It has the name RET_VAL (or ret_val) and is defined as the first output parameter. You declare the function value by assigning the name RET_VAL to the first output parameter in the declaration table.

As data type of the function value, all elementary data types as well as the data types DATE_AND_TIME, STRING, POINTER, ANY and user-defined data types UDT are permitted. ARRAY and STRUCT are not allowed.

As the first output parameter, the function value has no special role to play in the LAD or FBD programming language. It only gains significance in the SCL programming language, where you can use the block type FUNCTION as a "genuine" function. Here, a function FC can stand in place of an operand in a printout; the function value then represents the value of this function.

19.1.5 Initializing Block Parameters

When calling a block with block parameters, you initialize the block parameters with actual parameters. These can be constants, operands with absolute addresses, fully-addressed data operands or symbolically addressed variables. The actual parameter must be of the same data type as the block parameter.

You must initialize all of a function's block parameters at every call. In the case of function blocks, initialization of individual block parameters or all block parameters is optional.

19.2 Formal Parameters

In this chapter, you will learn how to access the block parameters within a block. You will learn that it is possible to access block parameters of elementary data type, components of an array or structure, and timers and counters without restriction. Section 19.4 shows you how you can 'pass on' block parameters to called blocks.

Access to parameters of complex data type and of parameter types POINTER and ANY is not supported by LAD or FBD. However, you can initialize acquired blocks or system blocks that have such parameters with the relevant variable. You can find examples of this in the program "Data Types" on the diskette accompanying the book.

Block parameters with data type BOOL

Block parameters of data type BOOL can be individual binary variables or binary components of arrays and structures. You can check input

19.2 Formal Parameters

Table 19.2 Accessing block parameters (general)

Data Types	Permitted for IN	Permitted for I_O	Permitted for OUT	Access in Block Possible
Elementary data types				
BOOL	x	x	x	yes
BYTE, WORD, DWORD, CHAR, INT, DINT, REAL, S5TIME, TOD, DATE	x	x	x	yes
Complex data types				
DT, STRING	x	x	x	no
ARRAY, STRUCT				
Individual binary components	x	x	x	yes
Individual digital components	x	x	x	yes
Complete variables	x	x	x	no
Parameter types				
TIMER	x	-	-	yes
COUNTER	x	-	-	yes
BLOCK_FC, BLOCK_FB	x	-	-	yes
BLOCK_DB	x	-	-	yes
BLOCK_SDB	x	-	-	no
POINTER, ANY	x	x	x [1]	no

[1] FC functions only

parameters and in/out parameters with contacts or with binary box inputs, and you can influence output parameters and in/out parameters with binary box outputs.

After the CPU has used the actual parameter specified as block parameter, it processes the functions as shown in the relevant chapters.

Block parameters of digital data type

Block parameters of digital data type occupy 8, 16 or 32 bits (all elementary data types except BOOL). They can be individual digital variables or digital components of arrays and structures.

You apply input and in/out parameters to digital box inputs, and you write output and in/out parameters to digital box outputs.

You exchange values between block parameters of different types or different sizes with the MOVE box as described in Chapter 6, "Transfer Functions".

Block parameters of data type DT and STRING

Direct access to block parameters of type DT and STRING is not possible. In function blocks, you can 'pass' input and output parameters of data type DT and STRING to parameters of called block.

Block parameters of data type ARRAY and STRUCT

Direct access to block parameters of type ARRAY and STRUCT is possible on a component-by-component basis, that is, you can access individual binary or digital components with the relevant operations.

Access to the complete variable (entire array or entire structure) is not possible and neither is access to individual components of complex or user-defined data type. In function blocks, you can 'pass' input and output parameters of data type ARRAY and STRUCT to parameters of called blocks.

221

Block parameters of user-defined data type

You handle block parameters of user-defined data type in the same way as block parameters of data type STRUCT.

Direct access to block parameters of data type UDT is possible on a component-by-component basis, that is, you can access individual binary or digital components with the relevant operations.

Access to the complete variable is not possible and neither is access to individual components of complex or user-defined data type. In function blocks, you can "pass on" input and output parameters of data type UDT to parameters of called blocks.

Block parameters of data type TIMER

You can use block parameters of data type TIMER with all functions as described in Chapter 7, "Timers". When a timer is started, the time value can also be a block parameter of data type S5TIME.

Block parameters of data type COUNTER

You can use block parameters of data type COUNTER with all functions as described in Chapter 8, "Counters". When setting a counter, the counter value can also be a block parameter of data type WORD.

Block parameters of data type BLOCK_DB

You can transfer a data block via a block parameter of data type BLOCK_DB. Call this data block with the OPN coil/box by labeling the OPN coil/box with the formal parameters. When opening a data block via a block parameter, the CPU always uses the global data block register (DB register).

Block parameters of data type BLOCK_FC

You can transfer an FC function via a block parameter of data type BLOCK_FC. Call this function with the CALL coil/box. You can use the CALL coil/box with a formal parameter and with or without any preceding logic operation if you are currently programming a function block. If you use the CALL coil/box with formal parameters in a function, a preceding logic operation is not permissible (absolute call only).

An FC function transferred via a block parameter must not have any block parameters.

Block parameters of data type BLOCK_FB

You can transfer an FB function block via a block parameter of data type BLOCK_FB. Direct access to block parameters of data type BLOCK_FB is not possible with LAD or FBD functions.

An FB function block transferred via a block parameter must not have any block parameters.

Block parameters of data type POINTER and ANY

Direct access to block parameters of data type POINTER and ANY is not possible with LAD or FBD functions.

19.3 Actual Parameters

When you call a block, you initialize its block parameters with constants, operands or variables with which it is to operate. These are the actual parameters. If you call the block often in your program, you usually use different actual parameters each time.

An actual parameter must agree in data type with the block parameter: You can only apply a binary actual parameter (for example, a memory bit) to a block parameter of type BOOL; you can only initialize a block parameter of type ARRAY with an identically dimensioned array variable. Table 19.3 gives an overview of which operands you can use as actual parameters with which data type.

When calling functions, you must initialize all block parameters with actual parameters.

When calling function blocks, it is not a mandatory requirement that you initialize block parameters with actual parameters. If you make no specification at a block parameter, the (old) values stored in the instance data block are used as the actual parameters. These can be, for example, the default values or the values of actual parameters from an earlier call. In/out parame-

19.3 Actual Parameters

Table 19.3 Initialization with Actual Parameters

Data Type of the Block Parameter	Permissible Actual Parameters
Elementary data type	▷ Simple operands, fully-addressed data operands, constants
	▷ Components of arrays or structures of elementary data type
	▷ Block parameters of the calling block
	▷ Components of block parameters of the calling block of elementary data type
Complex data type	▷ Variables or block parameters of the calling block
TIMER, COUNTER and BLOCK_xx	▷ Timers counters and blocks
POINTER	▷ Simple operands, fully-addressed data operands
	▷ Range pointer or DB pointer
ANY	▷ Variables of any data type
	▷ ANY pointer

ters of complex data type cannot be assigned default values, and neither can parameter types. You must provide these block parameters with actual parameters at least at the first call.

You can also use direct access to access the block parameters for the function block. Since they are located in a data block, you can handle the block parameters like data operands. Example: A function block with the instance data block "*Station_1*" controls a binary output parameter with the name "*Up*". Following processing in the function block (after its call), you can check the parameter under the symbolic address "*Station_1*".*Up*, without having initialized the output parameter.

Initializing block parameters with elementary data types

The actual parameters listed in Table 19.4 are permissible as actual parameters of elementary data type.

You can assign either absolute or symbolic addresses to input, output and bit memory operands. Input operands should be used only for input parameters and output operands for output parameters (however, this is not mandatory). Bit memory operands are suitable for all declaration types. You must apply peripheral inputs only to input parameters and peripheral outputs only to output parameters.

When you use (partially-addressed) data operands, you must make sure when you access the block parameter (in the called block) that the data block currently open is the "correct" one. Since the Program Editor may in certain circumstances change the data block when the block is called, partial addressing is not recommended for data operands. Use only fully-addressed data operands for this reason.

Temporary local data are usually symbolically addressed. They are located in the L stack of the calling block and are declared in the calling block.

If the calling block is a function block, you can also use its static local data as actual parameters (see Section 19.4, "Forwarding Block Parameters"). Static data are usually symbolically addressed; if you would like to assign absolute addresses via DI operands, please note the information in Section 18.2.4, "Special Points in Data Addressing".

With a block parameter of type BOOL, you can apply the constant TRUE (signal state "1") or FALSE (signal state "0"), and with block parameters of digital data type, you can apply all constants corresponding to the data type. Initialization with constants is permissible only for input parameters.

You can also initialize a block parameter of elementary data type with components of arrays and structures if the component is of the same data type as the block parameter.

19 Block Parameters

Table 19.4 Actual Parameters of Elementary Data Type

Operands	Permissible with IN	Permissible with I_O	Permissible with OUT	Binary operand or symbolic name	Digital operand or symbolic name
Inputs (process image)	x	x	x	I y.x	IB y, IW y, ID y
Outputs (process image)	x	x	x	Q y.x	QB y, QW y, QD y
Bit memory	x	x	x	M y.x	MB y, MW y, MD y
Peripheral inputs	x	-	-		PIB y, PIW y, PID y
Peripheral outputs	-	-	x		PQB y, PQW y, PQD y
Global data					
Partial addressing	x	x	x	DBX y.x	DBB y, DBW y, DBD y
Full addressing	x	x	x	DB z.DBX y.x	DB z.DBB y, etc.
Temporary local data	x	x	x	L y.x	LB y, LW y, LD y
Static local data	x	x	x	DIX y.x	DIB y, DIW y, DID y
Constants	x	-	-	TRUE, FALSE	All digital constants
Components of ARRAY or STRUCT	x	x	x	Complete component name	Complete component name

x = Bit number, y = Byte address, z = Data block number

Initializing block parameters of complex data type

Every block parameter can be of the complex type. Variables of the same type can be used as actual operands.

For initializing block parameters of type DT or STRING, individual variables or components of arrays or structures of the same data type are permissible.

STRING variables can be of variable length. If you have specified the STRING length when declaring an input or output parameter for a function block, the Editor reserves the specified space in the instance data block; if you have not specified any length, 256 bytes are reserved for the STRING variable. The maximum length of the STRING variable, which you specify in the declaration, must be the same for the actual parameter and the formal parameter. Exception: With FC functions, you specify either no length or the standard length of 254 bytes when declaring a STRING variable; here, you can use STRING variables of all lengths as actual parameters.

For initializing block parameters of type ARRAY or STRUCT, variables with exactly the same structure as the block parameters are permissible.

Initializing block parameters of user-defined data type

For complex or extensive data structures, the use of user-defined data types (UDTs) is recommended. First, you define the UDT and then you use it, for example, to generate the variable in the data block or to declare the block parameter. You can then use the variable when initializing the block parameter. Here, too, the actual parameter (the variable) must be of the same data type (the same UDT) as the block parameter.

Initializing block parameters of type TIMER, COUNTER and BLOCK_xx

You initialize a block parameter of type TIMER with a timer and a block parameter of type COUNTER with a counter. You can apply only blocks without their own parameters to block parameters of parameter types BLOCK_FC and BLOCK_FB. You initialize BLOCK_DBs with a data block.

Block parameters of types TIMER, COUNTER and BLOCK_xx must be input parameters.

Initializing block parameters of type POINTER

Pointers (constants) and operands are permissible for block parameters of parameter type POINTER. These pointers are either range pointers (32-bit pointers) or DB pointers (48-bit pointers). The operands are of elementary data type and can also be fully-addressed data operands.

Output parameters of type POINTER are not permitted for function blocks.

Initializing block parameters of type ANY

Variables of all data types are permissible for block parameters of parameter type ANY. The programming within the called block determines which variables (operands or data types) must be applied to the block parameters, or which variables are feasible. You can also specify a constant in the format of the ANY pointer "P#[data block.]Operand Datatype Number" and thus define an absolute-addressed area.

Output parameters of type ANY are not permissible for function blocks.

19.4 "Forwarding" Block Parameters

"Forwarding" (in other publications also called "passing on") block parameters is a special form of access and of initializing block parameters. The block parameters of the calling block are "forwarded" to the parameters of the called block. The formal parameter of the calling block then becomes the actual parameter of the called block.

In general, it is also the case here that the actual parameter must be of the same type as the formal parameter (that is, the relevant block parameters must agree in their data types). In addition, you can apply an input parameter of the calling blocks only to an input parameter of the called block, and similarly, an output parameter to an output parameter. You can apply an in/out parameter of the calling block to all declaration types of the called block.

There are restrictions with regard to data types caused by the variations in the storage of block parameter for functions and those for function blocks. Block parameters of elementary data type can be passed on without restriction in accordance with the information in the previous paragraph. Input and output parameters of complex type can only be forwarded if the calling block is a function block. Block parameters of parameter types TIMER, COUNTER and BLOCK_xx can only be passed on from one input parameter to another if the calling block is a function block. These statements are represented in Table 19.5.

19.5 Examples

19.5.1 Conveyor Belt Example

The example shows the transfer of signal states via block parameters. For this purpose, we use the function of a conveyor belt control system described in Chapter 5, "Memory Functions". The conveyor belt control system is to be programmed in a function block, and all inputs and outputs are to be routed through block parameters so that the function block can be called re-

Table 19.5 Permissible Combinations when forwarding Block Parameters

Calling → called	FC calls FC			FB calls FC			FC calls FB			FB calls FB		
Declaration type	E	C	P	E	C	P	E	C	P	E	C	P
Input → Input	x			x	x		x		x	x	x	x
Output → Output	x			x	x		x			x	x	
In/out → InpuT	x						x					
In/out → Output	x			x			x			x		
In/out → In/out	x			x						x		

E = Elementary data types
C = Complex data types
P = Parameter types TIMER, COUNTER and BLOCK_xx

19 Block Parameters

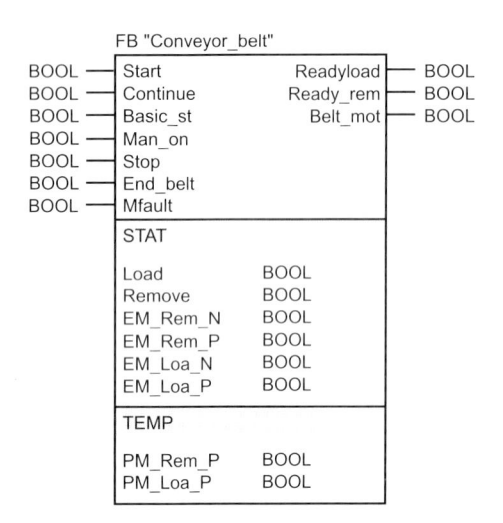

```
        FB "Conveyor_belt"
BOOL ──┤ Start          Readyload ├── BOOL
BOOL ──┤ Continue       Ready_rem ├── BOOL
BOOL ──┤ Basic_st       Belt_mot  ├── BOOL
BOOL ──┤ Man_on
BOOL ──┤ Stop
BOOL ──┤ End_belt
BOOL ──┤ Mfault
       │ STAT
       │
       │ Load            BOOL
       │ Remove          BOOL
       │ EM_Rem_N        BOOL
       │ EM_Rem_P        BOOL
       │ EM_Loa_N        BOOL
       │ EM_Loa_P        BOOL
       │
       │ TEMP
       │
       │ PM_Rem_P        BOOL
       │ PM_Loa_P        BOOL
```

Figure 19.3
Function Block for the Conveyor Belt Example

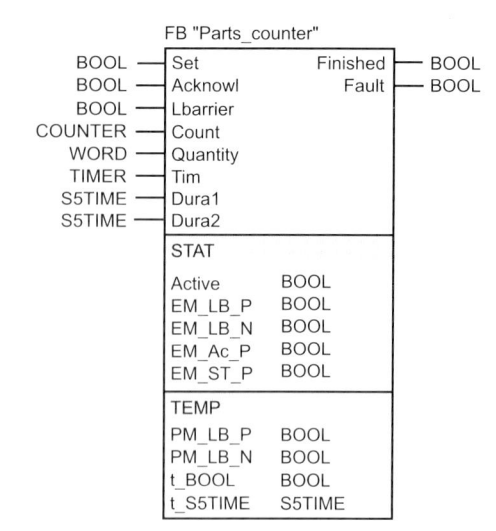

```
        FB "Parts_counter"
BOOL   ──┤ Set            Finished ├── BOOL
BOOL   ──┤ Acknowl        Fault    ├── BOOL
BOOL   ──┤ Lbarrier
COUNTER──┤ Count
WORD   ──┤ Quantity
TIMER  ──┤ Tim
S5TIME ──┤ Dura1
S5TIME ──┤ Dura2
         │ STAT
         │
         │ Active          BOOL
         │ EM_LB_P         BOOL
         │ EM_LB_N         BOOL
         │ EM_Ac_P         BOOL
         │ EM_ST_P         BOOL
         │
         │ TEMP
         │
         │ PM_LB_P         BOOL
         │ PM_LB_N         BOOL
         │ t_BOOL          BOOL
         │ t_S5TIME        S5TIME
```

Figure 19.4
Function Block for the Parts Counter Example

peatedly (for several conveyor belts). Figure 19.3 shows the input and output parameters for the function block as well as the static and temporary local data used.

Distributing the parameters is quite simple in this case: All binary operands that were inputs have become input parameters, all outputs have become output parameters, and all memory bits have become static local data. You will also have noticed that the names have been altered slightly because only letters, digits, and the underscore are permitted for block-local variables.

The function block "Conveyor_Belt" is to control two conveyor belts. For this purpose, it will be called twice; the first time with the inputs and outputs of conveyor belt 1 and the second time with those of conveyor belt 2. For each call, the function block requires an instance data block where it stores the data for the obtaining conveyor belt. The data block for conveyor belt 1 is to be called "BeltData1" and the data block for conveyor belt 2 is to be called "BeltData2".

You can find the executed programming example in the program on the "LAD_Book" or "FBD_Book" library under program "Conveyor Example" on the diskette accompanying the book. It shows the programming of function block FB 21 with input parameters, output parameters and static local data. You can use any data blocks as instance blocks; in the example, DB 21 is used for "Belt_data1" and DB 22 for "Belt_data2". In the Symbol Table, these data blocks have the data type of the function block (FB 21 in the example, if "Conveyor_belt" is the symbol for FB 21).

When you call the function block, you can use the inputs and outputs from the Symbol Table as actual parameters. In those cases where these global symbols contain special characters, you must place these symbols in quotation marks in the program. The Symbol Table is designed for all three examples in this chapter (Table 19.6).

19.5.2 Parts Counter Example

The example demonstrates the handling of block parameters of elementary data type. The "Parts Counter" example from Chapter 9, "Counters", is the basis of the function. The same function is implemented here as a function block, with all global variables declared either as block parameters or as static local data. The timers and counters are controlled here via individual elements.

Timers and counters are transferred via block parameters of type TIMER and COUNTER. These block parameters must be input parameters. The initial values of the counter (Quantity) and the timer (Dura1 and Dura2) can also be

19.5 Examples

transferred as block parameters; the data types of the block parameters correspond here to those of the actual parameters.

The edge memory bits are stored in the static local data and the pulse memory bits are stored in the temporary local data.

Table 19.6 Symbol Table for the Conveyor Belt, Parts Counter and Feed Examples

Symbol	Address	Data Type	Comment
Conveyor_belt	FB 21	FB 21	Conveyor belt control
Belt_data1	DB 21	FB 21	Data for conveyor belt 1
Belt_data2	DB 22	FB 21	Data for conveyor belt 2
Parts_counter	FB 22	FB 22	Counter control and monitor
CounterDat	DB 29	FB 22	Data for parts_counter
Feed	FB 20	FB 20	Feed with several belts
FeedDat	DB 20	FB 20	Data for Feed
Cycle	OB 1	OB 1	Main program, cyclic execution
Basic_st	I 0.0	BOOL	Set controller to the basic state
Man_on	I 0.1	BOOL	Switch on conveyor belt motors
/Stop	I 0.2	BOOL	Stop conveyor belt motors (zero active)
Start	I 0.3	BOOL	Start conveyor belt
Continue	I 0.4	BOOL	Acknowledgment that parts have been removed
Acknowl	I 0.6	BOOL	Acknowledge fault
Set	I 0.7	BOOL	Set counter, activate monitor
Lbarr1	I 1.0	BOOL	(Light barrier) "End of belt" sensor signal for conveyor belt 1
Lbarr2	I 1.1	BOOL	(Light barrier) "End of belt" sensor signal for conveyor belt 2
Lbarr3	I 1.2	BOOL	(Light barrier) "End of belt" sensor signal for conveyor belt 3
Lbarr4	I 1.3	BOOL	(Light barrier) "End of belt" sensor signal for conveyor belt 4
/Mfault1	I 2.0	BOOL	Motor protection switch conveyor belt 1 (zero-active)
/Mfault2	I 2.1	BOOL	Motor protection switch conveyor belt 2 (zero-active)
/Mfault3	I 2.2	BOOL	Motor protection switch conveyor belt 3 (zero-active)
/Mfault4	I 2.3	BOOL	Motor protection switch conveyor belt 4 (zero-active)
Readyload	Q 4.0	BOOL	Load new parts onto belt
Ready_rem	Q 4.1	BOOL	Remove parts from belt
Finished	Q 4.2	BOOL	Number of parts reached
Fault	Q 4.3	BOOL	Monitor activated
Belt_mot1	Q 5.0	BOOL	Switch on belt motor for conveyor belt 1
Belt_mot2	Q 5.1	BOOL	Switch on belt motor for conveyor belt 2
Belt_mot 3	Q 5.2	BOOL	Switch on belt motor for conveyor belt 3
Belt_mot 4	Q 5.3	BOOL	Switch on belt motor for conveyor belt 4
Quantity	MW 4	WORD	Number of parts
Dura1	MW 6	S5TIME	Monitoring time for light barrier covered
Dura2	MW 8	S5TIME	Monitoring time for light barrier not covered
Count	C 1	COUNTER	Counter for parts
Monitor	T 1	TIMER	Timer for monitor

19 Block Parameters

You can find the sample program on library "LAD_Book" or "FBD_Book" under "Conveyor Example" on the diskette accompanying the book. It contains function block FB 22 "Parts_counter" and the associated instance data block "CountDat". You can use the inputs, outputs, timer and counter from the Symbol Table (table in previous example) as actual parameters when calling the function block.

19.5.3 Feed Example

The same functions described in the two previous examples can also be called as local instances. In our example, this means that we program a function block called "Feed" that is to control four conveyor belts and count the conveyed parts. In this function block, FB "Conveyor Belt" is called four times and FB "Parts_counter" is called once. The call does not take place in each case with its own instance data block, but the called FBs are to store their data in the instance data block of function block "Feed".

Figure 19.5 shows how the individual conveyor belt controls are interconnected (FB "Parts_counter" is not represented here). The start signal is connected to the *Start* input of the controller for belt 1, the *ready_rem* output is connected to the *Start* input of belt 2, etc. Finally, the *ready_rem* output of belt 4 is connected to the *Remove* output of "Feed". The same signal sequence leads in the reverse direction from *Remove* over *Continue* and *Readyload* to *Load*.

Belt_mot, *Lbarr* and */Mfaultare* separate conveyor belt signals; *Reset*, *Startup* and *Stop* control all conveyor belts via *Basic_st*, *Man_on* and *Stop*.

The following program for function block "Feed" is designed in the same way. The input and output parameters of the function block can be taken from Figure 19.5. In addition, the digital values for the parts counter *Quantity*, *Dura1* and *Dura2* are designed as input parameters here. We declare the data of the individual conveyor belt controls and the data of the parts counter in the static local data in exactly the same way as for a user-defined data type, i.e. with name and data type. The variable "Belt1" is to receive the data structure of function block "Conveyor Belt", as is variable "Belt2", etc.; the variable "Check" receives the data structure of the function block "Parts_counter".

The program in the function block starts with the initialization of the signals common to all conveyor belts. Here, we make use of the fact that the block parameters of the function blocks called as local instances are static local data in the current block and can be treated as such. The block parameter Man_on in the current function block controls the *Man_on* input parameters of all four conveyor belt controls with a single assignment. We proceed in the same way with the signals *Stop* and *Reset*. And now the conveyor belt controls are initialized with the common signals. (You can, of course, also initialize these input parameters when the function block is called.)

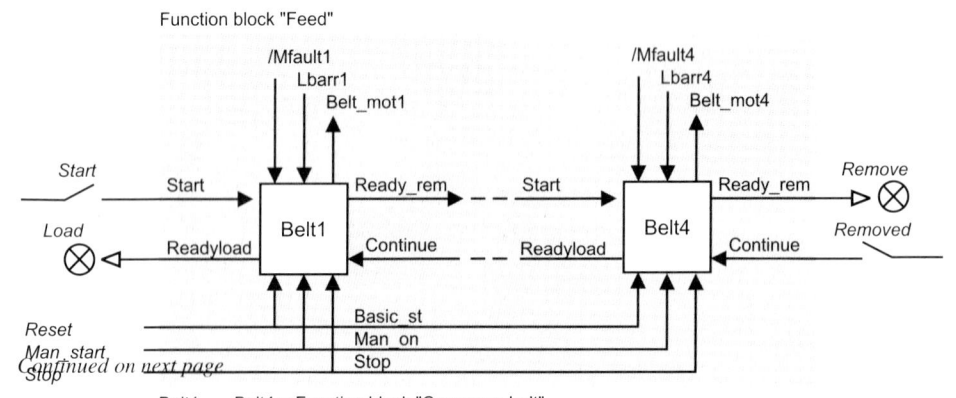

Continued on next page

Figure 19.5 Feed Programming Example

19.5 Examples

Table 19.7 Declaration Section of FB "Feed"

Address	Declaration	Name	Type	Initial Value	Comment
0.0	in	Start	BOOL	FALSE	Start conveyor belts
0.1	in	Removed	BOOL	FALSE	Parts have been removed from belt
0.2	in	Man_start	BOOL	FALSE	Start conveyor belts manually
0.3	in	Stop	BOOL	FALSE	Stop conveyor belts
0.4	in	Reset	BOOL	FALSE	Set control to the basic setting
2.0	in	Count	COUNTER		Counter for the parts
4.0	in	Quantity	WORD	W#16#200	Number of parts
6.0	in	Tim	TIMER		Timer for the monitor
8.0	in	Dura1	S5TIME	S5T#5s	Monitoring time for parts
10.0	in	Dura2	S5TIME	S5T#10s	Monitoring time for gap
12.0	out	Load	BOOL	FALSE	Load new parts onto belt
12.1	out	Remove	BOOL	FALSE	Remove parts from belt
	in_out				
14.0	stat	Belt1	Conveyor_belt		Control for belt 1
20.0	stat	Belt2	Conveyor_belt		Control for belt 2
26.0	stat	Belt3	Conveyor_belt		Control for belt 3
32.0	stat	Belt4	Conveyor_belt		Control for belt 4
38.0	stat	Check	Parts_counter		Control for counting and monitoring
	temp				

The subsequent calls of the function blocks for conveyor belt control contain only the block parameters for the individual signals for each conveyor belt and the connection to the block parameters of "Feed". The individual signals are the light barriers, the commands for the belt motor and the motor faults. (We make use here of the fact that when a function block is called, not all block parameters have to be initialized.) We program the connections between the individual belt controllers using assignments.

The FB "Parts_counter" is called as a local instance even if it has no closer connection with the signals of the conveyor belt controls. The instance data block for "Feed" stores the FB data.

The input parameters *Quantity*, *Dura1* and *Dura2 for* "Feed" need to be set only once. This can be done using default values (as in the example) or in the restart routine in OB 100 (for example, through direct assignment if these three parameters are treated as global data).

19 Block Parameters

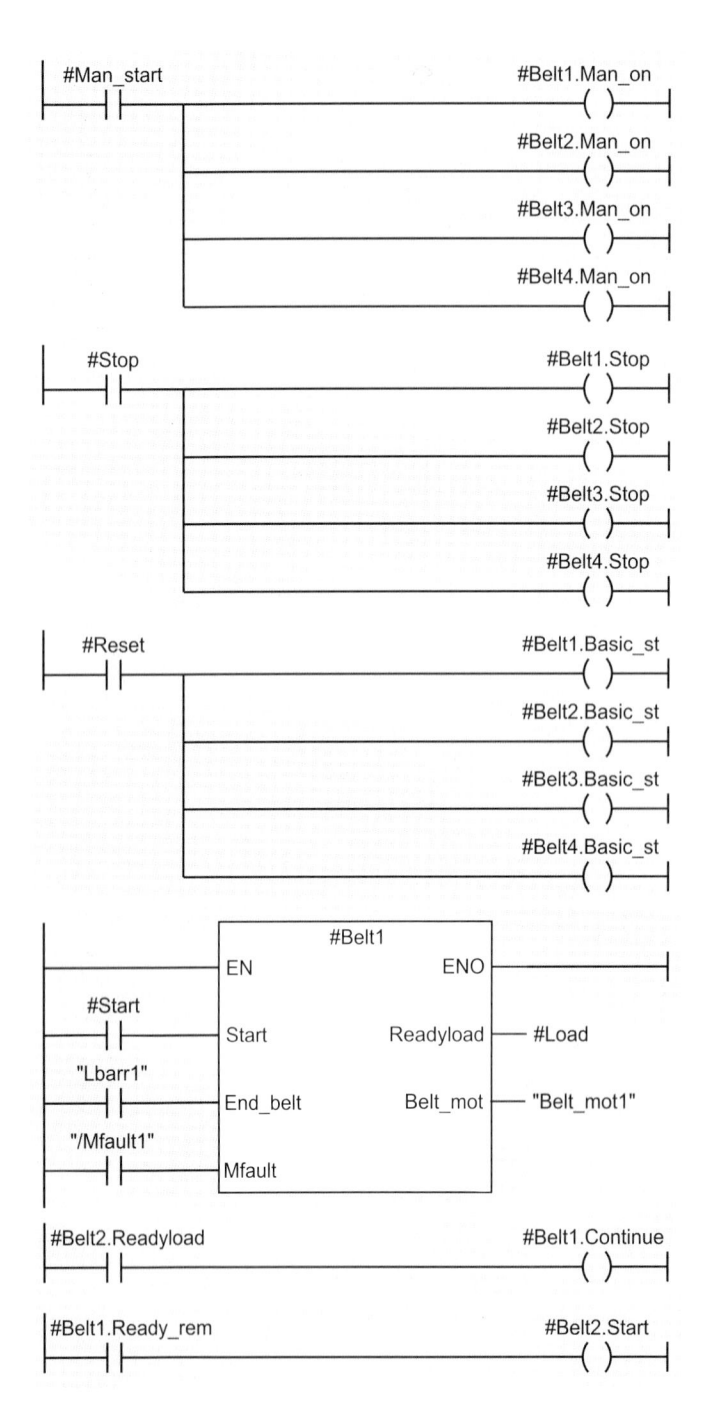

```
#Man_start                         #Belt1.Man_on
───┤ ├──────────────┬──────────────────( )───
                    │              #Belt2.Man_on
                    ├──────────────────( )───
                    │              #Belt3.Man_on
                    ├──────────────────( )───
                    │              #Belt4.Man_on
                    └──────────────────( )───


#Stop                              #Belt1.Stop
───┤ ├──────────────┬──────────────────( )───
                    │              #Belt2.Stop
                    ├──────────────────( )───
                    │              #Belt3.Stop
                    ├──────────────────( )───
                    │              #Belt4.Stop
                    └──────────────────( )───


#Reset                             #Belt1.Basic_st
───┤ ├──────────────┬──────────────────( )───
                    │              #Belt2.Basic_st
                    ├──────────────────( )───
                    │              #Belt3.Basic_st
                    ├──────────────────( )───
                    │              #Belt4.Basic_st
                    └──────────────────( )───


                         #Belt1
                    ┌─────────────────┐
              ──────┤EN            ENO├──────
   #Start          │                 │
   ──┤ ├───────────┤Start    Readyload├──── #Load
   "Lbarr1"        │                 │
   ──┤ ├───────────┤End_belt  Belt_mot├──── "Belt_mot1"
   "/Mfault1"      │                 │
   ──┤ ├───────────┤Mfault           │
                    └─────────────────┘

#Belt2.Readyload                   #Belt1.Continue
───┤ ├─────────────────────────────────( )───

#Belt1.Ready_rem                   #Belt2.Start
───┤ ├─────────────────────────────────( )───
```

Figure 19.6
Program for Feed example (LAD)

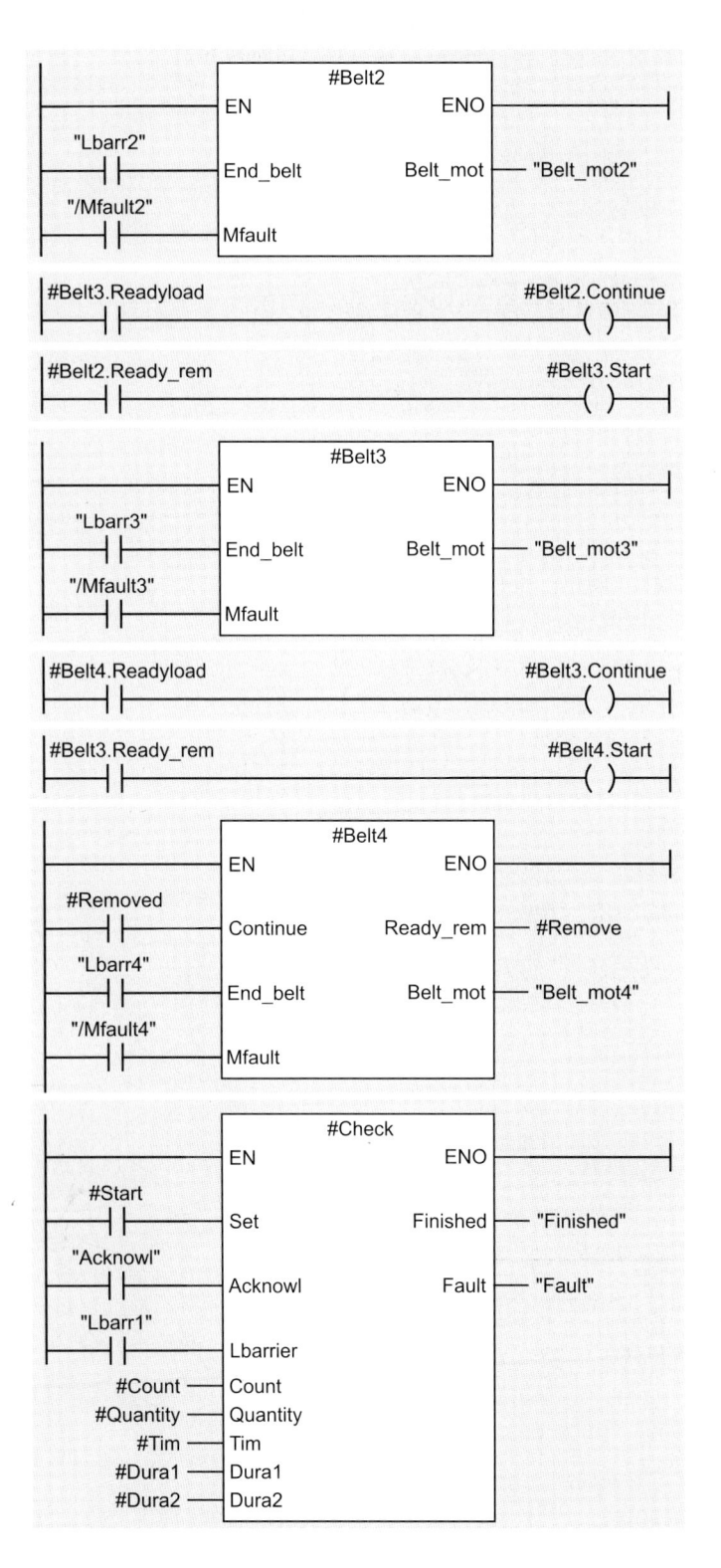

19.5 Examples

Continued

Programm for Feed Example (LAD)

```
           ┌─────────────────────┐
           │        #Belt2       │
       ────┤ EN              ENO ├────
  "Lbarr2" │                     │
     ──┤├──┤ End_belt   Belt_mot ├── "Belt_mot2"
 "/Mfault2"│                     │
     ──┤├──┤ Mfault              │
           └─────────────────────┘

  #Belt3.Readyload              #Belt2.Continue
     ──┤├────────────────────────────( )──

  #Belt2.Ready_rem              #Belt3.Start
     ──┤├────────────────────────────( )──

           ┌─────────────────────┐
           │        #Belt3       │
       ────┤ EN              ENO ├────
  "Lbarr3" │                     │
     ──┤├──┤ End_belt   Belt_mot ├── "Belt_mot3"
 "/Mfault3"│                     │
     ──┤├──┤ Mfault              │
           └─────────────────────┘

  #Belt4.Readyload              #Belt3.Continue
     ──┤├────────────────────────────( )──

  #Belt3.Ready_rem              #Belt4.Start
     ──┤├────────────────────────────( )──

           ┌─────────────────────┐
           │        #Belt4       │
       ────┤ EN              ENO ├────
  #Removed │                     │
     ──┤├──┤ Continue   Ready_rem├── #Remove
  "Lbarr4" │                     │
     ──┤├──┤ End_belt   Belt_mot ├── "Belt_mot4"
 "/Mfault4"│                     │
     ──┤├──┤ Mfault              │
           └─────────────────────┘

           ┌─────────────────────┐
           │        #Check       │
       ────┤ EN              ENO ├────
    #Start │                     │
     ──┤├──┤ Set         Finished├── "Finished"
 "Acknowl" │                     │
     ──┤├──┤ Acknowl        Fault├── "Fault"
  "Lbarr1" │                     │
     ──┤├──┤ Lbarrier            │
   #Count ─┤ Count               │
 #Quantity─┤ Quantity            │
      #Tim─┤ Tim                 │
    #Dura1─┤ Dura1               │
    #Dura2─┤ Dura2               │
           └─────────────────────┘
```

19 Block Parameters

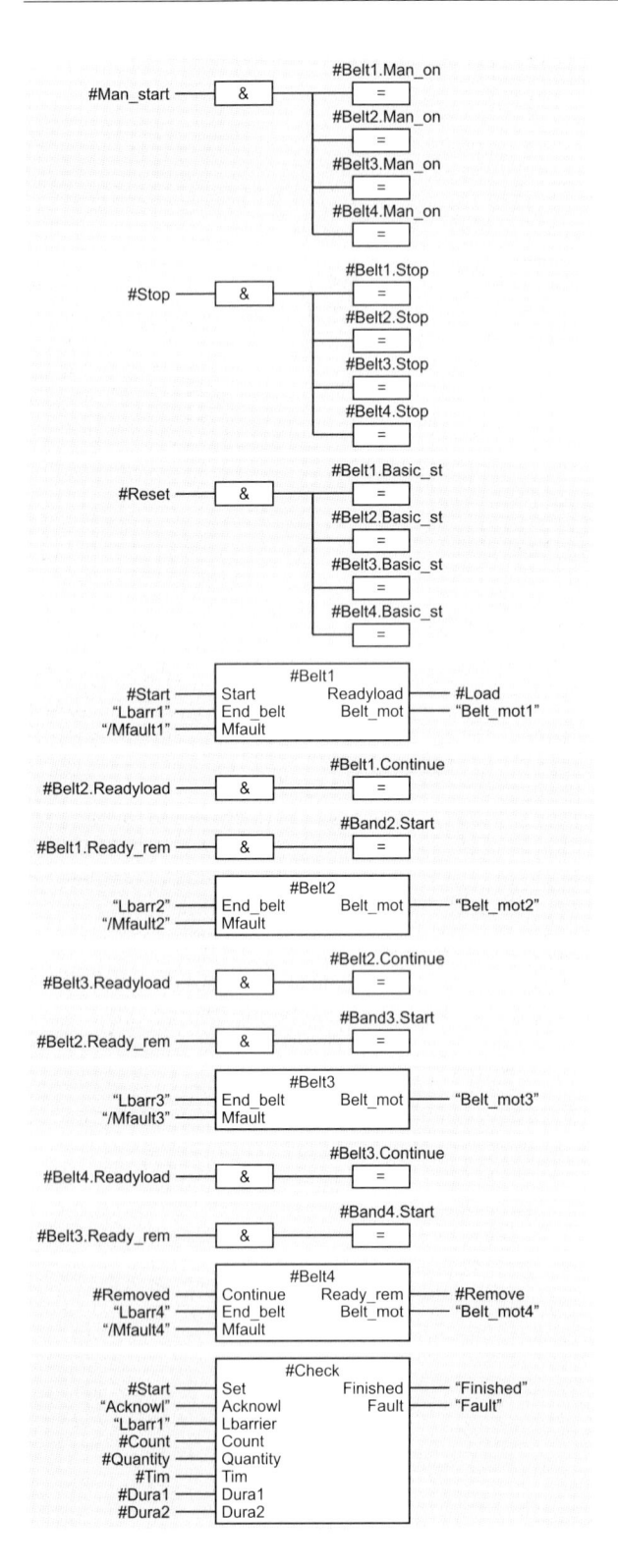

```
#Man_start ──┬─ & ──┬── #Belt1.Man_on
             │       │      =
             │       ├── #Belt2.Man_on
             │       │      =
             │       ├── #Belt3.Man_on
             │       │      =
             │       └── #Belt4.Man_on
             │              =

#Stop ──── & ──┬── #Belt1.Stop
                │      =
                ├── #Belt2.Stop
                │      =
                ├── #Belt3.Stop
                │      =
                └── #Belt4.Stop
                       =

#Reset ──── & ──┬── #Belt1.Basic_st
                 │      =
                 ├── #Belt2.Basic_st
                 │      =
                 ├── #Belt3.Basic_st
                 │      =
                 └── #Belt4.Basic_st
                        =
```

                              #Belt1
#Start ─────── Start   Readyload ─── #Load
"Lbarr1" ───── End_belt   Belt_mot ── "Belt_mot1"
"/Mfault1" ─── Mfault

                              #Belt1.Continue
#Belt2.Readyload ──── & ──── =

                              #Band2.Start
#Belt1.Ready_rem ──── & ──── =

                              #Belt2
"Lbarr2" ───── End_belt   Belt_mot ── "Belt_mot2"
"/Mfault2" ─── Mfault

                              #Belt2.Continue
#Belt3.Readyload ──── & ──── =

                              #Band3.Start
#Belt2.Ready_rem ──── & ──── =

                              #Belt3
"Lbarr3" ───── End_belt   Belt_mot ── "Belt_mot3"
"/Mfault3" ─── Mfault

                              #Belt3.Continue
#Belt4.Readyload ──── & ──── =

                              #Band4.Start
#Belt3.Ready_rem ──── & ──── =

                              #Belt4
#Removed ───── Continue   Ready_rem ── #Remove
"Lbarr4" ───── End_belt   Belt_mot ── "Belt_mot4"
"/Mfault4" ─── Mfault

                              #Check
#Start ─────── Set   Finished ── "Finished"
"Acknowl" ──── Acknowl   Fault   ── "Fault"
"Lbarr1" ───── Lbarrier
#Count ─────── Count
#Quantity ──── Quantity
#Tim ───────── Tim
#Dura1 ─────── Dura1
#Dura2 ─────── Dura2

Figure 19.7
Program for Feed Example (FBD)

Program Processing

This section of the book discusses the various methods of program processing.

The **main program** executes cyclically. After each program pass, the CPU returns to the beginning of the program and executes it again. This is the "standard" method of processing PLC programs.

Numerous system functions support the utilization of system services, such as controlling the real-time clock or communication via bus systems. In contrast to the static settings made when parameterizing the CPU, system functions can be used dynamically at program run time.

The main program can be temporarily suspended to allow **interrupt servicing**. The various types of interrupts (hardware interrupts, watchdog interrupts, time-of-day interrupts, time-delay interrupts, multiprocessor interrupts) are divided into priority classes whose processing priority you may yourself, to a large degree, determine. Interrupt servicing allows you to react quickly to signals from the controlled process or implement periodic control procedures independently of the processing time of the main program.

Before starting the main program, the CPU initiates a **start-up program** in which you can make specifications regarding program processing, define default values for variables, or parameterize modules.

Error handling is also part of program processing. STEP 7 distinguishes between synchronous errors, which occur during processing of a statement, and asynchronous errors, which can be detected independently of program processing. In both cases you can adapt the error routine to suit your needs.

20 **Main program**

Program structure; scan cycle control; response time; program functions; multicomputing operation; data exchange with system functions; start information

21 **Interrupt handling**

Hardware interrupts; watchdog interrupts; time-of-day interrupts; time-delay interrupts; multiprocessor interrupt; handling interrupt events

22 **Start-up characteristics**

Power-up, memory reset, retentivity; complete restart; warm restart; ascertain module address; parameterize modules

23 **Error handling**

Synchronous errors (programming errors, access errors); handling synchronous error events; asynchronous errors; system diagnostics

20 Main Program

The main program is the cyclically scanned user program; cyclic scanning is the "normal" way in which programs execute in programmable logic controllers. The large majority of control systems use only this form of program execution. If event-driven program scanning is used, it is in most cases only in addition to the main program.

The main program is invoked in organization block OB 1. It executes at the lowest priority level, and can be interrupted by all other types of program processing. The mode selector on the CPU's front panel must be at RUN or RUN-P. When in the RUN-P position, the CPU can be programmed via a programming device. In the RUN position, you can remove the key so that no one can change the operating mode without proper authorization; when the mode selector is at RUN, programs can only be read.

20.1 Program Organization

20.1.1 Program Structure

To analyze a complex automation task means to subdivide that task into smaller tasks or functions in accordance with the structure of the process to be controlled. You then define the individual tasks resulting from this subdividing process by determining the functions and stipulating the interface signals to the process or to other tasks. This breakdown into individual tasks can be done in your program. In this way, the structure of your program corresponds to the subdivision of the automation task.

A subdivided user program can be more easily configured, and can be programmed in sections (even by several people in the case of very large user programs). And finally, but not lacking in importance, subdividing the program simplifies both debugging and service and maintenance.

The structuring of the user program depends on its size and its function. A distinction is made between three different "methods" :

In a **linear program**, the entire main program is in organization block OB 1. Each current path is in a separate network. STEP 7 numbers the networks in sequence. When editing and debugging, you can reference every network directly by its number.

A **partitioned program** is basically a linear program which is subdivided into blocks. Reasons for subdividing the program might be because it is too long for organization block OB 1 or because you want to make it more readable. The blocks are then called in sequence. You can also subdivide the program in another block the same way you would the program in organization block OB 1. This method allows you to call associated process-related functions for processing from within one and the same block. The advantage of this program structure is that, even though the program is linear, you can still debug and run it in sections (simply by omitting or adding block calls).

A **structured program** is used when the conceptual formulation is particularly extensive, when you want to reuse program functions, or when complex problems must be solved. Structuring means dividing the program into sections (blocks) which embody self-contained functions or serve a specific functional purpose and which exchange the fewest possible number of signals with other blocks.

Assigning each program section a specific (process-related) function will produce easily readable blocks with simple interfaces to other blocks when programmed.

The LAD and FBD programming languages support structured programming through functions with which you can create "blocks" (self-contained program sections). Chapter 3, "SIMATIC S7 Program", discusses under the header "Blocks" the different kinds of blocks and their uses. You will find a detailed description of the functions for calling and ending blocks in Chapter 18, "Block Functions". The blocks receive the signals and data to be processed via the call interface (the block parameters), and forward the results over this same interface. The options for passing parameters are described in detail in Chapter 19, "Block Parameters".

20.1.2 Program Organization

Program organization determines whether and in what order the CPU will process the blocks which you have generated. To organize your program, you program block calls in the desired sequence in the supraordinate blocks. You should chose the order in which the blocks are called so that it mirrors the process-related or function-related subdivision of the controlled plant.

Nesting depth

The maximum depth applies for a priority class (for the program in an organization block), and is CPU-dependent. On the CPU 314, for example, the nesting depth is eight, that is, beginning with one organization block (nesting depth 1), you can add seven more blocks in the "horizontal" direction (this is called "nesting"). If more blocks are called, the CPU goes to STOP with a "Block overflow" error. Do not forget to include system function block (SFB) calls and system function (SFC) calls when calculating the nesting depth.

A data block call, which is actually only the opening or selecting of a data area, has no effect on the nesting depth of blocks, nor is the nesting depth affected by calling several blocks in succession (linear block calls).

Practice-related program organization

In organization block OB 1, you should call the blocks in the main program in such a way as to roughly organize your program. A program can be organized on either a process-related or function-related basis.

The following points of discussion can give only a rough, very general view with the intention of giving the beginner some ideas on program structuring and on translating his control task into reality. Advanced programmers normally have sufficient experience to organize a program to suit the special control task at hand.

A **process-related program structure** closely follows the structure of the plant to be controlled. The individual program sections correspond to the individual parts of the plant or of the process to be controlled. Subordinate to this rough structure are the scanning of the limit switches and operator panels and the control of the actuators and display devices (in different parts of the plant). Bit memory or global data are used for signal interchange between different parts of the plant.

A **function-related program structure** is based on the control function to be executed. Initially, this method of program structuring does not take the controlled plant into account at all. The plant structure first becomes apparent in the subordinate blocks when the control function defined by the rough structure is subdivided further.

In practice, a hybrid of these two concepts is normally used. Figure 20.1 shows an example: A functional structure is mirrored in the operating mode program and in the data processing program which goes above and beyond the plant itself. Program sections Feeding Conveyor 1, Feeding Conveyor 2, Process and Discharging Conveyor are process-related.

The example also shows the use of different types of blocks. The main program is in OB 1; it is in this program that the blocks for the operating modes, the various pieces of plant equipment, and for data processing are called. These blocks are function blocks with an instance data block as data store. Feeding Conveyor 1 and Feeding Conveyor 2 are identically structured; FB 20, with DB 20 as instance data block for Feeding Conveyor 1 and with DB 21 as instance data block for Feeding Conveyor 2, is used for control.

In the conveyor control program, function FC 20 processes the interlocks; it scans inputs or memory bits and controls FB 20's local data.

235

20 Main Program

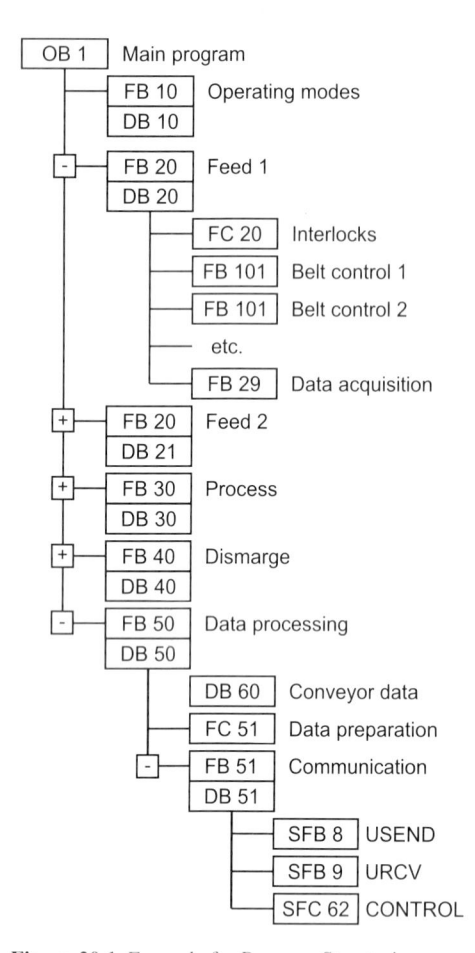

```
OB 1  Main program
 ├─ FB 10   Operating modes
 │  DB 10
 ├─ FB 20   Feed 1
 │  DB 20
 │   ├─ FC 20    Interlocks
 │   ├─ FB 101   Belt control 1
 │   ├─ FB 101   Belt control 2
 │   │  etc.
 │   └─ FB 29    Data acquisition
 ├─ FB 20   Feed 2
 │  DB 21
 ├─ FB 30   Process
 │  DB 30
 ├─ FB 40   Dismarge
 │  DB 40
 └─ FB 50   Data processing
    DB 50
     ├─ DB 60    Conveyor data
     ├─ FC 51    Data preparation
     └─ FB 51    Communication
        DB 51
         ├─ SFB 8    USEND
         ├─ SFB 9    URCV
         └─ SFC 62   CONTROL
```

Figure 20.1 Example for Program Structuring

Function block FB 101 contains the control program for a conveyor belt, and is called once for each belt. The call is a local instance, so that its local data are in instance data block DB 20. The same applies for the data acquisition program in FB 29.

The data processing program in FB 50, which uses DB 50, processes the data acquired with FB 29 (and other blocks), which are located in global data block DB 60. Function FC 51 prepares these data for transfer. The transfer is controlled by FB 51 (with DB 51), in which system blocks SFB 8, SFB 9 and SFB 62 are called. Here, too, the SFBs save their instance data in "supraordinate" data block DB 51.

20.2 Scan Cycle Control

20.2.1 Process Image Updating

The process image is part of the CPU's internal system memory (Chapter 1.4, "CPU Memory Areas"). It begins at I/O address 0 and ends at an upper limit stipulated by the CPU. On appropriately equipped CPUs, you can define this limit yourself.

Normally, all digital modules lie in the process image address area, while all analog modules have addresses outside this area. If the CPU has free address allocation, you can use the configuration table to direct any module over the process image or address it outside the process image area.

The process image consists of the process-image input table (inputs I) and the process-image output table (outputs Q).

After CPU restart and prior to the first execution of OB 1, the operating system transfers the signal states of the process-image output table to the output modules and accepts the signal states of the input modules into the process-image input table. This is followed by execution of OB 1 where normally the inputs are combined with each other and the outputs are controlled. Following termination of OB 1, a new cycle begins with the updating of the process image (Figure 20.2).

If an error occurs during automatic updating of the process image, e.g. because a module is no longer accessible, organization block OB 85 "Program Execution Errors" is called. If OB 85 is not available, the CPU goes to STOP.

Subprocess images

With appropriately equipped CPUs, you can subdivide the process image into up to 9 or 16 subprocess images. You make this subdivision during parameterization of the signal modules by defining the subprocess image via which the module is to be addressed when you assign addresses. You can separate the subdivision according to process-image input table and process-image output table.

All modules that you do not assign to one of the subprocess images 1 to 8 or 15 are stored in subprocess image 0. This subprocess image 0 is

20.2 Scan Cycle Control

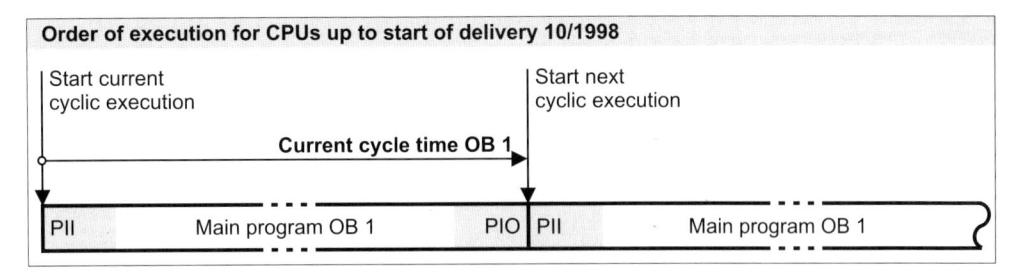

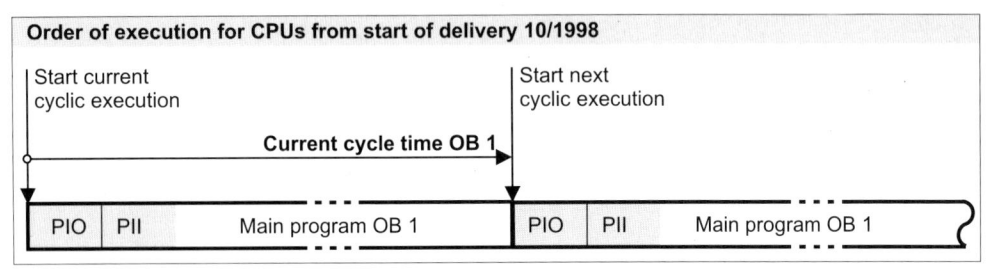

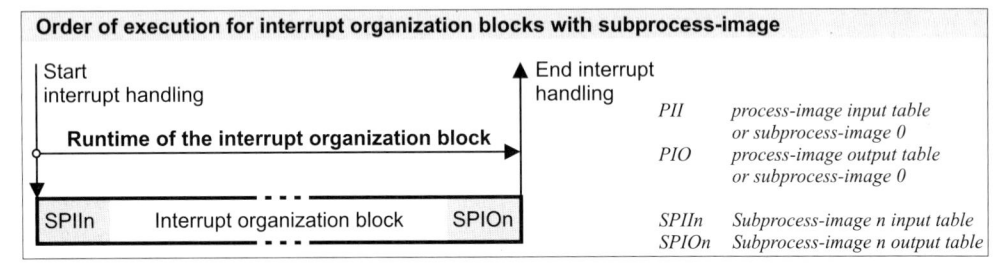

Figure 20.2 Updating the Process Image

updated automatically by the operating system of the CPU as part of cyclic execution.

With appropriately equipped CPUs, you can also assign the subprocess images to the interrupt organization blocks so that they are automatically updated when these OBs are called.

SFC 26 UPDAT_PI
SFC 27 UPDAT_PO

You update the subprocess images by calling a system function in the user program. For the subprocess-image input table, you use SFC 26 UPDAT_PI and for the subprocess-image output table, you use SFC 27 UPDAT_PO.

Table 20.1 shows the parameters of these SFCs. You can also update subprocess image 0 with these SFCs.

You can carry out updating of individual subprocess images by calling these SFCs at any time and at any location. For example, you can define a subprocess image for a priority class (a program execution level) and you can then cause this subprocess image to be updated at the start and at the end of the relevant organization block when this priority class is processed.

Updating of a process image can be interrupted by calling a higher priority class. If an error occurs during updating of a process image, e.g. because a module can no longer be accessed, this error is reported via the function value of the SFC.

20.2.2 Scan Cycle Monitoring Time

Program scanning in organization block OB 1 is monitored by the so-called "scan cycle mon-

237

20 Main Program

Table 20.1 Parameters for the SFCs for Process Image Updating

Parameter Name	SFC		Declaration	Data Type	Contents, Description
PART	26	27	INPUT	BYTE	Number of the subprocess image (0 to 15)
RET_VAL	26	27	OUTPUT	INT	Error information
FLADDR	26	27	OUTPUT	WORD	On an access error: the address of the first byte to cause the error

itor" or "scan cycle watchdog". The default value for the scan cycle monitoring time is 150 ms. You can change this value in the range from 1 ms to 6 s by parameterizing the CPU accordingly.

If the main program takes longer to scan than the specified scan cycle monitoring time, the CPU calls OB 80 ("Timeout"). If OB 80 has not been programmed, the CPU goes to STOP.

The scan cycle monitoring time includes the full scan time for OB 1. It also includes the scan times for higher priority classes which interrupt the main program (in the current cycle). Communication processes carried out by the operating system, such as GD communication or PG access to the CPU (block status!), also increase the runtime of the main program. The increase can be reduced in part by the way you parameterize the CPU ("Cyclic load from communication" on the "Cycle/Clock memory bits" tab). The CPU's cyclic memory test (S7-300) also increases the scan cycle time. You can limit or disable this test by parameterizing the CPU accordingly.

Cycle statistics

If you have an online connection from a programming device to an operating CPU, select PLC → MODULE INFORMATION to call up a dialog box that contains several tabs. The "Cycle Time" tab shows the current cycle time as well as the shortest and longest cycle time. The parameterized minimum cycle duration and the scan cycle monitoring time are also displayed.

The cycle time for the last cycle and the minimum and maximum cycle time since the PLC was last started up can also be read in the temporary local data in the start information of OB 1.

SFC 43 RE_TRIGR
Restarting the scan cycle monitoring time

An SFC 43 RE_TRIGR system function call restarts the scan cycle monitoring time; the timer restarts with the new value set via CPU parameterization. SFC 43 has no parameters.

Operating system run times

The scan cycle time also includes the operating system run times. These are composed of the following:

▷ System control of cyclic scanning ("no-load cycle"), fixed value

▷ Updating of the process image; dependent on the number of bytes to be updated

▷ Updating of the timers; dependent on the number of timers to be updated

▷ Communications load

Communications functions for the CPU include the transfer of user program blocks or data exchange between CPU modules using system functions. The time the CPU is to use for these functions can be limited by parameterizing the CPU.

All values at operating system runtime are properties of the relevant CPU.

20.2.3 Minimum Scan Cycle Time, Background Scanning

With appropriately equipped CPUs, you may specify a minimum scan cycle time. If the main program (including interrupts) takes less time, the CPU waits until the specified minimum scan cycle time has elapsed before beginning the next cycle by recalling OB 1.

20.2 Scan Cycle Control

The default value for the minimum scan cycle time is 0 ms, that is to say, the function is disabled. You can set a minimum scan cycle time of from 1 ms to 6 s in "Cycle/Clock memory bits" tab when you parameterize the CPU.

Background scanning OB 90

In the interval between the actual end of the cycle and expiration of the minimum cycle time, the CPU executes organization block OB 90 "Background scanning" (Figure 20.3). OB 90 is executed "in slices". When the operating system calls OB 1, execution of OB 90 is interrupted; it is then resumed at the point of interruption when OB 1 has terminated.

OB 90 can be interrupted after each statement, any system block called in OB 90, however, is first scanned in its entirety.

The length of a "slice" depends on the current scan cycle time of OB 1. The closer OB 1's scan time is to the minimum scan cycle time, the less time remains for executing OB 90. The program scan time is not monitored in OB 90.

OB 90 is scanned only in RUN mode. It can be interrupted by interrupt and error events, just like OB 1. The start information in the temporary local data (Byte 1) also tells which events cause OB 90 to execute from the beginning:

▷ B#16#91
After a CPU restart,

▷ B#16#92
After a block processed in OB 90 was deleted or replaced;

▷ B#16#93
After (re)loading of OB 90 in RUN mode;

▷ B#16#95
After the program in OB 90 was scanned and a new background cycle begins.

20.2.4 Response Time

If the user program in OB 1 works with the signal states of the process images, this results in a response time which is dependent on the program execution time (scan cycle time). The response time lies between one and two scan cycles, as the following example explains.

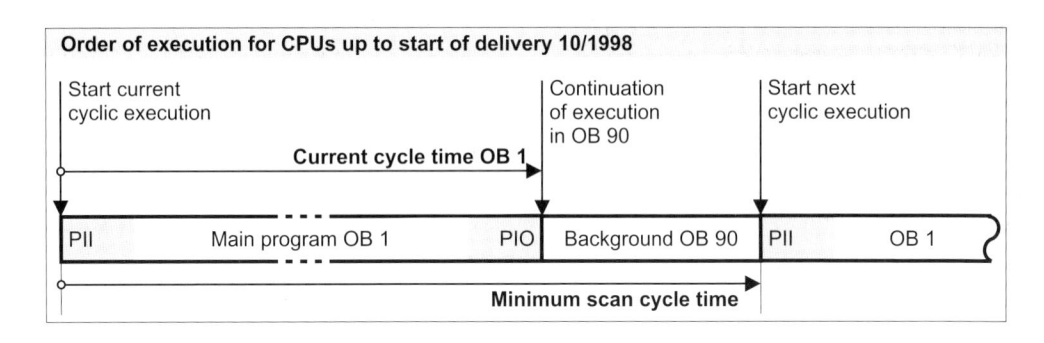

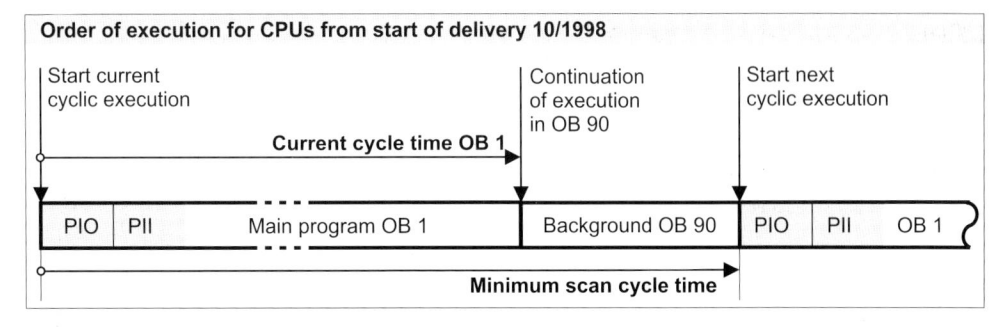

Figure 20.3 Minimum Cycle Duration and Background Scanning

When a limit switch is activated, for instance, it changes its signal state from "0" to "1". The programmable controller detects this change during the subsequent updating of the process image, and sets the inputs allocated to the limit switch to "1". The program evaluates this change by resetting an output, for example, in order to switch off the corresponding motor. The new signal state of the output that was reset is transferred at the end of the program scan; only then is the corresponding bit reset on the digital output module.

In a best-case situation, the process image is updated immediately following the change in the limit switch's signal. It would then take only one cycle for the relevant output to respond (Figure 20.4). In the worst-case situation, updating of the process image was just completed when the limit switch signal changed. It would then be necessary to wait approximately one cycle for the programmable controller to detect the signal change and set the input. After yet another cycle, the program can respond.

When so considered, the user program's execution time contains all procedures in one program cycle (including, for instance, the servicing of interrupts, the functions carried out by the operating system, such as updating timers, controlling the MPI interface and updating the process images).

The response time to a change in an input signal can thus be between one and two cycles. Added to the response time are the delays for the input modules, the switching times of contactors, and so on.

In some instances, you can reduce the response times by addressing the I/Os directly or calling program sections on an event-driven basis.

20.2.5 Start Information

The CPU's operating system forwards start information to organization block OB 1, as it does to every organization block, in the first 20 bytes of temporary local data. You can generate the declaration for the start information yourself or you can use information from standard library *Standard Library* unter *Organization Blocks*.

Table 20.2 shows this start information for the OB1, the default symbolic designation, and the data types. You can change the designation at any time and choose names more acceptable to you. Even if you don't use the start information, you must reserve the first 20 bytes of temporary local data for this purpose (for instance in the form of a 20-byte array).

In SIMATIC S7, all event messages have a fixed structure which is specified by the event class. The start information for OB 1, for instance, reports event B#16#11 as a standard OB call. From the contents of the next byte you can tell whether the main program is in the first cycle after power-up and is therefore calling, for instance, initialization routines in the cyclic program.

The priority and OB number of the main program are fixed. With three INT values, the start

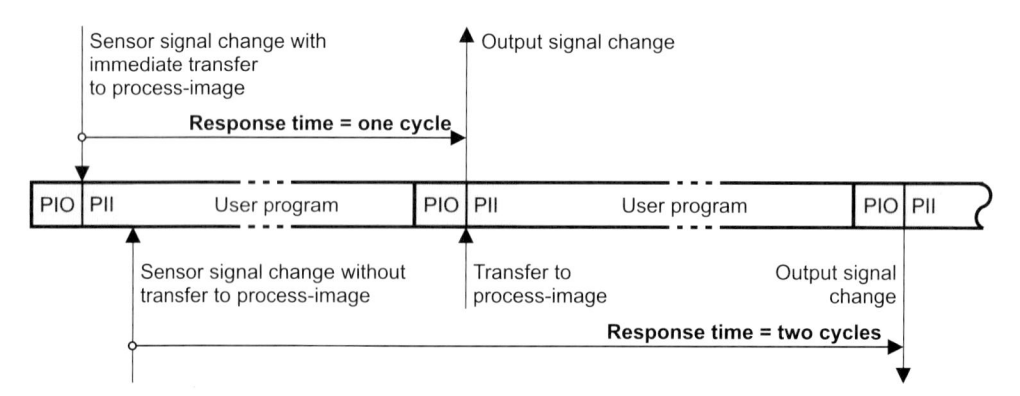

Figure 20.4 Response Times of Programmable Controllers

20.2 Scan Cycle Control

Table 20.2 Start Information for the OB 1

Name	Data Type	Description	Contents
OB1_EV_CLASS	BYTE	Event class	B#16#11 = Call standard OB
OB1_STRT_INFO	BYTE	Start information	B#16#01 = 1st cycle after complete restart B#16#02 = 1st cycle after warm restart B#16#03 = Every other cycle
OB1_PRIORITY	BYTE	Priority	B#16#01
OB1_OB_NUMBR	BYTE	OB Number	B#16#01
OB1_RESERVED_1	BYTE	Reserved	-
OB1_RESERVED_2	BYTE	Reserved	-
OB1_PREV_CYCLE	INT	Previous scan cycle time	in ms
OB1_MIN_CYCLE	INT	Minimum scan cycle time	in ms
OB1_MAX_CYCLE	INT	Maximum scan cycle time	in ms
OB1_DATE_TIME	DT	Event occurrence	Call time of the OB (cyclic)

information provides information on the cycle time of the last scan cycle and on the minimum and maximum cycle times since the last power-up. The last value, in DATE_AND_TIME format, indicates when the priority control program received the event for calling OB 1.

Note that direct reading of the start information for an organization block is possible only in that organization block because that information consists of temporary local data. If you require the start information in blocks which lie on deeper levels, call system function SFC RD_SINFO at the relevant location in the program.

SFC 6 RD_SINFO
Reading out start information

System function SFC 6 RD_SINFO makes the start information on the current organization block (that is, the OB at the top of the call tree) and on the start-up OB last executed available to you even at a deeper call level (Table 20.3).

Output parameter TOP_SI contains the first 12 bytes of start information on the current OB, output parameter START_UP_SI the first 12 bytes of start information on the last start-up OB executed. There is no time stamp in either case.

SFC 6 RD_SINFO can not only be called at any location in the main program but in every priority class, even in an error organization block or in the start-up routine. If the SFC is called in an interrupt organization block, for example, TOP_SI contains the start information of the interrupt OB. In the case of a call at restart, TOP_SI and START_UP_SI have the same contents.

Table 20.3 Parameters for SFC 6 RD_SINFO

SFC	Parameter Name	Declaration	Data Type	Contents, Description
6	RET_VAL	OUTPUT	INT	Error information
	TOP_SI	OUTPUT	STRUCT	Start information for the current OB (with the same structure as START_UP_SI)
	START_UP_SI	OUTPUT	STRUCT BYTE BYTE BYTE BYTE BYTE WORD DWORD	Start information for the last OB started: Event ID and event class Event number Execution priority (number of the execution level) ID of supplementary information 2_3 ID of supplementary information 1 Supplementary information 1 Supplementary information 2_3

20.3 Program Functions

In addition to parameterizing the CPU with the Hardware Configuration, you can also select a number of program functions dynamically at runtime via the integrated system functions.

20.3.1 Real-Time Clock

The following system functions can be used to control the CPU's real-time clock:

▷ SFC 0 SET_CLK
  Set date and time
▷ SFC 1 READ_CLK
  Read date and time
▷ SFC 48 SNC_RTCB
  Synchronize CPU clocks

You will find a list of system function parameters in Table 20.4.

When several CPUs are connected to one another in a subnetwork, initialize one of the CPU's clocks as "master clock". When parameterizing the CPU, also enter the synchronization interval after which all clocks in the subnetwork are to be automatically synchronized to the master clock.

Call SFC 48 SNC_RTCB in the CPU with the master clock. This call synchronizes all clocks in the subnetwork independently of automatic synchronization. When you set a master clock with SFC 0 SET_CLK, all other clocks in the subnetwork are automatically synchronized to this value.

20.3.2 Read System Clock

A CPU's system clock starts running on power-up or on a complete restart. The system clock keeps running as long as the CPU is executing the restart routine or is in RUN mode. When the CPU goes to STOP or HOLD, the current system time is "frozen".

If you initiate a warm restart on an S7-400 CPU, the system clock starts running again using the saved value as its starting time. Cold restart or complete restart (warm restart) reset the system time.

The system time has data format TIME, whereby it can assume only positive values

TIME#0ms to
TIME#24d20h31m23s647ms.

In the event of an overflow, the clock starts again at 0. A CPU3xx (except for CPU 318) updates the system clock every 10 milliseconds, CPU 318 and a CPU 4xx every millisecond.

SFC 64 TIME_TCK
Read system time

You can read the current system time with system function SFC 64 TIME_TCK. The RET_VAL parameter contains the system time in the TIME data format.

You can use the system clock, for example, to read out the current CPU runtime or, by computing the difference, to calculate the time between two SFC 64 calls. The difference between two values in TIME format is computed using DINT subtraction.

20.3.3 Run-Time Meter

A run-time meter in a CPU counts the hours. You can use the run-time meter for such tasks as determining the CPU runtime or ascertaining the runtime of devices connected to that CPU.

Table 20.4 SFC Parameters for the Real-Time Clock

SFC	Parameter Name	Declaration	Data Type	Contents, Description
0	PDT	INPUT	DT	Date and time (new)
	RET_VAL	OUTPUT	INT	Error information
1	RET_VAL	OUTPUT	INT	Error information
	CDT	OUTPUT	DT	Date and time (current)
48	RET_VAL	OUTPUT	INT	Error information

The number of run-time meters per CPU depends on the CPU. When the CPU is at STOP or HOLD, the run-time meter also stops running; when the CPU is restarted, the run-time meter begins again with the previous value.

When a run-time meter reaches 32767 hours, it stops and reports an overflow. A run-time meter can be set to a new value or reset to zero only via an SFC call.

The following system functions are available to control a run-time meter:

▷ SFC 2 SET_RTM
  Set run-time meter

▷ SFC 3 CTRL_RTM
  Start or stop run-time meter

▷ SFC 4 READ_RTM
  Read run-time meter

Table 20.5 shows the parameter for these system functions.

The NR parameter stands for the number of the run-time meter, and has the data type BYTE. It can be initialized using a constant or a variable (as can all input parameters of elementary data type). The PV parameter (data type INT) is used to set the run-time meter to an initial value. SFC 3's-S-parameter starts (with signal state "1") or stops (with signal state "0") the selected run-time meter. CQ indicates whether the run-time meter was running (signal state "1") or stopped (signal state "0") when scanned. The CV parameter records the hours in INT format.

20.3.4 Compressing CPU Memory

Multiple deletion and reloading of blocks, which often occur during online block modification, can result in gaps in the CPU's work memory and in the RAM load memory which decrease the amount of usable space in memory. When you call the "Compress" function, you start a CPU program which fills these gaps by pushing the blocks together. You can initiate the "Compress" function via a programming device connected to the CPU or by calling system function **SFC 25 COMPRESS**. The parameters for SFC 25 are listed in Table 20.6.

The compression procedure is distributed over several program cycles. The SFC returns BUSY = "1" to indicate that it is still in progress, and DONE = "1" to indicate that it has completed the compression operation. The SFC cannot compress when an externally initiated compression is in progress, when the "Delete Block" function is active, or when PG functions are accessing the block to be shifted (for instance the Block Status function).

Note that blocks of a particular CPU-specific maximum length cannot be compressed, so that gaps would still remain in CPU memory. Only the Compress function initiated via the PG while the CPU is at STOP closes all gaps.

20.3.5 Waiting and Stopping

The system function **SFC 47 WAIT** halts the program scan for a specified period of time.

Table 20.5 Parameters of the SFCs for the Run-Time Meter

SFC	Parameter	Declaration	Data Type	Contents, Description
2	NR	INPUT	BYTE	Number of the run-time meter (B#16#01 to B#16#08)
	PV	INPUT	INT	New value for the run-time meter
	RET_VAL	OUTPUT	INT	Error information
3	NR	INPUT	BYTE	Number of the run-time meter (B#16#01 to B#16#08)
	S	INPUT	BOOL	Start (with "1") or stop (with "0") run-time meter
	RET_VAL	OUTPUT	INT	Error information
4	NR	INPUT	BYTE	Number of the run-time meter (B#16#01 to B#16#08)
	RET_VAL	OUTPUT	INT	Error information
	CQ	OUTPUT	BOOL	Run-time meter running ("1") or stopped ("0")
	CV	OUTPUT	INT	Current value of the run-time meter

Table 20.6 Parameters for SFC 25 COMPRESS

SFC	Parameter	Declaration	Data Type	Contents, Description
25	RET_VAL	OUTPUT	INT	Error information
	BUSY	OUTPUT	BOOL	Compression still in progress (with "1")
	DONE	OUTPUT	BOOL	Compression completed (with "1")

SFC 47 WAIT has input parameter WT of data type INT in which you can specify the waiting time in microseconds (µs). The maximum waiting time is 32767 microseconds; the minimum waiting time corresponds to the execution time of the system function, which is CPU-specific. SFC 47 can be interrupted by higher-priority events. On an S7-300, this increases the waiting time by the scan time of the higher-priority interrupt routine.

The system function **SFC 46 STP** terminates the program scan, and the CPU goes to STOP. SFC 46 STP has no parameters.

20.3.6 Multiprocessing Mode

The S7-400 enables multiprocessing. As many as four appropriately designed CPUs can be operated in one rack on the same P bus and K bus.

An S7-400 station is automatically in multiprocessor mode if you arrange more than one CPU in the central rack in the Hardware Configuration. The slots are arbitrary; the CPUs are distinguished by a number assigned automatically in ascending order when the CPUs are plugged in. You can also assign this number yourself on the "Multicomputing" tab.

The configuration data for all the CPUs must be loaded into the PLC, even when you make changes to only one CPU. After assigning parameters to the CPUs, you must assign each module in the station to a CPU.

This is done by parameterizing the module in the "Addresses" tab under "CPU Allocation" (Figure 20.5). At the same time that you assign the module's address area, you also assign the module's interrupts to this CPU. With VIEW → FILTER → CPU No. x-MODULES, you can emphasize the modules assigned to a CPU in the configuration tables.

The CPUs in a multiprocessing network all have the same operating mode. This means

▷ They must all be parameterized with the same restart mode;
▷ They all go to RUN simultaneously;
▷ They all go to HOLD when you debug in single-step mode in one of the CPUs;
▷ They all go to STOP as soon as one of the CPUs goes to STOP.

When one rack in the station fails, organization block OB 86 is called in each CPU. The user programs in these CPUs execute independently of one another; they are not synchronized.

An SFC 35 MP_ALM call starts organization block OB 60 "Multiprocessor interrupt" in all CPUs simultaneously (see section 21.6, "Multiprocessor Interrupt").

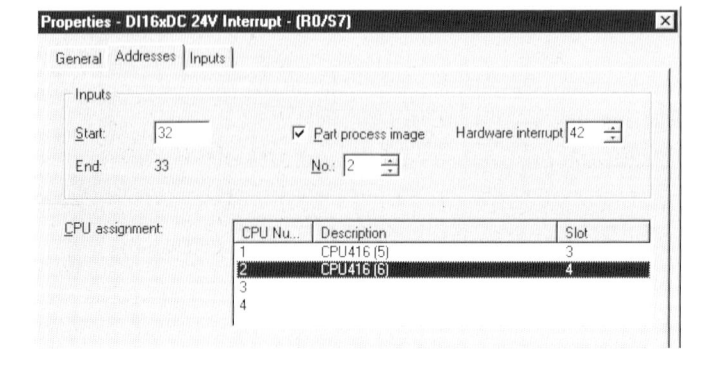

Figure 20.5
Module Assignments in Multiprocessor Mode

20.4 Communication via Distributed I/O

Similarly to the way in which centrally arranged modules are assigned to a CPU and are controlled from the CPU, distributed modules (stations, DP slaves) are assigned to a DP master. The DP master with all its DP slaves is referred to as a DP master system. One S7 station can contain several DP master systems.

Like centrally arranged modules, the DP slaves occupy addresses in the I/O area of the CPU (logic address area). The DP master is "transparent", as it were, for the addresses of the DP slaves; the CPU "sees" the addresses of the DP slaves, so the addresses of the DP slaves and those of the centrally arranged modules must not overlap. The addresses of the DP slaves must also not overlap with those of DP slaves in other DP master systems assigned to the CPU.

There are DP slaves that behave like centrally arranged modules: they have a user data area and a system data area and they can initiate process and diagnostics interrupts if appropriately equipped. These stations are referred to a "DP-S7 slaves". "DP standard slaves" comply with EN 50170, Volume 2, PROFIBUS. The essential difference between them is in the way they read and in the structure of the diagnostics data.

Configuring of the distributed I/O is also very similar to that of centralized modules. The starting point is the DP master with the graphically represented DP master system. The stations are "hung" onto this and then addressed and parameterized.

Special user data consistency requirements demand special system functions for distributed I/O; so, in spite of serial transfer on DP slaves, data consistency can go beyond the 4 bytes guaranteed by the S7 system and you can connect groups of DP slaves in such a way that they can supply or output data synchronously.

20.4.1 Addressing Distributed I/O

Every DP slave has three addresses in addition to the node address: a geographical address, a module starting address and a diagnostics address (Figure 20.6).

Node address

Every node on the PROFBIBUS subnetwork has a unique address, the node address (station number) in that subnetwork that distinguishes it from the other nodes on the subnetwork. The station is accessed on PROFIBUS with this node address.

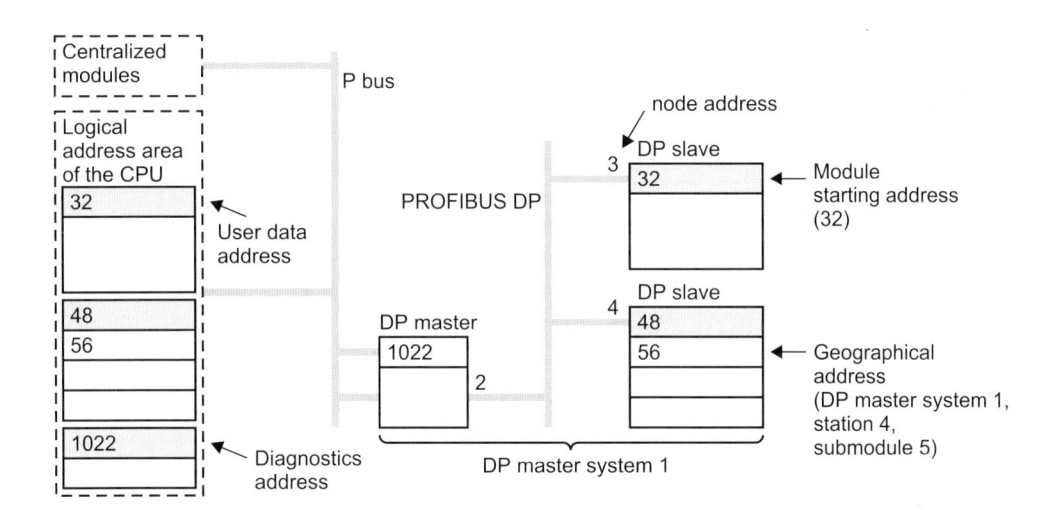

Figure 20.6 Addresses in a DP Master System

20 Main Program

Please note that there must be a gap of at least 1 between the addresses of the active bus nodes (e.g. in the case of DP master and nodes in cross traffic). STEP 7 takes this into account when assigning node addresses automatically.

Geographical address

The geographical address of a DP slave corresponds to the slot address of a centralized module. It consists of the DP master system ID (determined during configuring) and the node address on PROFIBUS (corresponding to the rack number).

In the case of modularly designed DP slaves, the slot number is added to this and in the case of modules with submodules, the submodule slot number is added (as with the S7-300, slot numbering begins at 4).

Module starting address, data consistency

Under the module starting address ("logic base address") you can access the user data of a compact DP slave or of a module in a modular DP slave. This address corresponds to the module starting address of a centralized module. If the addresses of a DP slave have a data consistency of 1, 2, or 4 bytes, you can access the user data with load and transfer statements or with the MOVE box. If the module starting address is in the process image, the user data are transferred in the course of transferring the process image. Assignment to a subprocess image is also possible.

If the data consistency is 3 bytes, or more than 4 bytes (up to a CPU-specific quantity), you require system functions SFC 14 DPRD_DAT and SFC 15 DPWR_DAT for reading or writing the user data. You cannot access such DP slaves with load and transfer statements or with the MOVE box, even if the module starting address is in the process image. (The SFCs bypass the process image to read or write the data areas created at the RECORD parameter, unless the process image is the data source or the data destination at the RECORD parameter).

Figure 20.7 illustrates the transfer of user data. The DP master transfers the user data of "its" DP slaves cyclically; in the example, it transfers from a station with the module starting address of 32 and a data consistency of 4 bytes (DP slave 1) and from a station with the module starting address 48 and a data consistency of 8 bytes (DP slave 2).

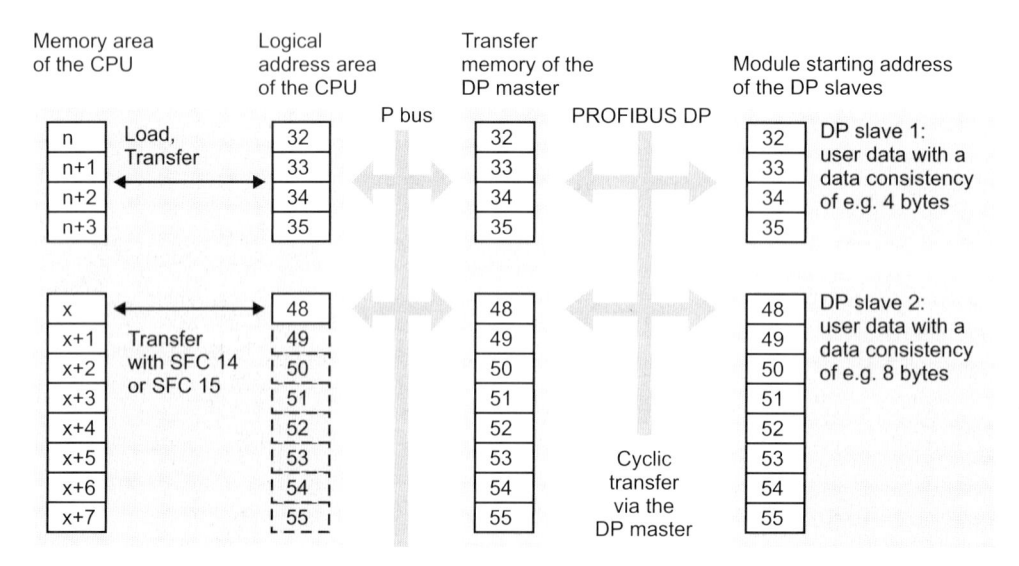

Figure 20.7 Transferring User Data in Distributed I/O

20.4 Communication via Distributed I/O

The user data bytes of DP slave 1 are located in the transfer area of the DP master on the P bus or the CPU and can be transferred to, for example, a data block in the memory area of the CPU with Load or Transfer (MOVE box), in the same way as a centralized module.

The user data bytes of DP slave 2 are also all located in the transfer area of the DP master, but only the module starting address (48 in the example) is accessible on the P bus.

The remaining addresses are (theoretically) freely assignable, but they are disabled by the Hardware Configuration for any other use. SFCs 14 and 15 address the user data of DP slave 2 via this module starting address and they exchange these data with a memory area of the CPU, for example, with a data block.

This address cannot be accessed with Load and Transfer or with the MOVE box, not even by the operating system when updating the process image; in the example, the process image is not updated between the addresses 48 and 55. However, you can specify to the system functions SFC 14/15 areas as source or destination and thereby still use these addresses.

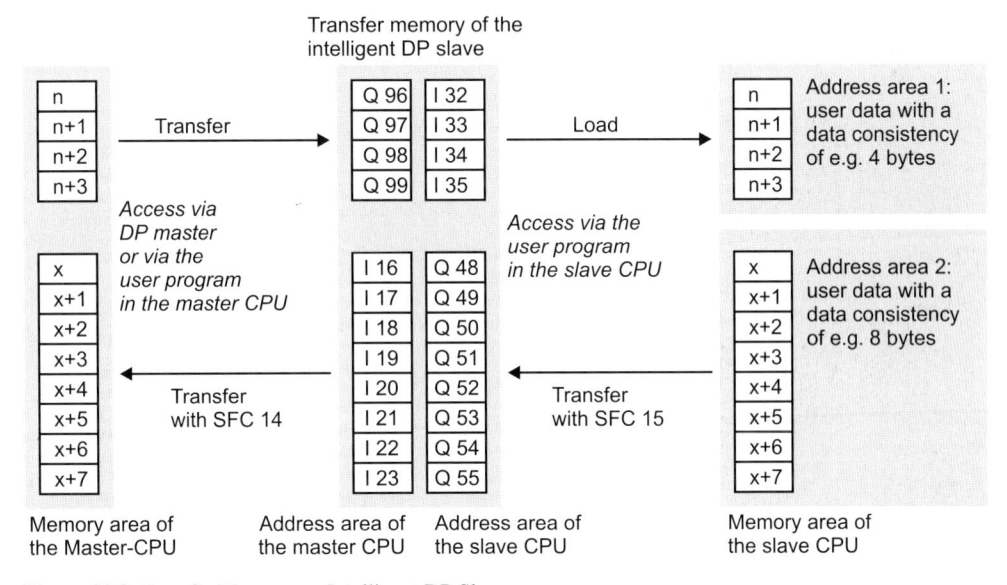

Transfer memory on intelligent DP slaves

In the case of compact and modular DP slaves, the addresses of the inputs and outputs are located in the I/O area of the centralized CPU (referred to below as the "master CPU").

In the case of intelligent slaves, the master CPU has no direct access to the input/output modules of the DP slave. Every intelligent DP slave therefore has a transfer memory that can be subdivided into several subsidiary areas of different length and data consistency. From the viewpoint of the master CPU, the intelligent DP slave then appears as a compact or modular DP slave, depending on the subdivision.

The size of the overall transfer memory depends on the DP slave. You define the addresses of the transfer memory at configuration: the addresses from the viewpoint of the slave CPU when configuring the intelligent DP slaves and the addresses from the viewpoint of the master CPU when inserting the intelligent DP slaves into the DP master system. Exception: if the CP 342-5DP forms the DP interface for the intelligent slave, the subdivision of its transfer memory is not configured until the slave is connected to the DP master system.

Figure 20.8 Transfer Memory on Intelligent DP Slaves

From the viewpoint of the DP master (more precisely: from the viewpoint of the centralized master CPU) the addresses of the transfer memory must not overlap with addresses of other modules in the (centralized) S7 station. From the viewpoint of the slave CPU, the addresses of the transfer memory must not overlap with those of the modules arranged in the intelligent DP slave.

The address areas of the transfer memory behave like individual modules with regard to user data access and data consistency, i.e. the lowest address of an address area is the "module starting address". According to the data consistency level set, you access the user data of an address area with Load/Transfer/MOVE or with SFC 14/15, both in the program of the master CPU and in the program of the slave CPU. In the program of the slave CPU, you can also use SFC 7 DP_PRAL to trigger a process interrupt in the master CPU for an address area.

The example in Figure 20.8 shows a transfer memory with two address areas. The master CPU sees address area 1 is an output module that can be written to with Transfer/MOVE (or also with binary memory operations if the addresses are in the process image). For the slave CPU, address area 1 is an input module that can be read with Load/MOVE, or with check operations if the addresses are in the process image. Address area 2 with a data consistency of 8 bytes can be written to by the slave CPU only with SFC 15, and read by the master CPU only with SFC 14.

Diagnostics address

Every DP master and every DP slave occupies an additional one byte of "diagnostics address". You read the diagnostics data via the diagnostics address.

On DP standard slaves, you use system function SFC 13 DPNRM_DG to read the diagnostics data, on DP S7 slaves you use SFC 59 RD_REC to read data record DS 1 that contains the diagnostics data. The diagnostics data read with SFC 123 DPNRM_DG have the structure defined in the standard (Figure 20.9). Data record DS 1 read with SFC 59 RD_REC contains the 4 bytes of module diagnostics information that

General structure of the diagnostics data of a DP standard slave

Byte	
0 ... 2	Station status 1, 2 and 3
3	Master station number
4 ... 5	Manufacturer's ID
6 ... n	further slave-specific diagnostics data

General structure of the diagnostics data of a DP S7 slave

Byte	
0 ... 3	Data record DS 0 (start information in OB 82)
4 ... n	further slave-specific diagnostics data

Figure 20.9
Structural Principle of the Standard Diagnostics Data and of the Diagnostics Data Record DS 1

are also available in, for example, the start information of diagnostics interrupt OB 82 (this corresponds to the diagnostics data record DS 0). The structure of the remaining diagnostics data is module-specific.

Modules or devices that have no user data of their own can also have a diagnostics address if they can provide diagnostics data, like for example, a DP master or a redundantly configurable power supply.

The diagnostics address occupies one byte of peripheral inputs. STEP 7 assigns the diagnostics address as default starting with the highest address in the I/O area of the CPU. The address overview in the Hardware Configuration indicates the diagnostics address with a star.

20.4.2 Configuring Distributed I/O

General procedure

You configure the distributed I/O essentially in the same way as the centralized modules. Instead of arranging modules in a mounting rack, you assign DP stations (PROFIBUS nodes) to a DP master system. The following order is recommended for the necessary actions:

1) Create a new project or open an existing one with the SIMATIC Manager.

20.4 Communication via Distributed I/O

2) Create a PROFIBUS subnetwork in the project with the SIMATIC Manager and, if required, set the bus profile

3) Use the SIMATIC Manager to create the master station in the project that is to accommodate the DP master, e.g. an S7-400 station.

If your system contains intelligent DP slaves, you also create the relevant slave stations at this point, e.g. S7-300 stations.

You start the Hardware Configuration by opening the master station.

4) With the Hardware Configuration, you place a DP master in the master station. This can be, for example, a CPU with integral DP interface. You assign the previously created PROFIBUS subnetwork to the DP interface and you then have a DP master system. You can also configure the remaining modules later. Save and compile the station.

5) If you have created an S7 station for an intelligent DP slave, you open this in the Hardware Configuration and you "plug in" the module with the desired DP interface, e.g. an S7-300 CPU with integral DP interface or an ET 200X basic module BM 147/CPU. If you set the DP interface as "DP slave", assign the previously created PROFIBUS subnetwork to the DP interface and configure the user data interface from the viewpoint of the DP slave (transfer memory). You can also configure the remaining modules later. Save and compile the station.

Proceed in the same way for the remaining stations intended for intelligent DP slaves.

6) Open the master station with the DP master system and use the mouse to drag the PROFIBUS nodes (compact and modular DP slaves) from the hardware catalog to the DP master system. Assign node addresses and, if necessary, set the module starting address and the diagnostics address.

7) If you have created intelligent DP slaves, drag the relevant icon (in the hardware catalog under "PROFIBUS-DP" and "Already configured stations") with the mouse to the DP master system.

Open the icon and assign the already configured DP slave ("Connect"), assign a node address and configure the user data interface from the viewpoint of the DP master (or from the viewpoint of the central master CPU). Proceed in the same way with every intelligent DP slave.

8) Save and compile all stations. The DP master system is now configured. You can now supplement the configuration with centralized modules or with further DP slaves.

You can also represent the DP master system configured in this way graphically with the Network Configuration tool. Open Network Configuration by, for example, double-clicking on a subnetwork. Select VIEW → DP SLAVES to display the slaves. You can also create a DP master system (or more precisely, assign the nodes to a PROFIBUS subnetwork) with the Network Configuration tool. You parameterize the stations after opening them with the Hardware Configuration. Here too, you must first set up an intelligent DP slave before you can integrate it into a DP master system.

Configuring the DP master

You must have created a project and an S7 station with the SIMATIC Manager. You open the S7 station and create a mounting rack (see Section 2.3 "Configuring Stations"). Now drag the DP master module from the Hardware Catalog to the configuration table of the mounting rack. You may already have selected a CPU with DP connection. In the line below, the DP master is displayed with a connection to a DP master system in the station window (broken black-and-white bar).

When placing the DP master module, you select in a window the PROFIBUS subnetwork to which the DP master system is to be assigned and the node address to be assigned to the DP master. You can also create a new PROFIBUS subnetwork in this window.

If there is no DP master system available (it may be that it is obscured behind an object or it is outside the visible area), create one by selecting the DP master in the configuration window and then selecting INSERT → DP MASTER SYSTEM. You can change the node address and the connection to the PROFIBUS subnetwork by selecting the module and then making your changes with the "PROFIBUS" button on the "General" tab under EDIT → OBJECT PROPERTIES.

CP 342-5DP as DP master

If a CP 342-5DP is the DP master, place it in the configuration table of the station, select it and then EDIT → OBJECT PROPERTIES. Set "DP Master" on the "Mode" tab.

The "Addresses" tab shows the user data address occupied by the CP in the address area of the CPU. From the viewpoint of the master CPU, the CP 342-5DP is an "analog module" with a module starting address and 16 bytes of user data.

Only DP standard slaves, or DP S7 slaves that behave like DP standard slaves, can be connected to a CP 342-5DP as DP master. You can find the suitable DP slaves in the hardware catalog under "PROFIBUS-DP" and "CP 342-5DP". Select the desired slave type and drag it to the DP master system.

The transfer memory as DP master has a maximum length of 240 bytes. It is transferred as one with the loadable blocks FC 1 DP_SEND and FC 2 DP_RECV (included in the *Standard Library* under the *Communication Blocks* program).

The data consistency covers the entire transfer memory.

You read the diagnostics data of the connected DP slaves with FC 3 DP_DIAG (e.g. station list, diagnostics data of a specific station). FC 4 DP_CTRL transfers control jobs to the CP 342-5DP (e.g. SYNC/FREEZE command, CLEAR command, set operating state of the CP 342-5DP).

If you select CPU or CP 342-5DP, VIEW → ADDRESS OVERVIEW shows you a list of the assigned addresses, inputs and/or outputs. You can also screen the existing address gaps.

Configuring compact DP slaves

The compact DP slaves are to be found in the hardware catalog under "PROFIBUS-DP" and the relevant sub-catalog, e.g. ET 200B. Click on the DP slave selected and drag it to the icon for the DP master system.

You will see the properties sheet of the station; here, you set the node address and the diagnostics address. Then the DP slave appears as an icon in the upper section of the station window and the lower section contains a configuration table for this station.

A double-click on the icon in the upper section of the station window opens a dialog box with one or more tabs in which you set the desired station properties. In the lower sub-window, you then see the input/output addresses. Double-clicking on an address line shows you a window where you can change the suggested addresses.

The lower sub-window shows optionally the configuration table of the selected DP slave or of the master system (toggle with the "arrow" button).

Configuring modular DP slaves

The modular DP slaves can be found in the hardware catalog under "PROFIBUS-DP" and the relevant sub-catalog, e.g. ET 200M.

Click on the selected interface module (basic module) and drag it to the icon for the DP master system. This screens the properties sheet for the station; here, you set the node address and the diagnostics address. Then the DP slave appears as an icon in the upper section of the station window and the lower section contains configuration table for this station.

Now place the modules that you can find in the hardware catalog *under the selected interface module (!)* in the configuration table. Double-clicking on the line opens the properties sheet of the module and allows you to parameterize the module.

Please note that both the address area per slave in the DP master and the address area of the DP slave interface module are limited. For example, the limit on the CPU 315-2DP as DP master is 122 bytes of inputs and 122 bytes of outputs per slave (eight 8-channel modules in the

ET 200M would exceed this limit: 8 x 16 = 128 bytes), and the limit on the ET 200X as DP slave is 104 bytes of inputs and 104 bytes of outputs.

Configuring a CPU with integral DP interface as an intelligent DP slave

With an appropriately equipped CPU, you can parameterize the station either as a DP master station or as a DP slave station. Before the station can be connected as a DP slave to a DP master system, it must be created. The procedure for doing this is exactly the same as that for a "normal" station; insert an S7 station into the project using the SIMATIC Manager and open the *Hardware* object. Drag a mounting rack to the window in the Hardware Configuration and place the desired modules. For configuring the DP slave, it is enough to place the CPU; you can add all other modules later.

When inserting the CPU, the properties sheet of the PROFIBUS interface is screened. Here, you must assign a subnetwork to the DP interface and you must assign an address. If the PROFIBUS subnetwork does not yet exist in the project, you can create a new one with the NEW button. This is the subnetwork to which the intelligent slave will later be connected.

You can open the properties sheet of the interface by selecting the DP interface and then EDIT → OBJECT PROPERTIES or by double-clicking on the interface. On the "Slave Configuration" tab, select "Use Controller as Slave". Now you can configure the user data interface from the viewpoint of the DP slave (Figure 20.10); Section 20.4.1 "Addressing Distributed I/O" provides information on the user data interface under "Transfer memory in intelligent DP slaves").

The size and structure of the transfer memory is CPU-specific. On the CPU 315-2DP, for example, you can divide the entire transfer memory into up to 32 address areas that you can access separately. Such an address area can be up to 32 bytes in size. The entire transfer memory can have up to 122 input addresses and 122 output addresses.

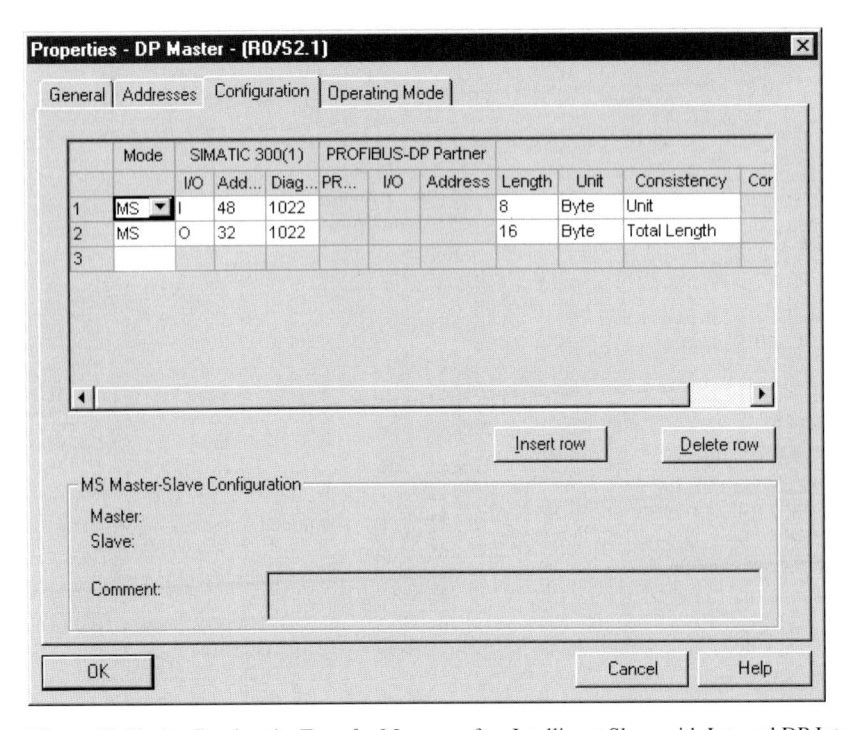

Figure 20.10 Configuring the Transfer Memory of an Intelligent Slave with Integral DP Interface

The addresses defined here are located in the address volume of the slave CPU. These addresses must not overlap with addresses of centralized or distributed modules in the DP slave station. The lowest address of an address area is the "module starting address".

The user program in the slave CPU gets diagnostics information from the DP master via the diagnostics address specified on this tab.

You terminate configuration of the intelligent DP slave with STATION → SAVE AND COMPILE. Connecting the intelligent DP slave into the DP master system is described below.

Configuring a BM 147/CPU as an intelligent DP slave

If you want to create an ET 200X as an intelligent DP slave, first insert a SIMATIC 300 station in the SIMATIC Manager under the project and open the *Hardware* object.

In the Hardware Configuration, drag the object *MB147/CPU* from the hardware catalog under "PROFIBUS-DP" and "ET200X" to the free window, or select it by double-clicking. Select the node address and the PROFIBUS subnetwork in the properties sheet of the DP interface that then appears (or create one and assign it to the DP interface). You get a configuration table as with a SIMATIC 300 station. The displayed CPU represents the intelligent BM 147 interface module or ET 200X station here. This CPU has no MPI address in the address table since it does not have an MPI interface (the BM 147/CPU is programmed via the MPI interface of the master station).

A double-click on the CPU line opens the window for the CPU properties; a double-click on the DP interface opens the properties window of the interface. Set the address areas for the user data interface, here from the viewpoint of the DP slave. The starting addresses are fixed at 128 on the BM147; the maximum size of the user data area is 32 bytes of inputs and 32 bytes of outputs. You can divide this area into eight sub-areas with different data consistency. The diagnostics address is set to 127; the slave program receives diagnostics information from the DP master via this address.

Further configuration of the ET 200X station is carried out in the same way as that for an S7-300 station with fixed slot addressing. You can only arrange the modules that can be found in the hardware catalog under "BM147/CPU".

You terminate configuration of the intelligent DP slave with STATION → SAVE AND COMPILE. Connecting the intelligent DP slave into the DP master system is described below.

Configuring an S7-300 station with CP 342-5DP as an intelligent slave

If you insert an S7-300 station in the SIMATIC Manager, open the *Hardware* object and configure a "normal" S7-300 station. Among other things, you arrange a CP 342-5DP communications processor in the configuration table.

When inserting the station, the properties sheet of the DP interface appears; the subnetwork to which the intelligent DP slave is later to be connected is to be assigned to the DP interface here and you must also assign the node address.

To open the properties window, select the CP 342-5DP and then EDIT → OBJECT PROPERTIES, or double-click on the CP 342-5DP. Select the option "DP Slave" on the "Mode" tab.

The "Addresses" tab shows the user data interface from the viewpoint of the slave CPU (starting address and 16 bytes in length). The size of the transfer memory on the CP 342-5DP as DP slave is up to 86 bytes, and you can subdivide this into different address areas after connection to the master system.

STATION → SAVE AND COMPILE terminates configuration of the intelligent DP slave.

Connecting an intelligent DP slave to a DP master

You must have created a project and configured a DP master station and the intelligent DP slave (in each case at least with the DP interface). The DP master and the DP slave must be configured for the same PROFIBUS subnetwork.

Open the master station; a DP master system (broken black-and-white bar) must exist, if not, create it with INSERT → DP MASTER SYSTEM. In the hardware catalog under "PROFIBUS DP" and "Already Configured Stations", you will find the objects that represent the intelligent slaves: "CP315-2DP" represents, for example, S7-300 stations with an integral DP slave, "X-BM147/CPU" represents ET200X stations with BM147/CPU and "S7-300 CP342-5 DP" represents S7-300 stations with CP 342-5 as DP slave interface module. Select the desired slave type and drag it to the DP master system.

CPU or ET200X as DP slave

A double-click on the DP slave opens the properties sheet. The slaves already configured for this PROFIBUS subnetwork are listed on the "Connection" tab. Select the desired slave and click on the "Connect" button. This causes the active connection to be carried out at the bottom of the same dialog box.

The option "Programming and Status/Modify" allows programming device access via the master CPU. The DP interface is then "open" for S7 functions in its capacity as active slave.

On the "General" tab, you set the diagnostics address of the DP slave from the viewpoint of the master station.

On the "Slave Configuration" tab, you now set the addresses of the user data interface from the viewpoint of the DP master. The output addresses on the master are the input addresses of the slave and vice versa. Section 20.4.1 "Addressing Distributed I/O" contains more information on the user data interface under "Transfer memory on intelligent DP slaves".

CP 342-5DP as DP slave

After dragging the DP slave to the master system, a window appears in which you select the already configured slave station. When the DP slave is selected, its configuration table is shown in the lower section of the station window.

Now you structure the transfer memory: select the "Universal submodule" from the hardware catalog (under "Already Configured Stations" and "S7-300 CP 342-5DP"), drag it to a line in the configuration table, or select a line and double-click on the "Universal submodule". You place one universal submodule for each individual (consistent) address area in the transfer memory; the maximum number is 32.

To open a window in which you define the properties of the address area, select the universal submodule and then EDIT → OBJECT PROPERTIES, or double-click on the line in the table: space, input or output area or both. Determine the starting address and the length of the area.

The addresses defined here are located in the address area of the master CPU. An area can be up to 64 bytes; the maximum overall size of the transfer area is 86 bytes.

If a CP 342-5DP is the DP master, structuring of the transfer memory can be omitted because the CP 342-5DP transfers the entire transfer area in one piece.

When subdividing the transfer memory, you arrange the address areas together without gaps starting from byte 0. You access the entire assigned transfer memory in the slave CPU with the loadable blocks FC 1 DP_SEND and FC 2 DP_RECV (included in the *Standard Library* under the *Communication Blocks* program).

The data consistency covers the entire transfer memory.

The option "Programming and Status/Modify" allows programming device access via the master CPU. The DP interface is then "open" for S7 functions in its capacity as active slave.

On the "General" tab, you set the diagnostics address of the DP slave from the viewpoint of the master station. The diagnostics data are read with FC 3 DP_DIAG (in the master station).

Section 20.4.1 "Addressing Distributed I/O" contains more information on the user data interface under "Transfer memory on intelligent DP slaves".

GSD files

You can "post install" DP slaves that are not included in the module catalog. For this purpose, you require the type file tailored to the slave (GSD file, device master data file). Select OPTIONS → INSTALL NEW GSD in the Hardware Configuration and specify the directory of the GSD file in the window that appears. Here, you can specify the pictogram that STEP 7 uses for graphical representation of the DP slave. STEP 7 accepts the GSD file and displays the slave in the hardware catalog under "Other Field Devices".

You can copy GSD files that already exist in another S7project to the current project with OPTIONS → IMPORT STATION GSD.

STEP 7 saves the GSD files in the directory ...\Step7\S7data\Gsd. The GSD files deleted when installing or importing at a later time are stored in the subdirectory ...\Gsd\Bkpx. From here, they can be restored with OPTIONS → INSTALL NEW GSD.

Configuring PROFIBUS PA

You require the SIMATIC PDM optional software for configuring a PROFIBUS-PA master system and for parameterizing the PA field devices. With the Hardware Configuration, you establish the connection to the DP master system with the DP/PA-Link: in the Hardware Catalog under "PROFIBUS-DP" and "DP/PA-Link", drag the IM 157 interface module to the DP master system. With the DP slave, a PA master system is simultaneously created in its own PROFIBUS subnetwork (45.45 kbits/s); it is indicated with a broken black-and-white bar.

The DP/PA interface transfers the data unmodified and uninterpreted between the bus systems; for this reason, it is not parameterized. The PA field devices are addressed by the DP master. They can be incorporated as DP standard slaves into the Hardware Configuration of STEP 7 via a GSD file. Following this, you will find the PA field devices in the hardware catalog under "PROFIBUS-DP" and "Other Field Devices".

Configuring the DP/AS-i-Link

You configure the DP/AS-Interface-Link like a modular DP slave. You will find the modules that you can drag to the DP master system, such as the DP/AS-i-Link 20 below, in the hardware catalog under "PROFIBUS-DP" and "DP/AS-i". In the windows that appear, establish first the setpoint configuration (16 to 20 bytes) and then the node address.

On the DP/AS-i-Link 20, you can define 16 bytes of inputs/outputs as the setpoint configuration, with an additional 4 bytes for control commands. In the latter case, the lower section of the window in the Hardware Configuration suggests 16 bytes of user data with addresses in the process image and 4 bytes of commands with addresses from, e.g. 512.

Select the DP slave and then EDIT → OBJECT PROPERTIES, or double-click on the DP slave, to open a window in which you can change the addresses suggested by the Hardware Configuration and also set the subprocess image, provided you have a suitable CPU.

Select the DP slave and then EDIT → OBJECT PROPERTIES, or double-click on the DP slave, to open the slave properties window. On the "Parameterize" tab, you set the parameters of the AS-i slaves, with 4 bits for each slave.

The AS-i master system with the AS-i slaves is not displayed by the Hardware Configuration.

Configuring SYNC/FREEZE groups

The SYNC control command causes the DP slaves combined as a group to output their output states simultaneously (synchronously). The FREEZE control command causes the DP slaves combined as a group to "freeze" the current input signal states simultaneously (synchronously), in order to allow them to be then fetched cyclically by the DP master. The UNSYNC and UNFREEZE control commands revoke the effect of SYNC and FREEZE respectively.

It is a requirement that the DP master and the DP slaves have the relevant functionality. From the object properties of a slave, you can see which command it supports (select DP slave, EDIT → OBJECT PROPERTIES, "General" tab under "SYNC/FREEZE Capabilities").

Per DP master system, you can form up to 8 SYNC/FREEZE groups that are to execute either the SYNC command or the FREEZE command or both. You can assign any DP slave to a group; on the CP 342-5DP of a specific version, one DP slave can be represented in up to 8 groups.

When you call SFC 11 DPSYC_FR, you cause the user program to output a command to a group (see Section 20.4.3 "Interface to the User Program" under "System functions for distributed I/O). The DP master then sends the relevant command simultaneously to all DP slaves in the specified group.

20.4 Communication via Distributed I/O

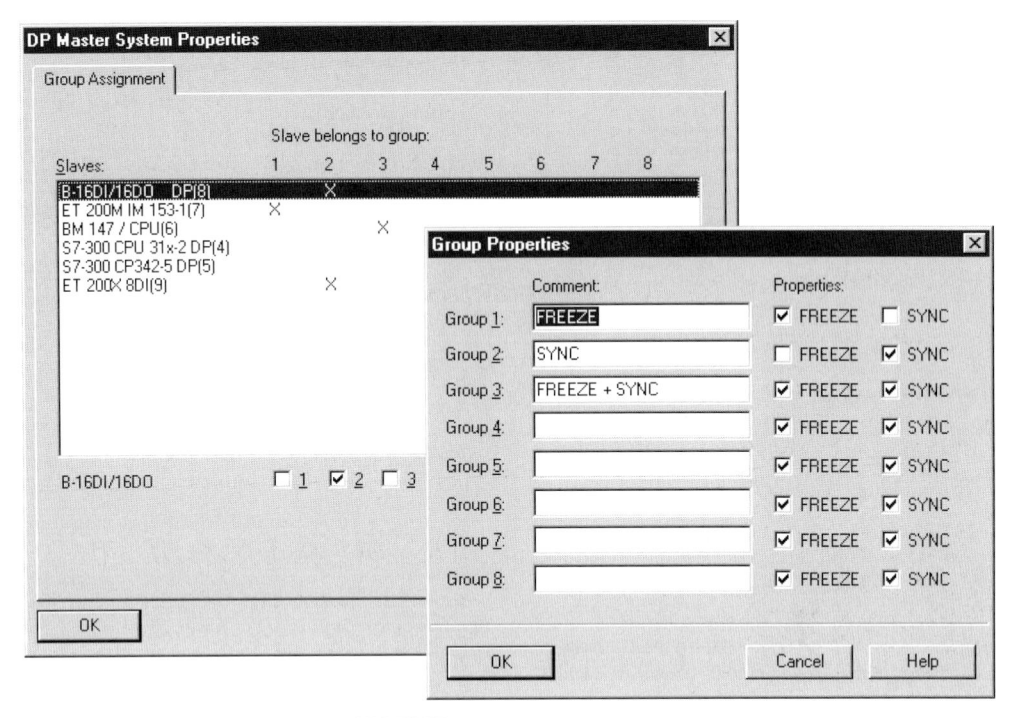

Figure 20.11 Configuring SYNC and FREEZE groups

You configure the SYNC/FREEZE group after configuring the DP master system (all DP slaves must exist in the DP master system). Select the DP master system (broken black-and-white bar) and select EDIT → OBJECT PROPERTIES. In the window that then appears, click first on the "Properties" button and define the commands of the groups that are to be executed (Figure 20.11). OK returns you to the properties window of the DP master system.

Here you select each of the DP slaves listed with its node number one after the other and select in each case the group to which it is to belong. If a DP slave cannot execute a specific command, such as FREEZE, the groups that contain this command cannot be selected, e.g. all groups with the FREEZE command. Terminate configuring of the SYNC/FREEZE groups with OK.

Please note that when configuring bus cycles of the same length (equidistant), groups 7 and 8 acquire a special meaning.

Configuring equidistant (equal length) bus cycles

Normally, the DP master controls the DP slaves assigned to it cyclically without a pause. With S7 Communication, such as when the programming device executes modify functions via the PROFIBUS subnetwork, this can result in variations in the time intervals. If, for example, the outputs are to be modified via the distributed I/O at a regular interval, you can set equidistant bus cycles with the appropriately equipped DP master. The DP master must be the only Class 1 master on the PROFIBUS subnetwork for this purpose.

In the properties window of the PROFIBUS subnetwork (e.g. with the "Properties" button under EDIT → OBJECT PROPERTIES with the DP master selected) select the "Network Settings" tab and the "Options" button. On the "Equidistance" tab, select the "Activate equidistant bus cycles" checkbox and the equidistant behavior ("Speed-optimized", "Standard" or "Failure-tolerant").

If you configure SYNC/FREEZE groups in addition to the equidistant behavior, please note the following:

▷ For DP slaves in group 7, the DP master automatically initiates a SYNC/FREEZE command in every bus cycle. Initiation per user program is prevented.

▷ Group 8 is used for the equidistance signal and is disabled for DP slaves. You cannot configure equidistant behavior if you have already configured slaves for group 8.

Configuring cross traffic

In a DP master system, the DP master controls "its" slaves exclusively. With appropriately equipped stations, another node (master or slave, referred to as the "receiver") can "monitor" the PROFIBUS subnetwork to learn which input data a DP slave ("sender") is sending to "its" master. This is called "cross traffic".

You configure cross traffic when all stations on the PROFIBUS subnetwork are connected. Select the receiver and then EDIT → OBJECT PROPERTIES, or double-click on this station. This opens a window in which you can specify the sender on the "Cross Traffic" tab ("New" button).

In the next window, you determine the I/O address of the "receiver", starting from which the data are to be stored, and the monitored data of the "sender" (starting address and length).

20.4.3 System Functions for Distributed I/O

You can use the following SFCs in conjunction with the distributed I/O:

▷ SFC 7 DP_PRAL
  Initiate process interrupt

▷ SFC 11 DPSYN_FR
  Send SYNC/FREEZE commands

▷ SFC 12 D_ACT_DP
  Activate/deactivate DP slave

▷ SFC 13 DPNRM_DG
  Read diagnostics data from a DP standard slave

▷ SFC 14 DPRD_DAT
  Read user data from a DP slave

▷ SFC 15 DPWR_DAT
  Write user data to a DP slave

Table 20.7 contains the parameters of these SFCs.

SFC 7 DP_PRAL
Initiate a process interrupt

With SFC 7 DP_PRAL, you initiate a process interrupt in the DP master associated with an intelligent slave from the user program of that slave.

At the parameter AL_INFO you transfer an interrupt ID defined by you that is transferred to the start information of the interrupt OB called in the DP master (variable OBxx_POINT_ADDR). The interrupt request is initiated with REQ = "1"; the parameters RET_VAL and BUSY indicate job status. The job is complete when the interrupt OB in the DP master has been executed.

The transfer memory between the DP master and the intelligent DP slave can be subdivided into individual address areas that represent individual modules from the viewpoint of the master CPU. The lowest address of an address area is taken as the module starting address. You can initiate a process interrupt in the master for each of these address areas ("virtual" modules).

You specify an address area at SFC 7 with the parameters IOID and LADDR from the viewpoint of the slave CPU (the I/O ID and the starting address of the slave side). The start information of the interrupt OB then contains the addresses of the "module" initiating the interrupt from the viewpoint of the master CPU.

SFC 11 DPSYN_FR
Send SYNC/FREEZE commands

With SFC 11 DPSYN_FR, you send the commands SYNC, UNSYNC, FREEZE and UNFREEZE to a SYNC/FREEZE group that you have configured with the Hardware Configuration. The SEND is initiated with REQ = "1" and is completed when BUSY = "0" is signaled.

In the parameter GROUP each group occupies one bit (from bit 0 = group 1 to bit 7 = group 8). The commands in the parameter MODE are also organized by bit:

20.4 Communication via Distributed I/O

▷ UNFREEZE, if bit 2 = "1"
▷ FREEZE, if bit 3 = "1"
▷ UNSYNC, if bit 4 = "1"
▷ SYNC, if bit 5 = "1"

In this way, SYNC mode and FREEZE mode on the DP slaves are first switched off. The inputs of the DP slaves are scanned in sequence by the DP master and the outputs of the DP slaves are modified; the DP slaves pass the received output signals immediately to the output terminals.

If you want to "freeze" the input signals of several DP slaves at a specific time, you output the command FREEZE to the relevant group. The input signals then read in sequence by the DP

Table 20.7 Parameters for SFCs Used to Reference the Distributed I/O

SFC	Parameter	Declaration	Data Type	Contents, Description
7	REQ	INPUT	BOOL	Request to initiate with REQ = "1"
	IOID	INPUT	BYTE	B#16#54 = input ID B#16#55 = output ID
	LADDR	INPUT	WORD	Starting address of an address area in the transfer memory
	AL_INFO	INPUT	DWORD	Interrupt ID (transfer of the start info of the interrupt OB)
	RET_VAL	OUTPUT	INT	Error information
	BUSY	OUTPUT	BOOL	No acknowledge yet from DP master when BUSY = "1"
11	REQ	INPUT	BOOL	Request to send with REQ = "1"
	LADDR	INPUT	WORD	Configured diagnostics address of the DP master
	GROUP	INPUT	BYTE	DP slave group (from the Hardware Configuration)
	MODE	INPUT	BYTE	Command (see test)
	RET_VAL	OUTPUT	INT	Error information
	BUSY	OUTPUT	BOOL	Job still running when BUSY = "1"
12	REQ	INPUT	BOOL	Request to activate/deactivate with REQ = "1"
	MODE	INPUT	BYTE	Function mode 0 Check for whether the DP slave is activated or deactivated 1 Activate DP slave 2 Deactivate DP slave 3 Abort activate/deactivate
	LADDR	INPUT	WORD	Diagnostics or module starting address of the DP slave
	RET_VAL	OUTPUT	INT	Result of check or error information
	BUSY	OUTPUT	BOOL	Job still running when BUSY = "1"
13	REQ	INPUT	BOOL	Read request with REQ = "1"
	LADDR	INPUT	WORD	Configured diagnostic address of the DP slave
	RET_VAL	OUTPUT	INT	Error information
	RECORD	OUTPUT	ANY	Destination area for the diagnostic data read
	BUSY	OUTPUT	BOOL	Read still in progress when BUSY = "1"
14	LADDR	INPUT	WORD	Configured start address (from the I area)
	RET_VAL	OUTPUT	INT	Error information
	RECORD	OUTPUT	ANY	Destination area for user data read
15	LADDR	INPUT	WORD	Configured start address (from the Q area)
	RECORD	INPUT	ANY	Source area for user data to be written
	RET_VAL	OUTPUT	INT	Error information

master have the signal states they had when "frozen". These input signals retain their value until you output another FREEZE command to cause the DP slaves to read in and hold the current input signals, or until you switch the DP slaves back to the "normal" mode with the UN-FREEZE command.

If you want to output the output signals of several DP slaves synchronously at a specific time, first output the SYNC command to the relevant group. The addressed DP slaves then hold the current signals at the output terminals. Now you can transfer the desired signal states to the DP slaves. Following the transfer, you output the SYNC command again; this causes the DP slaves to switch the received output signals simultaneously through to the output terminals. The accessed DP slaves then hold the signals at the output terminals until you switch through the new output signals with a new SYNC command, or until you switch the DP slaves back to their "normal" mode with the UNSYNC command.

SFC 12 D_ACT_DP
Activate/deactivate DP slave

With SFC 12 D_ACT_DP, you deactivate a configured (and existing) DP slave so that it can no longer be accessed by the DP master. The output terminals of the deactivated output slaves have zero or a substitute value.

A deactivated slave can be taken off the bus without an error message; it will not be reported as faulty or missing. Calls of the asynchronous error organization blocks OB 85 (program execution errors, when the user data of the deactivated slave are located in an automatically updated process image) and OB 86 (station failure) cease. After deactivation, you must no longer access the DP slave from the program, otherwise I/O access errors will be signaled.

With SFC 12 D_ACT_DP, you reactivate a deactivated DP slave. The DP slave is configured and parameterized by the DP master in the same way as during a station restore. On activation, the asynchronous error OBs 85 and 86 are not started. If the parameter BUSY has signal state "0" following activation, the DP slave can be accessed from the user program.

SFC 13 DPNRM_DG
Read diagnostic data

SFC 13 DPNRM_DG reads diagnostic data from a DP slave. The read procedure is initiated with REQ = "1", and is terminated when BUSY = "0" is returned. Function value RET_VAL then contains the number of bytes read. Depending on the slave, diagnostic data may comprise from 6 to 240 bytes. If there are more than 240 bytes, the first 240 bytes are transferred and the relevant overflow bit is then set in the data.

The parameter RECORD writes to the area in which the read data are stored. Variables of data type ARRAY and STRUCT or an ANY pointer of data type BYTE
(e.g. P#DBzDBXy.xBYTEnnn)
are permissible as actual parameters.

SFC 14 DPRD_DAT
Read user data

SFC 14 DPRD_DAT reads consistent user data with a length of 3 bytes or greater than 4 bytes from a DP slave. You specify the length of the data consistency when you parameterize the DP slave.

The parameter LADDR receives the module starting address of the DP slave (input area).

The parameter RECORD writes to the area in which the read data are stored. Variables of data type ARRAY and STRUCT or an ANY pointer of data type BYTE are permissible as actual parameters (e.g. P#DBzDBXy.xBYTEnnn).

SFC 15 DPWR_DAT
Write user data

SFC 15 DPWR_DAT writes consistent user data with a length of 3 bytes or greater than 4 bytes to a DP slave. You specify the length of the data consistency when you parameterize the DP slave.

The parameter LADDR receives the module starting address of the DP slave (input area).

The parameter RECORD writes to the area in which the read data are stored. Variables of data type ARRAY and STRUCT or an ANY pointer of data type BYTE (e.g. P#DBzDBXy.xBYTEnnn) are permissible as actual parameters.

20.5 Global Data Communication

20.5.1 Fundamentals

Global data communication (GD communication) is a communications service integrated into the operating system of the CPU and used for exchanging small volumes of non-time-critical data via the MPI bus. The transferrable global data include

▷ Inputs and outputs (process images)

▷ Memory bits

▷ Data in data blocks

▷ Timer and counter values as data to be sent.

It is a requirement that the CPUs are networked together via the MPI interface or connected via the K bus as in the S7-400 mounting rack. All CPUs must exist in the same STEP 7 project in order to be able to configure GD communication.

The cyclic GD communication service does not require an operating system: there are system functions available for event-driven GD communication on the S7-400.

Please note that a receiver CPU does not acknowledge receipt of global data. The sender therefore does not receive any response to tell it if a receiver has received data and if so, which one. However, you can screen the communication status between two CPUs as well as the overall status of all GD circles of a CPU.

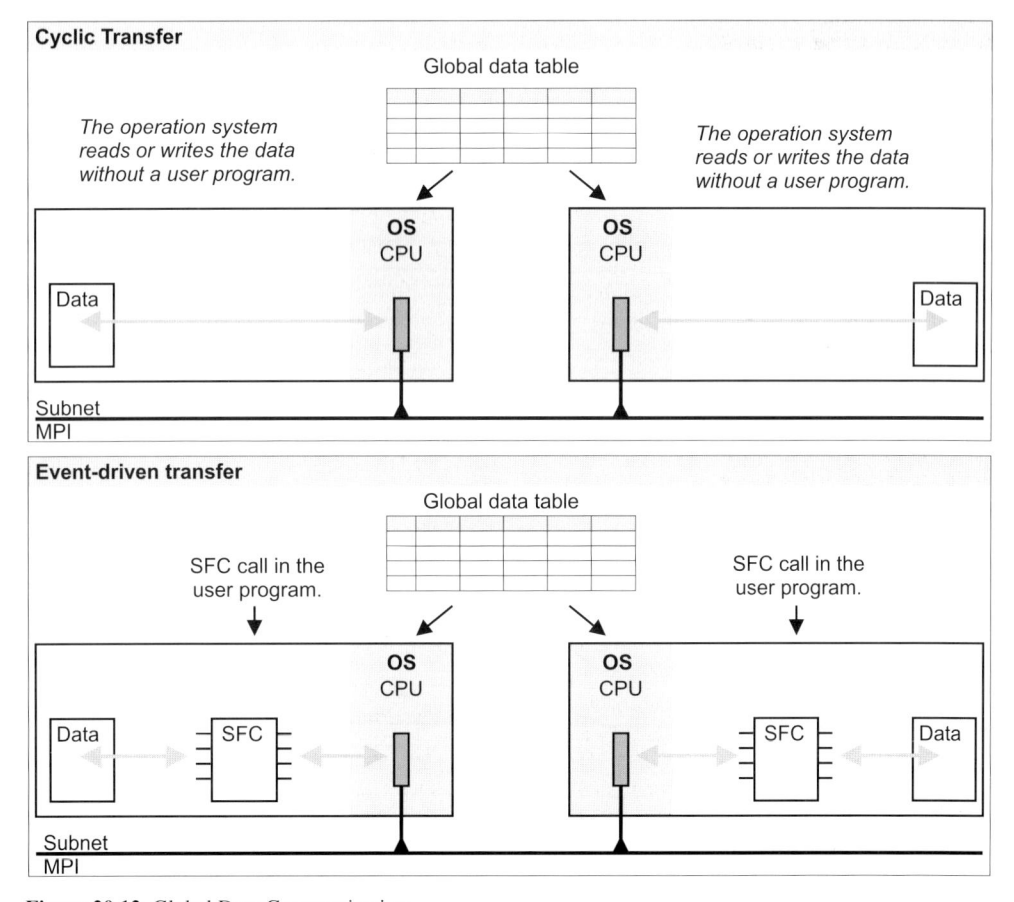

Figure 20.12 Global Data Communication

Sending and receiving global data is controlled with what are known as scan rates. These specify the number of (user program) cycles after which the CPU sends or receives the data. Sending and receiving takes place synchronously between the sender and the receiver at the cycle control point in each case, i.e. following cyclic program execution and before a new program cycle begins (like process image updating, for example).

Data is exchanged in the form of data packets (GD packets) between CPUs grouped into GD circles.

GD circle

The CPUs that exchange a shared GD packet form a GD circle. A GD circle can be any of the following

▷ The one-way connection of a CPU that sends a GD packet to several other CPUs that then receive that packet.

▷ The two-way connection between two CPUs where each of the two CPUs can send a GD packet to the other.

▷ The two-way connection between three CPUs where each of the three CPUs can send one GD packet to the other two CPUs (S7-400 CPUs only).

Up to 15 CPUs can exchange data with each other in one GD circle. One CPU can also belong to several GD circles.

See Table 20.8 for the resources of each individual CPU here.

GD packet

A GD packet comprises the packet header and one or more global data elements (GD elements):

▷ Packet header (8 bytes)
▷ ID of 1st GD element (2 bytes)
▷ User data of 1st GD element (x bytes)
▷ ID of 2nd GD element (2 bytes)
▷ User data of 2nd GD element (x bytes)
▷ etc.

Each GD element consists of 2 bytes of description and the actual net data. 3 bytes are required in the GD packet to transfer a memory byte, 4 bytes are required for a memory word, and 6 bytes for a memory doubleword. A Boolean variable occupies 1 byte of net data; it therefore requires the same space as a byte-sized variable. Timer and counter values with 2 bytes each occupy 4 bytes each in the GD packet.

A GD element can also be an address area. MB 0:15, for example, represents the area from memory byte MB 0 to MB 15, and DB20.DBW14:8 represents the data area located in DB 20 that starts from data word DBW 14 and comprises 8 data words.

The maximum size of a GD packet is 32 bytes on the S7-300 and 64 bytes on the S7-400. The maximum number of net data bytes per packet is reached with the transfer of only one GD element that contains up to 22 bytes on the S7-300 and up to 54 bytes on the S7-400.

Table 20.8 CPU Resources for Global Data Communication

GD resources Max. number of:	CPU 312 CPU 313 CPU 314	CPU 315 CPU 316	CPU 318	CPU 412 CPU 413 CPU 414	CPU 416 CPU 417
GD circles per CPU	4	4	8	8	16
Receive GD packets per CPU	4	4	16	16	32
Receive GD packets per circle	1	1	2	2	2
Send GD packets per CPU	4	4	8	8	16
Send GD packets per circle	1	1	1	1	1
Max. size of a GD packet	32 bytes	32 bytes	64 bytes	64 bytes	64 bytes
Max data consistency	8 bytes	8 bytes	32 bytes	16 bytes	32 bytes

20.5 Global Data Communication

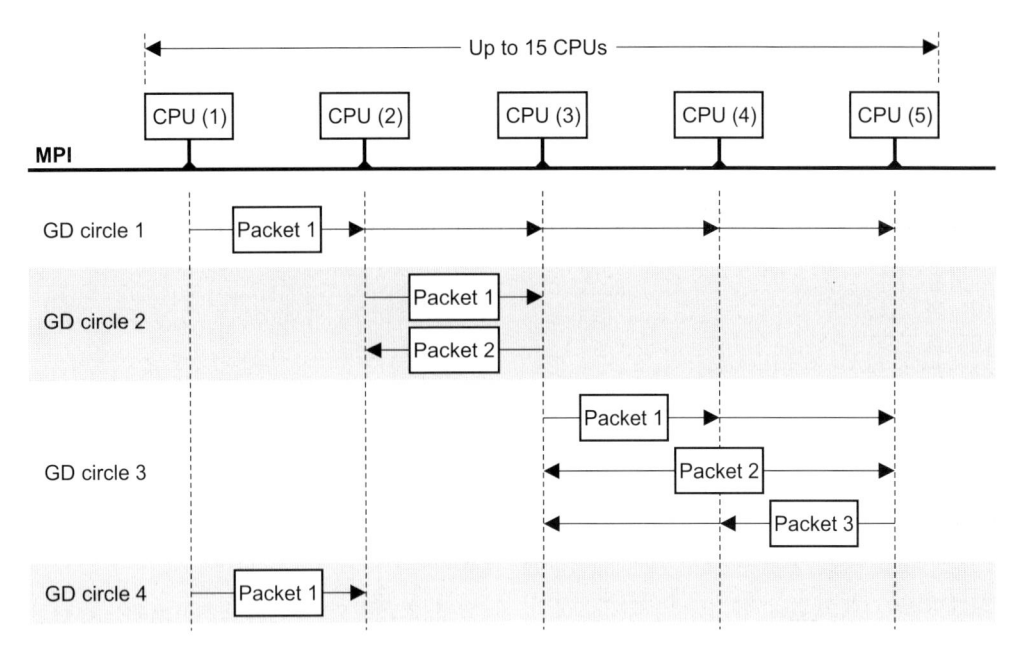

Figure 20.13 Example of GD Circles

Data consistency

The data consistency covers one GD element. If a GD element overwrites a CPU-specific variable, the areas specified in Table 20.8 apply.

If a GD element is greater than the length of the data consistency, blocks with consistent data of the relevant length are formed, starting with the first byte.

20.5.2 Configuring GD communication

Requirements

You must have created a project, there must be an MPI subnetwork available and you must have configured the S7 stations. The CPU, at least, must be available in the stations. Under the "Properties" button of the MPI interface, on the "General" tab of the properties window of the CPU (double-click on the CPU line in the Hardware Configuration or on the line with the MPI interface submodule), you can set the MPI address and select the MPI subnetwork with which the CPU is connected.

Global data table

You configure GD communication by filling out a table. With the icon for the MPI subnetwork selected in the SIMATIC Manager or in the Network Configuration, you can call up an empty table with OPTIONS → DEFINE GLOBAL DATA. Select a column and then EDIT → CPU. Select the station in the left half of the project selection window that then opens, and select the CPU in the right half. This CPU is accepted into the global data table with "OK".

Proceed in exactly the same way with the other CPUs participating in GD communication. A global data table can contain up to 15 CPU columns.

To configure data transfer between CPUs, select the first line under the send CPU and specify the address whose value is to be transferred (terminate with RETURN).

With EDIT → SENDER, you define this value as the value to be sent, indicated by a prefixed character ">" and shading. In the same line under Receiver CPU, you enter the address that is to accept the value (the property "Receiver" is set as default). You may use timer and counter functions only as senders; the receiver must be an address of word width for each timer or counter function.

A line can contain several receivers but only one sender (Table 20.9). After filling this in, you select GD TABLE → COMPILE.

After compiling (phase 1), the system data created are sufficient for global data communication. If you also configure the GD status (the status of the GD connection) and the scan rates, you must then compile the GD table for a second time.

GD ID

Following error-free compiling, STEP 7 completes the "GD ID" column. The GD ID shows you how the transferred data are structured into GD circles, GD packets and GD elements. For example, the GD ID "GD 2.1.3" corresponds to GD circle 2, GD packet 1, GD element 3. You can then find the resource assignment (number of GD circles) per CPU in the CPU column of the global data table.

GD status

Following compiling, you can enter the addresses for the communication status into the global data table with VIEW → GD STATUS. The overall status (GST) shows the status of all communications connections in the table. The status (GDS) shows the status of a communications connection (a transmitted GD packet). The status is shown in a doubleword in each case.

Scan rates

The GD communication service requires a significant portion of execution time in the CPU operating system and demands transmission time on the MPI bus. To keep this "communications load" to a minimum, it is possible to specify a "scan rate". A scan rate specifies the number of program cycles after which the data (or more precisely, a GD packet) are to be sent or received.

Since the data are not updated in every program cycle with a scan rate, you should avoid sending time-critical data via this form of communication.

After the first (error-free) compilation, you can use VIEW → SCAN RATES to define the scan rates (SRs) yourself for each GD packet and each CPU. The scan rate is set as standard in such a way that with an "empty" CPU (no user program) the GD packets are sent and received approximately every 10 ms. If a user program is then loaded, the time interval increases.

You can enter the scan rates in the area between 1 and 255. Please note, that as the scan rates decrease, the communications load on the CPU increases. To keep the communications load within tolerable limits, set the scan rate in the

Table 20.9 Example of a GD Table with Status and Scan Rates

GD-Identifier	Station 417 \ CPU417 (3)	Station 417 \ CPU414 (4)	Station 416\ CPU 416 (5)	Station 315 Slave\ CPU315 (7)	Station 314CP\ CPU314 (10)
GST	MD100	MD100	MD100	DB10.DBD200	DB10.DBD200
GDS 1.1	DB9.DBD0		MD92	DB10.DBD204	DB10.DBD204
SR 1.1	44	0	44	8	8
GD 1.1.1	>DB9.DBW10		MW90	DB10.DBW208	DB10.DBW208
GDS 2.1	MD96	MD96			
SR 2.1	44	23	0	0	0
GD 2.1.1	>Z10:10	DB3.DBW20:10			
GDS 3.1			MD96		
SR 3.1	0	0	44	8	8
GD 3.1.1			>MW98	DB10.DBW220	DB10.DBW210

20.6 SFC Communication

Table 20.10 SFC Parameters for GD Communication

Parameter	Present in SFC		Declaration	Data Type	Contents, Description
CIRCLE_ID	60	61	INPUT	BYTE	Number of the GD circle
BLOCK_ID	4	61	INPUT	BYTE	Number of the GD packet to be sent or to be received
RET_VAL	60	61	OUTPUT	INT	Error information

send CPU in such a way that the product of scan rate and cycle time on the S7-300 is greater than 60 ms and on the S7-400 greater than 10 ms. In the receive CPU, this product must be less than that in the send CPU to avoid the loss of any GD packets.

With scan rate 0, you switch off data exchange of the relevant GD packet if you only want to send or receive it event-driven with SFCs.

After configuring the GD status and the scan rates, you must compile the GD table for a second time. Then STEP 7 enters the compiled data in the *System data* object. GD communication becomes effective when you transfer the GD table to the connected CPUs with PLC → DOWNLOAD → TO MODULE.

GD communication also becomes effective when the *System data* object, that contains all hardware settings and parameter settings, is transferred.

20.5.3 System Functions for GD Communication

In S7-400 systems, you can also control GD communication in your program. Additionally or alternatively to the cyclic transfer of global data, you can send or receive a GD packet with the following SFCs:

▷ SFC 60 GD_SND
   Send GD packet

▷ SFC 61 GD_RCV
   Receive GD packet

The parameters for these SFCs are listed in Table 20.10. The prerequisite for the use of these SFCs is a configured global data table. After compiling this table, STEP 7 shows you, in the "GD Identifier" column, the numbers of the GD circles and GD packets which you need for parameter assignments.

SFC 60 GD_SND enters the GD packet in the system memory of the CPU and initiates transfer; SFC 61 GD_RCV fetches the GD packet from the system memory. If a scan rate greater than 0 has been specified for the GD packet in the GD table, cyclic transfer also takes place.

If you want to ensure data consistency for the entire GD packet when transferring with SFCs 60 and 61, you must disable or delay higher-priority interrupts and asynchronous errors on both the Send and Receive side during processing of SFC 60 or SFC 61.

The SFCs need not be called in pairs; "mixed" operation is also possible. For example, you can use SFC 60 GD_SND to have event-driven transmission of GD packets but then receive cyclically.

20.6 SFC Communication

20.6.1 Station-Internal SFC Communication

Fundamentals

With station-internal SFC communication, you can exchange data between programmable modules within a SIMATIC station. The communication functions required here are SFCs in the operating system of the CPU. These SFCs establish the communication connections themselves, if necessary. For this reason, these station-internal connections are not configured via the connection table ("Communication via non-configured connections", basic communication).

Station-internal SFC communication can take place, for example, parallel to cyclic data exchange via PROFIBUS-DP between the master CPU and the slave CPU, with data transfer being event-driven (Figure 20.14).

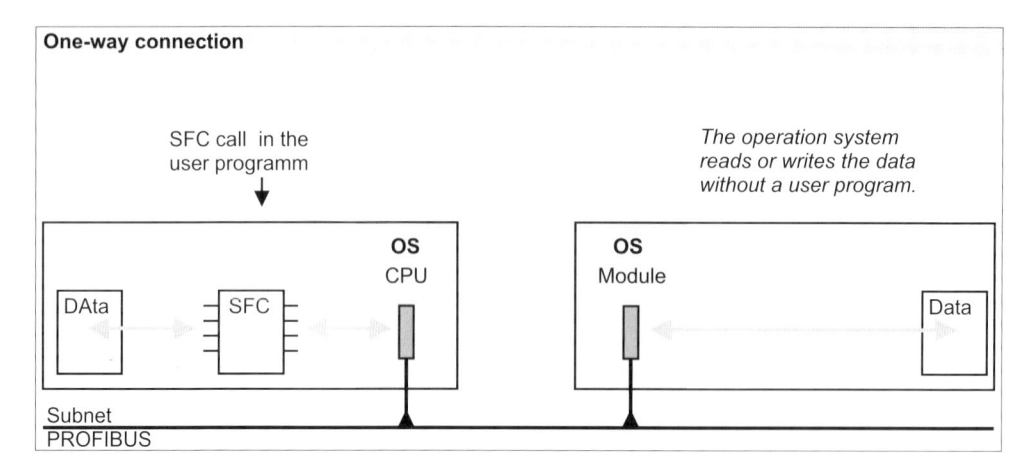

Figure 20.14 Station-Internal SFC Communication

Addressing the nodes, connections

Node identification is derived from the I/O address: at the LADDR parameter, you specify the module starting address and at the IOID parameter, you specify whether this address is in the input area or the output area.

These system functions establish the necessary communication connections dynamically and they clear the connections down again following completion of the job (programmable). If a connection buildup cannot be executed due to lack of resources either in the sending device or in the receiving device, "Temporary lack of resources" is signaled. Transfer must then be reinitiated. There can only be one connection between two communication partners in each direction.

You can use one system function for different communication connections by modifying the block parameters at runtime. An SFC cannot interrupt itself. A program section in which one of these SFCs is used can only be modified in the STOP mode; following this, a complete restart is executed.

User data, data consistency

These SFCs transfer up to 76 bytes as user data. Regardless of the direction of transfer, the operating system of a CPU arranges the user data in blocks that are consistent within themselves. On the S7-300, these blocks have a length of 8 bytes, on the CPU 412/413, they have a length of 16 bytes and on the CPU 414/416, they have a length of 32 bytes. If two CPUs exchange data, the block size of the "passive" CPU is decisive for data consistency.

Configuring Station-internal SFC Communication

Station-internal SFC communication is a special case in that it requires no configuring since data transfer is handled via dynamic connections. You simply use an existing PROFIBUS subnetwork or you create one either in the SIMATIC Manager (select the *Project* object and then INSERT → SUBNETWORK → PROFIBUS) or in the Network Configuration (see Section 2.4 "Configuring the Network").

Example: you have configured distributed I/O with a CPU 315-2DP as master. You use another CPU 315-2DP as an "intelligent" slave. You can now use station-internal SFC communication from both controllers to read and write data.

20.6.2 System Functions for Data Interchange Within a Station

The following system functions handle data transfers between two CPUs in the same station:

▷ SFC 72 I_GET
  Read data

▷ SFC 73 I_PUT
  Write data

▷ SFC 74 I_ABORT
  Disconnect

The parameters for these SFCs are listed in Table 20.11.

SFC 72 I_GET
Read data

A job is initiated with REQ = "1" and BUSY = "0" ("first call"). While the job is in progress, BUSY is set to "1". Changes to the REQ parameter no longer have any effect. When the job is completed, BUSY is reset to "0". If REQ is still "1", the job is immediately restarted.

When the read procedure has been initiated, the operating system in the partner CPU assembles and sends the requested data. An SFC call transfers the Receive data to the destination area. RET_VAL then shows the number of bytes transferred.

If CONT is = "0", the communication link is broken. If CONT is = "1", the link is maintained. The data are also read when the communication partner is in STOP mode.

The RD and VAR_ADDR parameters describe the area from which the data to be transferred are to be read or to which the receive data are to be written. Actual parameters may be addresses, variables or data areas addressed with an ANY pointer. The Send and Receive data are not checked for identical data types.

SFC 73 I_PUT
Write data

A job is initiated with REQ = "1" and BUSY = "0" ("first call"). While the job is in progress, BUSY is set to "1". Changes to the REQ parameter no longer have any effect. When the job is completed, BUSY is reset to "0". If REQ is still "1", the job is immediately restarted.

When the write procedure has been initiated, the operating system transfers all data from the source area to an internal buffer on the first call, and sends them to the partner in the link. There, the receiver writes the data into data area VAR_ADDR. BUSY is then set to "0". The data are also written when the receiving partner is at STOP.

The SD and VAR_ADDR parameters describe the area from which the data to be transferred are to be read or to which the receive data are to be written. Actual parameters may be addresses, variables or data areas addressed with an ANY pointer. The Send and Receive data are not checked for identical data types.

Table 20.11 SFC Parameters for Internal Station Communication

Parameter	For SFC			Declaration	Data Type	Contents, Description
REQ	72	73	74	INPUT	BOOL	Initiate job with REQ = "1"
CONT	72	73	-	INPUT	BOOL	CONT = "1": Connection remains intact after job terminates
IOID	72	73	74	INPUT	BYTE	B#16#54 = Input area B#16#55 = Output area
LADDR	72	73	74	INPUT	WORD	Module start address
VAR_ADDR	72	73	-	INPUT	ANY	Data area in partner CPU
SD	-	73	-	INPUT	ANY	Data area in own CPU which contains the Send data
RET_VAL	72	73	74	OUTPUT	INT	Error information
BUSY	72	73	74	OUTPUT	BOOL	Job in progress when BUSY = "1"
RD	72	-	-	OUTPUT	ANY	Data area in own CPU which will take the Receive data

SFC 74 I_ABORT
Disconnect

REQ = "1" breaks a connection to the specified communication partner. With I_ABORT, you can break only those connections established in the same station with I_GET or I_PUT.

While the job is in progress, BUSY is set to "1". Changes to the REQ parameter no longer have any effect. When the job is completed, BUSY is reset to "0". If REQ is still "1", the job is immediately restarted.

20.6.3 Station-External SFC Communication

Fundamentals

With station-external SFC communication, you can have event-driven data exchange between SIMATIC S7 stations. The stations must be connected to each other via an MPI subnetwork. The communications functions required for this are SFCs in the operating system of the CPU. These SFCs establish the communication connections themselves, if necessary. For this reason, these station-external connections are not configured via the connection table ("Communication via non-configured connections", basic communication).

Station-external SFC communication can execute event-driven data transfer, for example, parallel to cyclic global data communication.

Addressing the nodes, connections

These functions address nodes that are on the same MPI subnet. The node identification is derived from the MPI address (DEST_ID parameter).

These system functions set up the required communication links dynamically and - if specified - break them when the job has been executed. If a connection cannot be established because of a lack of resources in either the sender or the receiver, "temporary lack of resources" is reported. The transfer must then be retried. Between two communication partners, there can be only one connection in each direction.

On a transition from RUN to STOP, all active connections (all SFCs except X_RECV) are cleared.

By modifying the block parameters at run time, you can utilize a system function for different communication links. An SFC may not interrupt itself. You may modify a program section in which one of these SFCs is used only in STOP mode; a complete restart must then be executed.

User data, data consistency

These SFCs transfer a maximum of 76 bytes of user data. A CPU's operating system combines the user data into blocks consistent within themselves, without regard to the direction of transfer. In S7-300 systems, these blocks have a length of 8 bytes, in systems with a CPU 412/413 a length of 16 bytes, and in systems with a CPU 414/416 a length of 32 bytes.

If two CPUs exchange data via X_GET or X_PUT, the block size of the "passive" CPU is decisive to data consistency of the transferred data. In the case of a SEND/RECEIVE connection, all data of a call are consistent.

Configuring station-external SFC communication

Station-external SFC communication is a special case in that it requires no configuring since data transfer is handled via dynamic connections. You simply use an existing PROFIBUS subnetwork or you create one.

Example: you have a divided S7-400 mounting rack with one CPU 416 in each section. In addition, you connect an S7-300 station with a CPU 314 via an MPI cable to one of the S7-400s. You configure all three CPUs in the Hardware Configuration, for example, as "networked" via an MPI subnetwork. You can now use station-external SFC communication from all three controllers to exchange data.

20.6.4 System Functions for Station-External SFC Communication

The following system functions handle data transfers between partners in different stations:

▷ SFC 65 X_SEND
  Send data

▷ SFC 66 X_RCV
  Receive data

▷ SFC 67 X_GET
  Read data

▷ SFC 68 X_PUT
  Write data

20.6 SFC Communication

▷ SFC 69 X_ABORT
  Disconnect

The parameters for these SFCs are listed in Table 20.12.

SFC 65 X_SEND
Send data

A job is initiated with REQ = "1" and BUSY = "0" ("first call"). While the job is in progress, BUSY is set to "1"; changes to the REQ parameter now no longer have any effect. When the job terminates, BUSY is set back to "0". If REQ is still "1", the job is immediately restarted.

On the first call, the operating system transfers all data from the source area to an internal buffer, then transfers the data to the partner CPU.

BUSY is "1" for the duration of the send procedure. When the partner has signaled that it has fetched the data, BUSY is set to "0" and the send job terminated.

If CONT is = "0", the connection is broken and the respective CPU resources are available to other communication links. If CONT is = "1", the connection is maintained. The REQ_ID parameter makes it possible for you to assign an ID to the Send data which you can evaluate with SFC X_RCV.

The SD parameter describes the area from which the data to be sent are to be read. Actual parameters may be addresses, variables, or data areas addressed with an ANY pointer. Send and Receive data are not checked for matching data types.

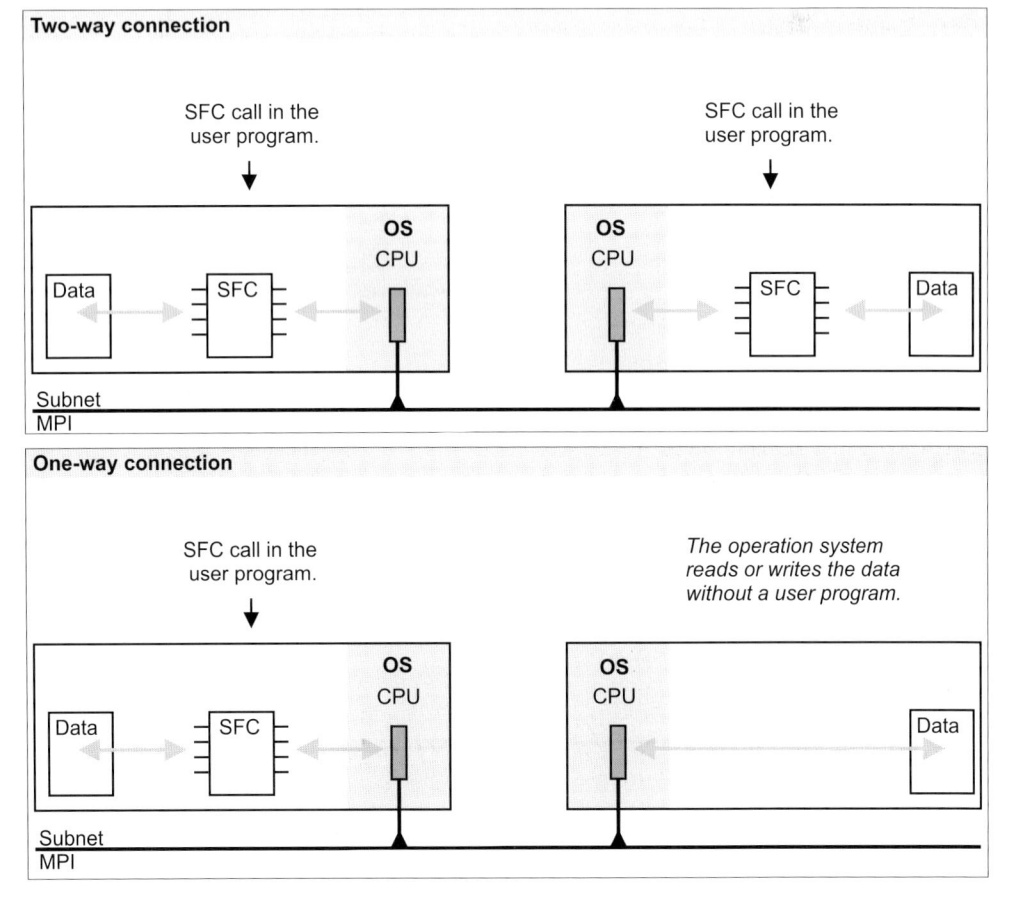

Figure 20.15 Station-External SFC Communication

20 Main Program

Table 20.12 SFC Parameters for External Station Communication

Parameter	For SFC					Declaration	Data Type	Contents, Description
REQ	65	-	67	68	69	INPUT	BOOL	Job initiation with REQ = "1"
CONT	65	-	67	68	-	INPUT	BOOL	CONT = "1": Connection is maintained when job is completed
DEST_ID	65	-	67	68	69	INPUT	WORD	Partner's node identification (MPI address)
REQ_ID	65	-	-	-	-	INPUT	DWORD	Job identification
VAR_ADDR	-	-	67	68	-	INPUT	ANY	Data area in partner CPU
SD	65	-	-	68	-	INPUT	ANY	Data area in own CPU which contains the Send data
EN_DT	-	66	-	-	-	INPUT	BOOL	If "1": Accept Receive data
RET_VAL	65	66	67	68	69	OUTPUT	INT	Error information
BUSY	65	-	67	68	69	OUTPUT	BOOL	Job in progress when BUSY = "1"
REQ_ID	-	66	-	-	-	OUTPUT	DWORD	Job identification
NDA	-	66	-	-	-	OUTPUT	BOOL	When "1": Data received
RD	-	66	67	-	-	OUTPUT	ANY	Data area in own CPU which will accept the Receive data

SFC 66 X_RCV
Receive data

The Receive data are placed in an internal buffer. Multiple packets can be put in a queue in the chronological order of their arrival.

Use EN_DT = "0" to check whether or not data were received; if so, NDA is "1", RET_VAL shows the number of bytes of Receive data, and REQ_ID is the same as the corresponding parameter in SFC 65 X_SEND. When EN_DT is = "1", the SFC transfers the first (oldest) packet to the destination area; NDA is then "1" and RET_VAL shows the number of bytes transferred. If EN_DT is "1" but there are no data in the internal queue, NDA is "0". On a complete restart, all data packets in the queue are rejected.

In the event of a broken connection or a restart, the oldest entry in the queue, if already "queried" with EN_DT = "0", is retained; otherwise, it is rejected like the other queue entries.

The RD parameter describes the area to which the Receive data are to be written. Actual parameters may be addresses, variables, or data areas addressed with an ANY pointer.

Send and Receive data are not checked for matching data types. When the Receive data are irrelevant, a "blank" ANY pointer (NIL ponter) as RD parameter in X_RCV is permissible.

SFC 67 X_GET
Read data

A job is initiated with REQ = "1" and BUSY = "0" ("first call"). While the job is in progress, BUSY is set to "1"; changes to the REQ parameter now no longer have any effect.

When the job terminates, BUSY is set back to "0". If REQ is still "1", the job is immediately restarted.

When the read procedure has been initiated, the operating system in the partner CPU assembles and sends the data required under VAR_ADDR. On an SFC call, the Receive data are entered in the destination area specified at the RD parameter. RET_VAL then shows the number of bytes transferred.

If CONT is "0", the communication link is broken. If CONT is "1", the connection is maintained. The data are then read even when the communication partner is in STOP mode.

The RD and VAR_ADDR parameters describe the area from which the data to be sent are to be read or to which the Receive data are to be written. Actual parameters may be addresses, variables, or data areas addressed with an ANY pointer. Send and Receive data are not checked for matching data types.

SFC 68 X_PUT
Write data

A job is initiated with REQ = "1" and BUSY = "0" ("first call"). While the job is in progress, BUSY is set to "1"; changes to the REQ parameter now no longer have any effect.

When the job terminates, BUSY is set back to "0". If REQ is still "1", the job is immediately restarted.

When the write procedure has been initiated, the operating system transfers all data from the source area specified at the SD parameter to an internal buffer on the first call, then sends the data to the partner CPU. There, the partner CPU's operating system writes the Receive data to the data area specified at the VAR_ADDR parameter. BUSY is then set to "0".

The data are written even if the communication partner is in STOP mode.

The RD and VAR_ADDR parameters describe the area from which the data to be sent are to be read or to which the Receive data are to be written. Actual parameters may be addresses, variables, or data areas addressed with an ANY pointer. Send and Receive data are not checked for matching data types.

SFC 69 X_ABORT
Disconnect

REQ = "1" breaks an existing connection to the specified communication partner. The SFC X_ABORT can be used to break only those connections established in the CPU's own station with the SFCs X_SEND, X_GET or X_PUT.

20.7 SFB Communication

20.7.1 Fundamentals

With SFB communication, you transfer larger volumes of data between SIMATIC S7 stations. The stations are connected to each other via a subnetwork; this can be an MPI subnetwork, a PROFIBUS subnetwork or an Ethernet subnetwork. The communications connections are static; they are configured in the connection table ("Communication via configured connections", extended communication).

The communications functions are system function blocks SFBs integrated in the operating system of the S7-400 CPUs. The associated instance data block is located in the user memory. If you want to use SFB communication, copy the interface description of the SFBs from the *Standard Library* under *System Function Blocks* to the *Blocks* container, generate an instance data block for each call and call the SFB with the associated instance data block. With incremental input, you can also select the SFB from the program element catalog and have the instance data block generated automatically.

Configuring SFB Communication

The prerequisite for communication via system function blocks is a configured connection table in which the communication links are defined.

A communication link is specified by a connection ID for each communication partner. STEP 7 assigns the connection IDs when it compiles the connection table. Use the "local ID" to initialize the SFB in the local or "own" module and the "remote ID" to initialize the SFB in the partner module.

The same logical connection can be used for different Send/Receive requests. To distinguish between them, you must add a job ID to the connection ID in order to define the relationship between the Send block and Receive block.

Initialization

SFB communication must be initialized at restart so that the connection to the communication partner can be established. Initialization takes place in the CPU that receives the attribute "Active connection buildup = Yes" in the connection table. You call the communication SFBs used in cyclic operation in a restart OB and initialize the parameters (provided they are available) as follows:

▷ REQ = FALSE

▷ ID = local connection ID from the connection table (data type WORD W#16#xxxx)

▷ PI_NAME = variable with the contents 'P_PROGRAM' in ASCII coding (e.g. ARRAY[1..9] OF CHAR).

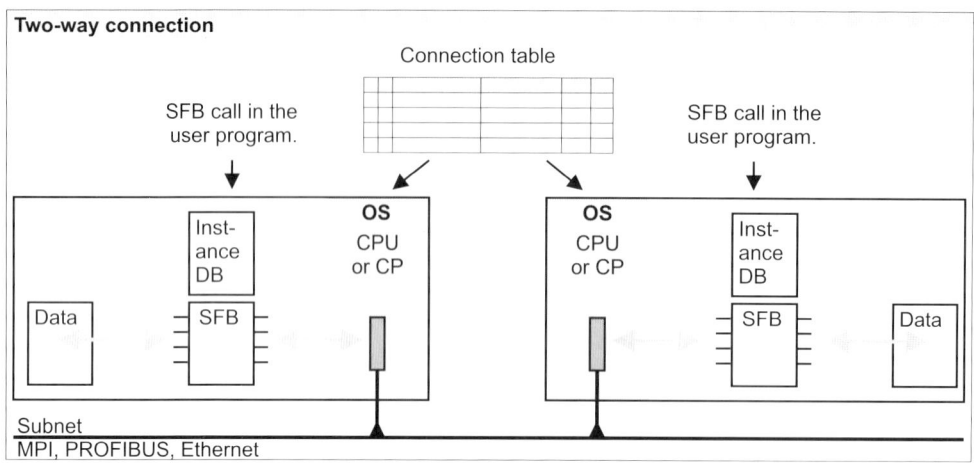

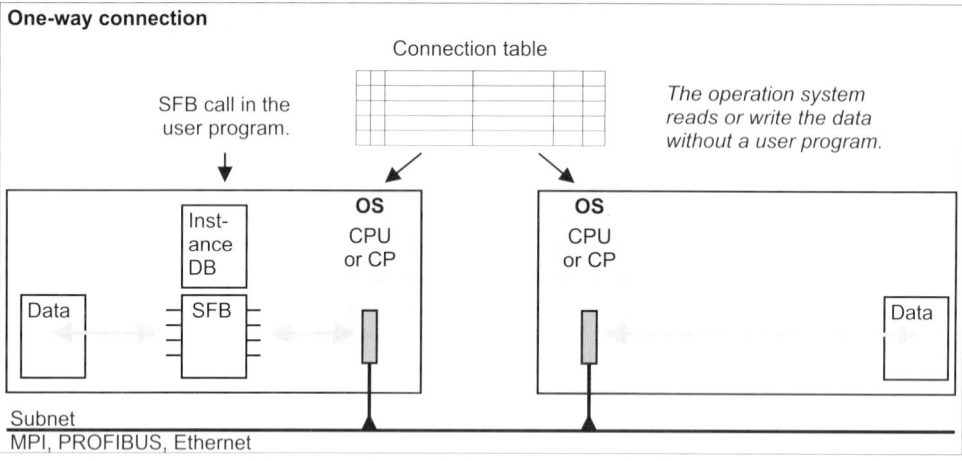

Figure 20.16 SFB Communication

The SFBs must continue to be called in a program loop until the DONE parameter has signal state "1". The parameters ERROR and STATUS give information concerning the errors that have occurred and the job status. You do not need to switch the data areas at restart (concerns the parameters ADDR_x, RD_x and SD_x).

In cyclic operation, you call the communication SFBs absolutely and you control data transfer via the parameters REQ and EN_R.

20.7.2 Two-Way Data Exchange

For two-way data exchange, you require one SEND block and one RECEIVE block each at the ends of a connection. Both blocks carry the connection IDs that are located in the connection table in the same line. You can also use several "block pairs" which are then distinguished from each other by the job ID.

The following SFBs are available for two-way data interchange:

▷ SFB 8 USEND
Uncoordinated sending of a data packet of CPU-specific length

▷ SFB 9 URCV
Uncoordinated receiving of a data packet of CPU-specific length

▷ SFB 12 BSEND
Sending of a data block of up to 64 Kbytes in length

▷ SFB 13 BRCV
Receiving of a data block of up to 64 Kbytes in length

SFB 8 and SFB 9 or SFB 12 and SFB 13 must always be used as a pair.

The parameters for these SFBs are listed in Table 20.13.

SFB 8 USEND and SFB 9 URCV
Uncoordinated sending and receiving

The SD_x and RD_x parameters are used to specify the variable or the area you want to transfer. Send area SD_x must correspond to the respective Receive area RD_x. Use the parameters without gaps, beginning with 1. No values need be specified for unneeded parameters (like an FB, not all SFB parameters need be assigned values).

The first time SFB 9 is called, a Receive mailbox is generated; on all subsequent calls, the Receive must fit into this mailbox.

A positive edge at the REQ (request) parameter starts the data exchange, a positive edge at the R (reset) parameter aborts it. A "1" in the EN_R (enable receive) parameter signals that the partner is ready to receive data.

Initialize the ID parameter with the connection ID, which STEP 7 enters in the connection table for both the local and the partner (the two IDs may differ). R_ID allows you to choose a specifiable but unique job ID which must be identical for the Send and Receive block. This allows several pairs of Send and Receive blocks to share a single logical connection (as each has a unique ID).

The block transfers the actual values of the ID and R_ID parameters to its instance data block on the first call. The first call establishes the communication relationship (for this instance) until the next complete restart.

With signal state "1" in the DONE or NDR parameter, the block signals that the job terminated without error. An error, if any, is flagged in the ERROR parameter. A value other than zero in the STATUS parameter indicates either a warning (ERROR = "0") or an error (ERROR = "1"). You must evaluate the DONE, NDR, ERROR and STATUS parameters after every block call.

Table 20.13 SFB Parameters for Sending and Receiving Data

Parameter	For SFB				Declaration	Data Type	Contents, Description
REQ	8	-	12	-	INPUT	BOOL	Start data exchange
EN_R	-	9	-	13	INPUT	BOOL	Receive ready
R	-	-	12	-	INPUT	BOOL	Abort data exchange
ID	8	9	12	13	INPUT	WORD	Connection ID
R_ID	8	9	12	13	INPUT	DWORD	Job ID
DONE	8	-	12	-	OUTPUT	BOOL	Job terminated
NDR	-	9	-	13	OUTPUT	BOOL	New data fetched
ERROR	8	9	12	13	OUTPUT	BOOL	Error occurred
STATUS	8	9	12	13	OUTPUT	WORD	Job status
SD_1	8	-	12	-	IN_OUT	ANY	First Send area
SD_2	8	-	-	-	IN_OUT	ANY	Second Send area
SD_3	8	-	-	-	IN_OUT	ANY	Third Send area
SD_4	8	-	-	-	IN_OUT	ANY	Fourth Send area
RD_1	-	9	-	13	IN_OUT	ANY	First Receive area
RD_2	-	9	-	-	IN_OUT	ANY	Second Receive area
RD_3	-	9	-	-	IN_OUT	ANY	Third Receive area
RD_4	-	9	-	-	IN_OUT	ANY	Fourth Receive area
LEN	-	-	12	13	IN_OUT	WORD	Length of data block in bytes

SFB 12 BSEND and SFB 13 BRCV
Block-oriented sending and receiving

Enter a pointer to the first byte of the data area in parameter SD_1 or RD_1 (the length is not evaluated); the number of bytes of Send or Receive data is in the LEN parameter.

Up to 64 Kbytes may be transferred; the data are transferred in blocks (sometimes called frames), and the transfer itself is asynchronous to the user program scan.

A positive edge at the REQ (request) parameter starts the data exchange, a positive edge at the R (reset) parameter aborts it. A "1" in the EN_R (enable receive) parameter signals that the partner is ready to receive data. Initialize the ID parameter with the connection ID, which STEP 7 enters in the connection table for both the local and the partner (the two IDs may differ).

R_ID allows you to choose a specifiable but unique job ID which must be identical for the Send and Receive block. This allows several pairs of Send and Receive blocks to share a single logical connection (as each has a unique ID).

The block transfers the actual values of the ID and R_ID parameters to its instance data block on the first call. The first call establishes the communication relationship (for this instance) until the next complete restart.

With signal state "1" in the DONE or NDR parameter, the block signals that the job terminated without error. An error, if any, is flagged in the ERROR parameter. A value other than zero in the STATUS parameter indicates either a warning (ERROR = "0") or an error (ERROR = "1"). You must evaluate the DONE, NDR, ERROR and STATUS parameters after *every* block call.

20.7.3 One-Way Data Exchange

In one-way data exchange, the communication SFB call is located in only one CPU. In the partner CPU, the operating system handles the necessary communication functions.

Table 20.14 SFB Parameters for Reading and Writing Data

Parameter	For SFB		Declaration	Data Type	Contents, Description
REQ	14	15	INPUT	BOOL	Start data exchange
ID	14	15	INPUT	WORD	Connection ID
NDR	14	-	OUTPUT	BOOL	New data fetched
DONE	-	15	OUTPUT	BOOL	Job terminated
ERROR	14	15	OUTPUT	BOOL	Error occurred
STATUS	14	15	OUTPUT	WORD	Job status
ADDR_1	14	15	IN_OUT	ANY	First data area in partner CPU
ADDR_2	14	15	IN_OUT	ANY	Second data area in partner CPU
ADDR_3	14	15	IN_OUT	ANY	Third data area in partner CPU
ADDR_4	14	15	IN_OUT	ANY	Fourth data area in partner CPU
RD_1	14	-	IN_OUT	ANY	First Receive area
RD_2	14	-	IN_OUT	ANY	Second Receive area
RD_3	14	-	IN_OUT	ANY	Third Receive area
RD_4	14	-	IN_OUT	ANY	Fourth Receive area
SD_1	-	15	IN_OUT	ANY	First Send area
SD_2	-	15	IN_OUT	ANY	Second Send area
SD_3	-	15	IN_OUT	ANY	Third Send area
SD_4	-	15	IN_OUT	ANY	Fourth Send area

The following SFBs are available for one-way data interchange:

▷ SFB 14 GET
Read data up to a CPU-specific maximum length

▷ SFB 15 PUT
Write data up to a CPU-specific maximum length

Table 20.14 lists the parameters for these SFBs.

The operating system in the partner CPU collects the data read with SFB 14; the operating system in the partner CPU distributes the data written with SFB 15. A Send or Receive (user) program in the partner CPU is not required.

A positive edge at parameter REQ (request) starts the data interchange. Set the ID parameter to the connection ID entered by STEP 7 in the connection table.

With a "1" in the DONE or NDR parameter, the block signals that the job terminated without error. An error, if any, is flagged with a "1" in the ERROR parameter.

A value other than zero in the STATUS parameter is indicative of either a warning (ERROR = "0") or an error (ERROR = "1"). You must evaluate the DONE, NDR, ERROR and STATUS parameters after *every* block call.

Use the ADDR_n parameter to specify the variable or the area in the partner CPU from which you want to fetch or to which you want to send the data. The areas in ADDR_n must coincide with the areas specified in SD_n or RD_n. Use the parameters without gaps, beginning with 1. Unneeded parameters need not be specified (as in an FB, an SFB does not have to have values for all parameters).

20.7.4 Transferring Print Data

SFB 16 PRINT allows you to transfer a format description and data to a printer via a CP 441 communications processor. Table 20.15 lists the parameters for this SFB.

A positive edge at the REQ parameter starts the data exchange with the printer specified by the ID and PRN_NR parameters. The block signals an error-free transfer by setting DONE to "1". An error, if any, is flagged by a "1" in the ERROR parameter. A value other than zero in the STATUS parameter is indicative of either a warning (ERROR = "0") or an error (ERROR = "1"). You must evaluate the DONE, ERROR and STATUS parameters after *every* block call.

Enter the characters to be printed in STRING format in the FORMAT parameter. You can integrate as many as four format descriptions for variables in this string, defined in parameters SD_1 to SD_4. Use the parameters without gaps, beginning with 1; do not specify values for unneeded parameters. You can transfer up to 420 bytes (the sum of FORMAT and all variables) per print request.

Table 20.15 Parameters for SFB 16 PRINT

Parameter	Declaration	Data Type	Contents, Description
REQ	INPUT	BOOL	Start data exchange
ID	INPUT	WORD	Connection ID
DONE	OUTPUT	BOOL	Job terminated
ERROR	OUTPUT	BOOL	Error occurred
STATUS	OUTPUT	WORD	Job status
PRN_NR	IN_OUT	BYTE	Printer number
FORMAT	IN_OUT	STRING	Format description
SD_1	IN_OUT	ANY	First variable
SD_2	IN_OUT	ANY	Second variable
SD_3	IN_OUT	ANY	Third variable
SD_4	IN_OUT	ANY	Fourth variable

20.7.5 Control Functions

The following SFBs are available for controlling the communication partner:

▷ SFB 19 START
  Execute a complete restart in the partner controller
▷ SFB 20 STOP
  Switch the partner controller to STOP
▷ SFB 21 RESUME
  Execute a warm restart in the partner controller

These SFBs are for one-way data exchange; no user program is required in the partner device for this purpose. The parameters for them are listed in Table 20.16.

A positive edge at the REQ parameter starts the data exchange. Enter as ID parameter the connection ID which STEP 7 entered in the connection table.

With a "1" in the DONE parameter, the block signals that the job terminated without error. An error, if any, is flagged by a "1" in the ERROR parameter. A value other than zero in the STATUS parameter is indicative of either a warning (ERROR = "0") or an error (ERROR = "1"). You must evaluate the DONE, ERROR and STATUS parameters after *every* block call.

Specify as PI_NAME an array variable with the contents "P_PROGRAM" (ARRAY [1..9] OF CHAR). The ARG and IO_STATE parameters are currently irrelevant, and need not be assigned a value.

SFB 19 START executes a complete restart of the partner CPU. Prerequisite is that the partner CPU is at STOP and that the mode selector is positioned to either RUN or RUN-P.

SFB 20 STOP sets the partner CPU to STOP. Prerequisite for error-free execution of this job request is that the partner CPU is not at STOP when the request is submitted.

SFB 21 RESUME executes a warm restart of the partner CPU. Prerequisite is that the partner CPU is at STOP, that the mode selector is set to either RUN or RUN-P, and that a warm restart is permissible at this time.

20.7.6 Monitoring Functions

The following system blocks are available for monitoring functions:

▷ SFB 22 STATUS
  Check partner status
▷ SFB 23 USTATUS
  Receive partner status
▷ SFC 62 CONTROL
  Check status of an SFB instance

Table 20.17 lists the parameters for the SFBs, Table 20.18 those for SFC 62.

The following applies for these system blocks: an error is indicated with "1" at the ERROR parameter. If the STATUS parameter has a value not equal to zero, this indicates either a warning (ERROR = "0") or an error (ERROR = "1").

Table 20.16 SFB Parameters for Partner Controller

Parameter	For SFB			Declaration	Data Type	Contents, Description
REQ	19	20	21	INPUT	BOOL	Start data exchange
ID	19	20	21	INPUT	WORD	Connection ID
DONE	19	20	21	OUTPUT	BOOL	Job terminated
ERROR	19	20	21	OUTPUT	BOOL	Error occurred
STATUS	19	20	21	OUTPUT	WORD	Job status
PI_NAME	19	20	21	IN_OUT	ANY	Program name (P_PROGRAM)
ARG	19	-	21	IN_OUT	ANY	Irrelevant
IO_STATE	19	20	21	IN_OUT	BYTE	Irrelevant

Table 20.17 SFB Parameters for Querying Status

Parameter	For SFB		Declaration	Data Type	Contents, Description
REQ	22	-	INPUT	BOOL	Start data exchange
EN_R	-	23	INPUT	BOOL	Ready to receive
ID	22	23	INPUT	WORD	Connection ID
NDR	22	23	OUTPUT	BOOL	New data fetched
ERROR	22	23	OUTPUT	BOOL	Error occurred
STATUS	22	23	OUTPUT	WORD	Job status
PHYS	22	23	IN_OUT	ANY	Physical status
LOG	22	23	IN_OUT	ANY	Logical status
LOCAL	22	23	IN_OUT	ANY	Status of an S7 CPU as partner

SFB 22 STATUS
Check the status of the partner device

SFB 22 STATUS fetches the status of the partner CPU and displays it in the PHYS (physical status), LOG (logical status) and LOCAL (operating status if the partner is an S7 CPU) parameters.

A positive edge at the REQ (request) parameter starts the query. Enter as ID parameter the connection ID which STEP 7 entered in the connection table.

With a "1" in the NDR parameter, the block signals that the job terminated without error. You must evaluate the NDR, ERROR and STATUS parameters after *every* block call.

SFB 23 USTATUS
Receive the status of the partner device

SFB 23 USTATUS receives the status of the partner, which it sends, unbidden, in the event of a change. The device status is displayed in the PHYS, LOG and LOCAL parameters.

A "1" in the EN_R (enable receive) parameter signals that the partner is ready to receive data. Initialize the ID parameter with the connection ID, which STEP 7 enters in the connection table.

With a "1" in the NDR parameter, the block signals that the job terminated without error. You must evaluate the NDR, ERROR and STATUS parameters after *every* block call.

Table 20.18 Parameters for SFC 62 CONTROL

Parameter	Declaration	Data Type	Contents, Description
EN_R	INPUT	BOOL	Ready to receive
I_DB	INPUT	BLOCK_DB	Instance data block
OFFSET	INPUT	WORD	Number of the local instance
RET_VAL	OUTPUT	INT	Error information
ERROR	OUTPUT	BOOL	Error detected
STATUS	OUTPUT	WORD	Status word
I_TYP	OUTPUT	BYTE	Block type identifier
I_STATE	OUTPUT	BYTE	Current status identifier
I_CONN	OUTPUT	BOOL	Connection status ("1" = connection exists)
I_STATUS	OUTPUT	WORD	STATUS parameter for SFB instance

SFC 62 CONTROL
Check the status of an SFB instance

SFC 62 CONTROL determines the status of an SFB instance and the associated connection in the local controller. Enter the SFB's instance data block in the I_DB parameter. If the SFB is called as local instance, specify the number of the local instance in the OFFSET parameter (zero when no local instance, 1 for the first local instance, 2 for the second, and so on).

A "1" in the EN_R (enable receive) parameter signals that the partner is ready to receive data.

Initialize the ID parameter with the connection ID, which STEP 7 enters in the connection table.

With a "1" in the NDR parameter, the block signals that the job terminated without error. You must evaluate the NDR, ERROR and STATUS parameters after *every* block call.

The parameters I_TYP, I_STATE, I_CONN and I_STATUS provide information concerning the status of the local SFB instance.

21 Interrupt Handling

Interrupt handling is always event-driven. When such an event occurs, the operating system interrupts scanning of the main program and calls the routine allocated to this particular event. When this routine has executed, the operating system resumes scanning of the main program at the point of interruption. Such an interruption can take place after every operation (statement).

Applicable events may be interrupts and errors. The order in which virtually simultaneous interrupt events are handled is regulated by a priority scheduler. Each event has a particular servicing priority. Several interrupt events can be combined into priority classes.

Every routine associated with an interrupt event is written in an organization block in which additional blocks can be called. A higher-priority event interrupts execution of the routine in an organization block with a lower priority. You can affect the interruption of a program by high-priority events using system functions.

21.1 General Remarks

SIMATIC S7 provides the following interrupt events (interrupts):

▷ Hardware interrupt
  Interrupt from a module, either via an input derived from a process signal or generated on the module itself

▷ Watchdog interrupt
  An interrupt generated by the operating system at periodic intervals

▷ Time-of-day interrupt
  An interrupt generated by the operating system at a specific time of day, either once only or periodically

▷ Time-delay interrupt
  An interrupt generated after a specific amount of time has passed; a system function call determines the instant at which this time period begins

▷ Multiprocessor interrupt
  An interrupt generated by another CPU in a multiprocessor network

Other interrupt events are the synchronous errors which may occur in conjunction with program scanning and the asynchronous errors, such as diagnostic interrupts. The handling of these events is discussed in Chapter 23, "Error Handling".

Priorities

An event with a higher priority interrupts a program being processed with lower priority because of another event. The main program has the lowest priority (priority class 1), asynchronous errors the highest (priority class 26), apart from the start-up routine. All other events are in the intervening priority classes. In S7-300 systems, the priorities are fixed; in S7-400 systems, you can change the priorities by parameterizing the CPU accordingly.

An overview of all priority classes, together with the default organization blocks for each, is presented in Section 3.1.2, "Priority Classes".

Disabling interrupts

The organization blocks for event-driven program scanning can be disabled and enabled with system functions SFC 39 DIS_IRT and SFC 40 EN_IRT and

delayed and enabled with SFC 41 DIS_AIRT and SFC 42 EN_AIRT (see Section 21.7, "Interrupt Handling").

Current signal states

In an interrupt handling routine, one of the requirements is that you work with the current

21 Interrupt Handling

signal states of the I/O modules (and not with the signal states of the inputs that were updated at the start of the main program) and write the fetched signal states direct to the I/O (not waiting until the process-image output table is updated at the end of the main program).

In the case of a few inputs and outputs for the interrupt handling routine, it is enough to access the I/O modules direct with load and transfer operations or with the MOVE box. You are recommended here to maintain a strict separation between the main program and the interrupt handling routine with regard to the I/O signals.

If you want to process many input and output signals in the interrupt handling routine, the solution on the S7-400 CPUs is to use subprocess images. When assigning addresses, you assign each module to a subprocess image. With SFC 26 UPDAT_PI and SFC 27 UPDAT_PO, you update the subprocess images in the user program (see also Section 20.2.1 "Process Image Updating").

On new S7-400 CPUs, you can assign an input and an output subprocess image to each interrupt organization block (each interrupt priority class) and so cause the process images to be updated automatically when the interrupt occurs.

Start Information, temporary local data

Table 21.1 provides an overview of the start information for the event-driven organization blocks. In S7-300 systems, the available temporary local data have a fixed length of 256 bytes. In S7-400 systems, you can specify the length per priority class by parameterizing the CPU accordingly (parameter block "local data"), whereby the total may not exceed a CPU-specific maximum. Note that the minimum number of bytes for temporary local data for the priority class used must be 20 bytes so as to be able to accommodate the start information. Specify zero for unused priority classes.

21.2 Hardware Interrupts

Hardware interrupts are used to enable the immediate detection in the user program of events in the controlled process, making it possible to respond with an appropriate interrupt handling routine. STEP 7 provides organization blocks OB 40 to OB 47 for servicing hardware interrupts; which of these eight organization blocks are actually available, however, depends on the CPU.

Table 21.1 Start Information for Interrupt Organization Blocks

Byte	Multiprocessor Interrupt	Hardware Interrupts	Watchdog Interrupts	Time-Delay Interrupts	Time-of-Day Interrupts
	OB 60	OB 40 to OB 47	OB 30 to OB 38	OB 20 to OB 23	OB 10 to OB 17
0	Event class	Event class	Event class	Event class	Event class
1	Start event	Start event	Start event	Start event	Start event
2	Priority class	Priority class	Priority class	Priority class	Priority class
3	OB number	OB number	OB number	OB number	OB number
4	-	-	-	-	-
5	-	Address identifier	-	-	-
6..7	Job identifier (INT)	Module start address (WORD)	Phase offset in ms (WORD)	Job identifier (WORD)	Interval (WORD)
8..9	-	Hardware interrupt	-	Expired delay	-
10..11	-	information (DWORD)	Time cycle in ms (INT)	(TIME)	-
12..19	Event instant (DT)	Event instant (DT)	Event instant (DT)	Event instant (DT)	Event instant (DT)

Hardware interrupt handling is programmed in the hardware configuration data. With system functions SFC 55 WR_PARM, SFC 56 WR_DPARM and SFC 57 PARM_MOD, you can (re)parameterize the modules with hardware interrupt capability even in RUN mode.

21.2.1 Generating a Hardware Interrupt

A hardware interrupt is generated on the modules with this capability. This could, for example, be a digital input module that detects a signal from the process or a function module that generates a hardware interrupt because of an activity taking place on the module.

By default, hardware interrupts are disabled. A parameter is used to enable servicing of a hardware interrupt (static parameter), and you can specify whether the hardware interrupt should be generated for a coming event, a leaving event, or both (dynamic parameter). Dynamic parameters are parameters which you can modify at runtime using SFCs.

In an intelligent DP slave equipped for this purpose, you can initiate a process interrupt in the master CPU with SFC 7 DP_PRAL.

The hardware interrupt is acknowledged on the module when the organization block containing the service routine for that interrupt has finished executing.

Resolution on the S7-300

If an event occurs during execution of a hardware interrupt OB which itself would trigger generation of the same hardware interrupt, that hardware interrupt will be lost when the event that triggered it is no longer present following acknowledgment.

It makes no difference whether the event comes from the module whose hardware interrupt is currently being serviced or from another module.

A diagnostic interrupt can be generated while a hardware interrupt is being serviced. If another hardware interrupt occurs on the same channel between the time the first hardware interrupt was generated and the time that interrupt was acknowledged, the loss of the latter interrupt is reported via a diagnostic interrupt to system diagnostics.

Resolution on the S7-400

If during execution of a hardware interrupt OB an event occurs on the same channel on the same module which would trigger the same hardware interrupt, that interrupt is lost. If the event occurs on another channel on the same module or on another module, the operating system restarts the OB as soon as it has finished executing.

21.2.2 Servicing Hardware Interrupts

Querying interrupt information

The start address of the module that triggered the hardware interrupt is in bytes 6 and 7 of the hardware interrupt OB's start information. If this address is an input address, byte 5 of the start information contains B#16#54; otherwise it contains B#16#55. If the module in question is a digital input module, bytes 8 to 11 contain the status of the inputs; for any other type of module, these bytes contain the interrupt status of the module.

Interrupt handling in the start-up routine

In the start-up routine, the modules do not generate hardware interrupts. Interrupt handling begins with the transition to RUN mode. Any hardware interrupts pending at the time of the transition are lost.

Error handling

If a hardware interrupt is generated for which there is no hardware interrupt OB in the user program, the operating system calls OB 85 (program execution error).

The hardware interrupt is acknowledged. If OB 85 has not been programmed, the CPU goes to STOP.

Hardware interrupts deselected when the CPU was parameterized cannot be serviced, even when the OBs for these interrupts have been programmed. The CPU goes to STOP.

Disabling, delaying and enabling

Calling of the hardware interrupt OBs can be disabled and enabled with system functions SFC 39 DIS_IRT and SFC 40 EN_IRT, and delayed and enabled with SFC 41 DIS_AIRT and SFC 42 EN_AIRT.

21.2.3 Configuring Hardware Interrupts with STEP 7

Hardware interrupts are programmed in the hardware configuration data. Open the selected CPU with EDIT → OBJECT PROPERTIES and choose the "Interrupts" tab in the dialog box.

In S7-300 systems (except for the CPU 318), the default priority for OB 40 is 16, and cannot be changed. In S7-400 systems and in CPU 318, you can choose a priority between 2 and 24 for every possible OB (on a CPU-specific basis); priority 0 deselects execution of an OB. You should never assign the same priority twice because interrupts can be lost when more than 12 interrupt events with the same priority occur simultaneously.

You must also enable the triggering of hardware interrupts on the respective modules. To this purpose, these modules are parameterized much the same as the CPU.

When it saves the hardware configuration, STEP 7 writes the compiled data to the *System Data* object in offline user program *Blocks;* from here, you can load the parameterization data into the CPU while the CPU is in STOP mode. The parameterization data for the CPU go into force immediately following loading; the parameter assignment data for the modules take effect after the next start-up.

21.3 Watchdog Interrupts

A watchdog interrupt is an interrupt which is generated at periodic intervals and which initiates execution of a watchdog interrupt OB. A watchdog interrupt allows you to execute a particular program periodically, independently of the processing time of the cyclic program.

In STEP 7, organization blocks OB 30 to OB 38 have been set aside for watchdog interrupts; which of these nine organization blocks are actually available depends on the CPU used.

Watchdog interrupt handling is set in the hardware configuration data when the CPU is parameterized.

21.3.1 Handling Watchdog Interrupts

Triggering watchdog interrupts in an S7-300

In an S7-300, the organization block for servicing watchdog interrupts is OB 35, which has the priority 12. You can set the interval in the range from 1 millisecond to 1 minute, in 1-millisecond increments, by parameterizing the CPU accordingly.

Triggering watchdog interrupts in an S7-400

You define a watchdog interrupt when you parameterize the CPU. A watchdog interrupt has three parameters: the interval, the phase offset, and the priority. You can set all three. Specifiable values for interval and phase offset are from 1 millisecond to 1 minute, in 1-millisecond increments; the priority may be set to a value between 2 and 24 or to zero, depending on the CPU (zero means the watchdog interrupt is not active).

STEP 7 provides the organization blocks listed in Table 21.2, in their maximum configurations.

Phase offset

The phase offset can be used to stagger the execution of watchdog interrupt handling routines despite the fact that these routines are timed to a multiple of the same interval. Use of the phase offset achieves a higher interval accuracy.

The start time of the time interval and the phase offset is the instant of transition from START-UP to RUN. The call instant for a watchdog interrupt OB is thus the time interval plus the

Table 21.2 Defaults for Watchdog Interrupts

OB	Time Interval	Phase	Priority
30	5 s	0 ms	7
31	2 s	0 ms	8
32	1 s	0 ms	9
33	500 ms	0 ms	10
34	200 ms	0 ms	11
35	100 ms	0 ms	12
36	50 ms	0 ms	13
37	20 ms	0 ms	14
38	10 ms	0 ms	15

phase offset. Figure 21.1 shows an example of this. No phase offset is set for time interval 1, time interval 2 is twice as long as time interval 1. Because of time interval 2's phase offset, the OBs for time interval 2 and those for time interval 1 are not called simultaneously. The lower-priority OB thus need not wait, and can precisely maintain its time interval.

Performance characteristics during startup

Watchdog interrupts cannot be serviced in the start-up OB. The time intervals do not begin until a transition is made to RUN mode.

Performance characteristics on error

When the same watchdog interrupt is generated again while the associated watchdog interrupt handling OB is still executing, the operating system calls OB 80 (timing error). If OB 80 has not been programmed, the CPU goes to STOP.

The operating system saves the watchdog interrupt that was not serviced, servicing it at the next opportunity. Only one unserviced watchdog interrupt is saved per priority class, regardless of how many unserviced watchdog interrupts accumulate.

Watchdog interrupts that were deselected when the CPU was parameterized cannot be serviced, even when the corresponding OB is available. The CPU goes to STOP in this case.

Disabling, delaying and enabling

Calling of the watchdog interrupt OBs can be disabled and enabled with system functions SFC 39 DIS_IRT and SFC 40 EN_IRT and delayed and enabled with SFC 41 DIS_AIRT and SFC 42 EN_AIRT.

21.3.2 Configuring Watchdog Interrupts with STEP 7

Watchdog interrupts are configured via the hardware configuration data. Simply open the selected CPU with EDIT → OBJECT PROPERTIES and choose the "Cyclic Interrupt" tab from the dialog box.

In S7-300 controllers (except for CPU 318), the processing priority is permanently set to 12. In S7-400 controllers and in CPU 318, you may set a priority between 2 and 24 for each possible OB (CPU-specific); priority 0 deselects the OB to which it is assigned. You should not assign a priority more than once, as interrupts might be lost if more than 12 interrupt events with the same priority occur simultaneously.

The interval for each OB is selected under "Execution", the delayed call instant under "Phase Offset".

When it saves the hardware configuration, STEP 7 writes the compiled data to the *System Data* object in the offline user program *Blocks*. From here, you can load the parameter assignment data into the CPU while the CPU is at STOP; the data take effect immediately.

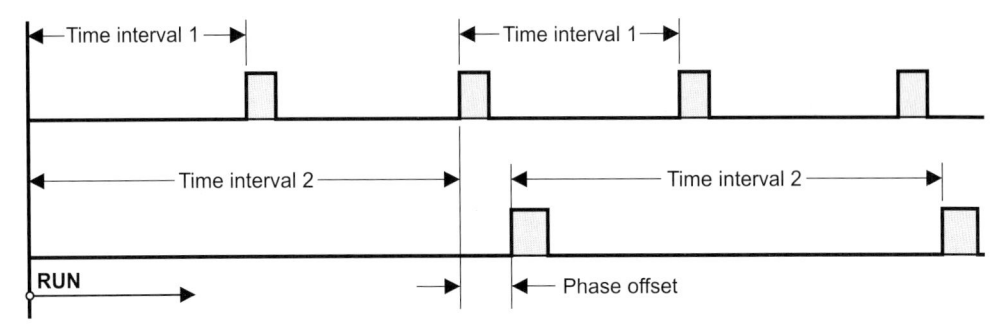

Figure 21.1 Example of Phase Offset for Watchdog Interrupts

21.4 Time-of-Day Interrupts

Time-of-day interrupts are used when you want to run a program at a particular time, either once only or periodically, for instance daily. In STEP 7, organization blocks OB 10 to OB 17 are provided for servicing time-of-day interrupts; which of these eight organization blocks are actually available depends on the CPU used.

You can configure the time-of-day interrupts in the hardware configuration data or control them at runtime via the program using system functions. The prerequisite for proper handling of the time-of-day interrupts is a correctly set real-time clock on the CPU.

21.4.1 Handling Time-of-Day Interrupts

General remarks

To start a time-of-day interrupt, you must first set the start time, then activate the interrupt. You can perform the two activities separately via the hardware configuration data or using SFCs. Note that when activated via the hardware configuration data, the time-of-day interrupt is started automatically following parameterization of the CPU.

You can start a time-of-day interrupt in two ways:

▷ Single-shot: the relevant OB is called once only at the specified time, or

▷ Periodically: depending on the parameter assignments, the relevant OB is started every minute, hourly, daily, weekly, monthly or yearly.

Following a single-shot time-of-day interrupt OB call, the time-of-day interrupt is canceled. You can also cancel a time-of-day interrupt with SFC 29 CAN_TINT.

If you want to once again use a canceled time-of-day interrupt, you must set the start time again, then reactivate the interrupt.

You can query the status of a time-of-day interrupt with SFC 31 QRY_TINT.

Performance characteristics during startup

During a cold restart or complete restart, the operating system clears all settings made with SFCs. Settings made via the hardware configuration data are retained. On a warm restart, the CPU resumes servicing of the time-of-day interrupts in the first complete scan cycle of the main program.

You can query the status of the time-of-day interrupts in the start-up OB by calling SFC 31, and subsequently cancel or re-set and reactivate the interrupts. The time-of-day interrupts are serviced only in RUN mode.

Performance characteristics on error

If a time-of-day interrupt OB is called but was not programmed, the operating system calls OB 85 (program execution error). If OB 85 was not programmed, the CPU goes to STOP.

Time-of-day interrupts that were deselected when the CPU was parameterized cannot be serviced, even when the relevant OB is available. The CPU goes to STOP.

If you activate a time-of-day interrupt on a single-shot basis, and if the start time has already passed (from the real-time clock's point of view), the operating system calls OB 80 (timing error). If OB 80 is not available, the CPU goes to STOP.

If you activate a time-of-day interrupt on a periodic basis, and if the start time has already passed (from the real-time clock's point of view), the time-of-day interrupt OB is executed the next time that time period comes due.

If you set the real-time clock ahead, whether for the purpose of correction or synchronization, thus skipping over the start time for the time-of-day interrupt, the operating system calls OB 80 (timing error). The time-of-day interrupt OB is then executed precisely once.

If you set the real-time clock back, whether for the purpose of correction or synchronization, an activated time-of-day interrupt OB will no longer be executed at the instants which are already past.

If a time-of-day interrupt OB is still executing when the next (periodic) call occurs, the operating system invokes OB 80 (timing error). When OB 80 and the time-of-day interrupt OB have executed, the time-of-day interrupt OB is restarted.

Disabling, delaying and enabling

Time-of-day interrupt OB calls can be disabled and enabled with SFC 39 DIS_IRT and SFC 40 EN_IRT, and delayed and enabled with SFC 41 DIS_AIRT and SFC 42 EN_AIRT.

21.4.2 Configuring Time-of-Day Interrupts with STEP 7

The time-of-day interrupts are configured via the hardware configuration data. Open the selected CPU with EDIT → OBJECT PROPERTIES and choose the "Time-of-Day" tab from the dialog box.

In S7-300 controllers (except for the CPU 318), the processing priority is permanently set to 2. In S7-400 controllers and in CPU 318, you can set a priority between 2 and 24, depending on the CPU, for each possible OB; priority 0 deselects an OB. You should not assign a priority more than once, as interrupts might be lost when more than 12 interrupt events with the same priority occur simultaneously.

The "Active" option activates automatic starting of the time-of-day interrupt. The "Execution" option screens a list which allows you to choose whether you want the OB to execute on a single-shot basis or at specific intervals. The final parameter is the start time (date and time).

When it saves the hardware configuration, STEP 7 writes the compiled data to the *System Data* object in the offline user program *Blocks*. From here, you can load the parameter assignment data into the CPU while the CPU is at STOP; these data then go into force immediately.

21.4.3 System Functions for Time-of-Day Interrupts

The following system functions can be used for time-of-day interrupt control:

▷ SFC 28 SET_TINT
  Set time-of-day interrupt

▷ SFC 29 CAN_TINT
  Cancel time-of-day interrupt

▷ SFC 30 ACT_TINT
  Activate time-of-day interrupt

▷ SFC 31 QRY_TINT
  Query time-of-day interrupt

The parameters for these system functions are listed in Table 21.3.

Table 21.3 SFC Parameters for Time-of-Day Interrupts

SFC	Parameter	Declaration	Data Type	Contents, Description
28	OB_NR	INPUT	INT	Number of the OB to be called at the specified time on a single-shot basis or periodically
	SDT	INPUT	DT	Start date and start time in the format DATE_AND_TIME
	PERIOD	INPUT	WORD	Period on which start time is based: W#16#0000 = Single-shot W#16#0201 = Every minute W#16#0401 = Hourly W#16#1001 = Daily W#16#1201 = Weekly W#16#1401 = Monthly W#16#2001 = Last in the month((412)) W#16#1801 = Yearly
	RET_VAL	OUTPUT	INT	Error information
29	OB_NR	INPUT	INT	Number of the OB whose start time is to be deleted
	RET_VAL	OUTPUT	INT	Error information
30	OB_NR	INPUT	INT	Number of the OB to be activated
	RET_VAL	OUTPUT	INT	Error information
31	OB_NR	INPUT	INT	Number of the OB whose status is to be queried
	RET_VAL	OUTPUT	INT	Error information
	STATUS	OUTPUT	WORD	Status of the time-of-day interrupt

SFC 28 SET_TINT
Set time-of-day interrupt

You determine the start time for a time-of-day interrupt by calling system function SFC 28 SET_TINT. SFC 28 sets only the start time; to start the time-of-day interrupt OB, you must activate the time-of-day interrupt with SFC 30 ACT_TINT. Specify the start time in the SDT parameter in the format DATE_AND_TIME, for instance DT#1997-06-30-08:30. The operating system ignores seconds and milliseconds and sets these values to zero. Setting the start time will overwrite the old start time value, if any.

An active time-of-day interrupt is canceled, that is, it must be reactivated.

SFC 30 ACT_TINT
Activate time-of-day interrupt

A time-of-day interrupt is activated by calling system function SFC 30 ACT_TINT. When a TOD interrupt is activated, it is assumed that a time has been set for the interrupt. If, in the case of a single-shot interrupt, the start time is already past, SFC 30 reports an error. In the case of a periodic start, the operating system calls the relevant OB at the next applicable time. Once a single-shot time-of-day interrupt has been serviced, it is, for all practical purposes, canceled. You can re-set and reactivate it (for a different start time) if desired.

SFC 29 CAN_TINT
Cancel time-of-day interrupt

You can delete a start time, thus deactivating the time-of-day interrupt, with system function SFC 29 CAN_TINT. The respective OB is no longer called. If you want to use this same time-of-day interrupt again, you must first set the start time, then activate the interrupt.

SFC 31 QRY_TINT
Query time-of-day interrupt

You can query the status of a time-of-day interrupt by calling system function SFC 31 QRY_TINT. The required information is returned in the STATUS parameter.

When the bits have signal state "1", they have the following meanings:

0	TOD interrupt disabled by operating system
1	New TOD interrupt rejected
2	TOD interrupt not activated and not expired
3	(- Reserved -)
4	TOD interrupt OB loaded
5	No disable
6	(and following – reserved -)

21.5 Time-Delay Interrupts

A time-delay interrupt allows you to implement a delay timer independently of the standard timers. In STEP 7, organization blocks OB 20 to OB 23 are set aside for time-delay interrupts; which of these four organization blocks are actually available depends on the CPU used.

The priorities for time-delay interrupt OBs are programmed in the hardware configuration data; system functions are used for control purposes.

21.5.1 Handling Time-Delay Interrupts

General remarks

A time-delay interrupt is started by calling SFC 32 SRT_DINT; this system function also passes the delay interval and the number of the selected organization block to the operating system. When the delay interval has expired, the OB is called.

You can cancel servicing of a time-delay interrupt, in which case the associated OB will no longer be called.

You can query the status of a time-delay interrupt with SFC 34 QRY_DINT.

Performance characteristics during startup

On a cold restart or complete restart, the operating system deletes all programmed settings for time-delay interrupts. On a warm restart, the settings are retained until processed in RUN mode, whereby the "residual cycle" is counted as part of the start-up routine.

You can start a time-delay interrupt in the start-up routine by calling SFC 32. When the delay interval has expired, the CPU must be in RUN mode in order to be able to execute the relevant organization block. If this is not the case, the CPU waits to call the organization block until the start-up routine has terminated, then calls the time-delay interrupt OB before the first network in the main program.

Performance characteristics on error

If no time-delay interrupt OB has been programmed, the operating system calls OB 85 (program execution error). If there is no OB 85 in the user program, the CPU goes to STOP.

If the delay interval has expired and the associated OB is still executing, the operating system calls OB 80 (timing error) or goes to STOP if there is no OB 80 in the user program.

Time-delay interrupts which were deselected during CPU parameterization cannot be serviced, even when the respective OB has been programmed. The CPU goes to STOP.

Disabling, delaying and enabling

The time-delay interrupt OBs can be disabled and enabled with system functions SFC 39 DIS_IRT and SFC 40 EN_IRT, and delayed and enabled with SFC 41 DIS_AIRT and SFC 42 EN_AIRT.

21.5.2 Configuring Time-Delay Interrupts with STEP 7

Time-delay interrupts are configured in the hardware configuration data. Simply open the selected CPU with EDIT → OBJECT PROPERTIES and choose the "Interrupts" tab from the dialog box.

In S7-300 controllers (except for the CPU 318), the priority is permanently preset to 3. In S7-400 controllers and in CPU 318, you can choose a priority between 2 and 24, depending on the CPU, for each possible OB; choose priority 0 to deselect an OB. You should not assign a priority more than once, as interrupts could be lost if more than 12 interrupt events with the same priority occur simultaneously.

When it saves the hardware configuration, STEP 7 writes the compiled data to the

System Data object in the offline user program *Blocks*. From here, you can transfer the parameter assignment data while the CPU is at STOP; the data take effect immediately.

21.5.3 System Functions for Time-Delay Interrupts

A time-delay interrupt can be controlled with the following system functions:

▷ SFC 32 SRT_DINT
  Start time-delay interrupt

▷ SFC 33 CAN_DINT
  Cancel time-delay interrupt

▷ SFC 34 QRY_DINT
  Query time-delay interrupt

The parameters for these system functions are listed in Table 21.4.

Table 21.4 SFC Parameters for Time-Delay Interrupts

SFC	Parameter	Declaration	Data Type	Contents, Description
32	OB_NR	INPUT	INT	Number of the OB to be called when the delay interval has expired
	DTIME	INPUT	TIME	Delay interval; permissible: T#1ms to T#1m
	SIGN	INPUT	WORD	Job identification in the respective OB's start information when the OB is called (arbitrary characters)
	RET_VAL	OUTPUT	INT	Error information
33	OB_NR	INPUT	INT	Number of the OB to be canceled
	RET_VAL	OUTPUT	INT	Error information
34	OB_NR	INPUT	INT	Number of the OB whose status is to be queried
	RET_VAL	OUTPUT	INT	Error information
	STATUS	OUTPUT	WORD	Status of the time-delay interrupt

SFC 32 SRT_DINT
Start time-delay interrupt

A time-delay interrupt is started by calling system function SFC 32 SRT_DINT. The SFC call is also the start time for the programmed delay interval. When the delay interval has expired, the CPU calls the programmed OB and passes the time delay value and a job identifier in the start information for this OB. The job identifier is specified in the SIGN parameter for SFC 32; you can read the same value in bytes 6 and 7 of the start information for the associated time-delay interrupt OB. The time delay is set in increments of 1 ms. The accuracy of the time delay is also 1 ms. Note that execution of the time-delay interrupt OB may itself be delayed when organization blocks with higher priorities are being processed when the time-delay interrupt OB is called. You can overwrite a time delay with a new value by recalling SFC 32. The new time delay goes into force with the SFC call.

SFC 33 CAN_DINT
Cancel time-delay interrupt

You can call system function SFC 33 CAN_DINT to cancel a time-delay interrupt, in which case the programmed organization block is not called.

SFC 34 QRY_DINT
Query time-delay interrupt

System function SFC 34 QRY_DINT informs you about the status of a time-delay interrupt. You select the time-delay interrupt via the OB number, and the status information is returned in the STATUS parameter.

When the bits have signal state "1", they have the following meanings:

0   Time-delay interrupt disabled by operating system
1   New time-delay interrupt rejected
2   Time-delay interrupt activated and not expired
3   (- Reserved -)
4   Time-delay OB loaded
5   No disable
6   (and following: - reserved -)

21.6 Multiprocessor Interrupt

The multiprocessor interrupt allows a synchronous response to an event in all CPUs in multiprocessor mode. A multiprocessor interrupt is triggered using SFC 35 MP_ALM. Organization block OB 60, which has a fixed priority of 25, is the OB used to service a multiprocessor interrupt.

General remarks

An SFC 35 MP_ALM call initiates execution of the multiprocessor interrupt OB. If the CPU is in single-processor mode, OB 60 is started immediately. In multiprocessor mode, OB 60 is started simultaneously on all participating CPUs, that is to say, even the CPU in which SFC 35 was called waits before calling OB 60 until all the other CPUs have indicated that they are ready.

The multiprocessor interrupt is not programmed in the hardware configuration data; it is already present in every CPU with multicomputing capability. Despite this fact, however, a sufficient number of local data bytes (at least 20) must still be reserved in the CPU's "Local Data" tab under priority class 25.

Performance characteristics during startup

The multiprocessor interrupt is triggered only in RUN mode. An SFC 35 call in the start-up routine terminates after returning error 32 929 (W#16#80A1) as function value.

Performance characteristics on error

If OB 60 is still in progress when SFC 35 is recalled, the system function returns error code 32 928 (W#16#80A0) as function value. OB 60 is not started in any of the CPUs.

The unavailability of OB 60 in one of the CPUs at the time it is called or the disabling or delaying of its execution by system functions has no effect, nor does SFC 35 report an error.

Disabling, delaying and enabling

The multiprocessor OB can be disabled and enabled with system functions SFC 39 DIS_IRT and SFC 40 EN_IRT, and delayed and enabled with SFC 41 DIS_AIRT and SFC 42 EN_AIRT.

Table 21.5 Parameters for SFC 35 MP_ALM

Parameter	Declaration	Data Type	Contents, Description
JOB	INPUT	BYTE	Job identification in the range B#16#00 to B#16#0F
RET_VAL	OUTPUT	INT	Error information

SFC 35 MP_ALM
Multiprocessor interrupt

A multiprocessor interrupt is triggered with system function SFC 35 MP_ALM. Its parameters are listed in Table 21.5.

The JOB parameter allows you to forward a job identifier. The same value can be read in bytes 6 and 7 of OB 60's start information in all CPUs.

21.7 Handling Interrupts

The following system functions are available for handling interrupts and asynchronous errors:

▷ SFC 39 DIS_IRT
   Disable interrupts
▷ SFC 40 EN_IRT
   Enable disabled interrupts
▷ SFC 41 DIS_AIRT
   Delay interrupts
▷ SFC 42 EN_AIRT
   Enable delayed interrupts

Table 21.6 lists the parameters for these system functions.

These system functions affect all interrupts and all asynchronous errors. System functions SFC 36 to SFC 38 are provided for handling synchronous errors.

SFC 39 DIS_IRT
Disabling interrupts

System function SFC 39 DIS_IRT disables servicing of new interrupts and asynchronous errors. All new interrupts and asynchronous errors are rejected. If an interrupt or asynchronous error occurs following a Disable, the organization block is not executed; if the OB does not exist, the CPU does not go to STOP.

The Disable remains in force for all priority classes until it is revoked with SFC 40 EN_IRT. After a complete restart, all interrupts and asynchronous errors are enabled.

The MODE and OB_NR parameters are used to specify which interrupts and asynchronous errors are to be disabled. MODE = B#16#00 disables all interrupts and asynchronous errors. MODE = B#16#01 disables an interrupt class whose first OB number is specified in the OB_NR parameter.

For example, MODE = B#16#01 and OB_NR = 40 disables all hardware interrupts; OB = 80 would disable all asynchronous errors. MODE = B#16#02 disables the interrupt or asynchronous error whose OB number you entered in the OB_NR parameter.

Regardless of a Disable, the operating system enters each new interrupt or asynchronous error in the diagnostic buffer.

SFC 40 EN_IRT
Enabling disabled interrupts

System function SFC 40 EN_IRT enables the interrupts and asynchronous errors disabled with SFC 39 DIS_IRT. An interrupt or asynchronous error occurring after the Enable will be serviced by the associated organization block; if that organization block is not in the user program, the CPU goes to STOP (except in the case of OB 81, the organization block for power supply errors).

The MODE and OB_NR parameters specify which interrupts and asynchronous errors are to be enabled. MODE = B#16#00 enables all interrupts and asynchronous errors. MODE = B#16#01 enables an interrupt class whose first OB number is specified in the OB_NR parameter. MODE = B#16#02 enables the interrupt or asynchronous error whose OB number you entered in the OB_NR parameter.

Table 21.6 SFC Parameters for Interrupt Handling

SFC	Parameter	Declaration	Data Type	Contents, Description
39	MODE	INPUT	BYTE	Disable mode (see text)
	OB_NR	INPUT	INT	OB number (see text)
	RET_VAL	OUTPUT	INT	Error information
40	MODE	INPUT	BYTE	Enable mode (see text)
	OB_NR	INPUT	INT	OB number (see text)
	RET_VAL	OUTPUT	INT	Error information
41	RET_VAL	OUTPUT	INT	(New) number of delays
42	RET_VAL	OUTPUT	INT	Number of delays remaining

SFC 41 DIS_AIRT
Delaying Interrupts

System function SFC 41 DIS_AIRT delays the servicing of higher-priority new interrupts and asynchronous errors. Delay means that the operating system saves the interrupts and asynchronous errors which occurred during the delay and services them when the delay interval has expired. Once SFC 41 has been called, the program in the current organization block (in the current priority class) will not be interrupted by a higher-priority interrupt; no interrupts or asynchronous errors are lost.

A delay remains in force until the current OB has terminated its execution or until SFC 42 EN_AIRT is called.

You can call SFC 41 several times in succession. The RET_VAL parameter shows the number of calls. You must call SFC 42 precisely the same number of times as SFC 41 in order to reenable the interrupts and asynchronous errors.

SFC 42 EN_AIRT
Enabling delayed interrupts

System function SFC 42 EN_AIRT reenables the interrupts and asynchronous errors delayed with SFC 41. You must call SFC 42 precisely the same number of times as you called SFC 41 (in the current OB). The RET_VAL parameter shows the number of delays still in force; if RET_VAL is = 0, the interrupts and asynchronous errors have been reenabled.

If you call SFC 42 without having first called SFC 41, RET_VAL contains the value 32896 (W#16#8080).

22 Restart Characteristics

22.1 General Remarks

22.1.1 Operating Modes

Before the CPU begins processing the main program following power-up, it executes a restart routine. START-UP is one of the CPU's operating modes, as is STOP or RUN. This chapter describes the CPU's activities on a transition from and to START-UP and in the restart routine itself.

Following power-up ①, the CPU is in the STOP mode (Figure 22.1). If the keyswitch on the CPU's front panel is at RUN or RUN-P, the CPU switches to START-UP mode ②, then to RUN mode ③. If an "unrecoverable" error occurs while the CPU is in START-UP or RUN mode or if you position the keyswitch to STOP, the CPU returns to the STOP mode ④ ⑤.

The user program is tested with breakpoints in single-step operation in the HOLD mode. You can switch to this mode from both RUN and START-UP, and return to the original mode when you abort the test ⑥ ⑦. You can also set the CPU to the STOP mode from the HOLD mode ⑧.

When you parameterize the CPU, you can define restart characteristics with the "Restart" tab such as the maximum permissible amount of time for the Ready signals from the modules following power-up or whether the CPU is to start up when the configuration data do not coincide with the actual configuration or in what mode the CPU restart is to be in.

SIMATIC S7 has three restart modes, namely *cold restart*, *complete restart* and *warm restart*. On a cold restart or complete restart, the main program is always processed from the beginning. A warm restart resumes the main program at the point of interruption, and "finishes" the cycle.

S7 CPUs supplied before 10/98 have complete restart and warm restart. The complete restart corresponds in functionality to the warm restart.

You can scan a program on a single-shot basis in START-UP mode. STEP 7 provides organization blocks OB 102 (cold restart), OB 100 (complete restart or warm restart) and OB 101 (warm restart) expressly for this purpose. Sample applications are the parameterization of modules unless this was already taken care of by the CPU, and the programming of defaults for your main program.

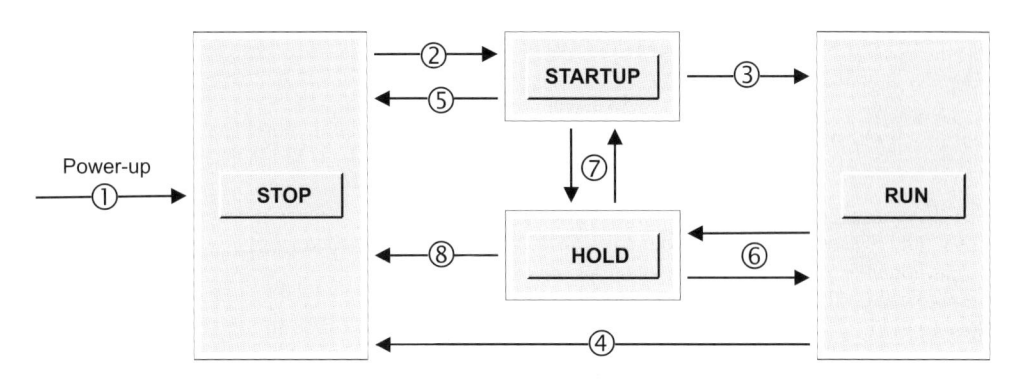

Figure 22.1 CPU Operating Modes

22.1.2 HOLD Mode

The CPU changes to the HOLD mode when you test the program with breakpoints (in "single-step mode"). The STOP LED then lights up and the RUN LED blinks.

In HOLD mode, the output modules are disabled. Writing to the modules affects the module memory, but does not switch the signal states "out" to the module outputs. The modules are not reenabled until you exit the HOLD mode.

In HOLD mode, everything having to do with timing is discontinued. This includes, for example, the processing of timers, clock memory and run-time meters, cycle time monitoring and minimum scan cycle time, and the servicing of time-of-day and time-delay interrupts. Exception: the real-time clock continues to function normally.

Every time the progression is made to the next statement in test mode, the timers for the duration of the single step run a little further, thus simulating a dynamic behavior similar to "normal" program scanning.

In HOLD mode, the CPU is capable of passive communication, that is, it can receive global data or take part in the unilateral exchange of data.

If the power fails while the CPU is in HOLD mode, battery-backed CPUs go to STOP on power recovery. CPUs without backup batteries execute an automatic complete restart.

22.1.3 Disabling the Output Modules

In the STOP and HOLD modes, modules are disabled (OD (output disable) signal). Disabled output modules output a zero signal or, if they have the capability, the replacement value. Via a variable table, you can control outputs on the modules with the "Isolate PQ" function, even in STOP mode.

During restart, the output modules remain disabled. Only when the cyclic scan begins are the output modules enabled. The signal states in a module's memory (not the process image!) are then applied to the outputs.

While the output modules are disabled, module memory can be set either with direct access (transfer statements or MOVE box with address area PQ) or via the transfer of the process-image output table. If the CPU removes the Disable signal, the signal states in module memory are applied to the external outputs.

On a cold restart (OB 102) and complete restart (OB 100), the process images and the module memory are cleared. If you want to scan inputs in OB 102 or in OB 100, you must load the signal states from the module using direct access. You can then set the inputs (transfer them, for instance, with load statements or with the MOVE box from address area PI to address area I), then work with the inputs. If you want certain outputs to be set on a transition from a complete restart to the cyclic program (prior to calling OB 1), you must use direct access to address the output modules. It is not enough to just set the outputs (in the process image), as the process-image output table is not transferred at the end of the complete restart routine.

On a warm restart, the "old" process-image input and process-image output tables, which were valid prior to power-down or STOP, are used in OB 101 and in the remainder of the cycle. At the end of that cycle, the process-image output table is transferred to module memory (but not yet switched through to the external outputs, since the output modules are still disabled).

You now have the option of parameterizing the CPU to clear the process-image output table and the module memory at the end of the warm restart. Before switching to OB 1, the CPU revokes the Disable signal so that the signal states in the module memory are applied to the external outputs.

22.1.4 Restart Organization Blocks

On a cold restart, the CPU calls organization block OB 102; on a complete restart, it calls organization block OB 100. In the absence of OB 100 or OB 102, the CPU begins cyclic program execution immediately.

On a warm restart, the CPU calls organization block OB 101 on a single-shot basis before processing the main program. If there is no OB 101, the CPU begins scanning at the point of interruption.

The start information in the temporary local data has the same format for the restart organization blocks; Table 22.1 shows the start information for OB 100. The reason for the restart is shown in the restart request (Byte 1):

▷ B#16#81 Manual complete restart (OB 100)
▷ B#16#82 Automatic complete restart (OB 100)
▷ B#16#83 Manual warm restart (OB 101)
▷ B#16#84 Automatic warm restart (OB 101)
▷ B#16#84 Manual cold restart (OB 102)
▷ B#16#84 Automatic cold restart (OB 102)

The number of the stop event and the additional information define the restart more precisely (tells you, for example, whether a manual complete restart was initiated via the mode selector). With this information, you can develop an appropriate event-related restart routine.

22.2 Power-Up

22.2.1 STOP Mode

The CPU goes to STOP in the following instances:

▷ When the CPU is switched on
▷ When the mode selector is set from RUN to STOP
▷ When an "unrecoverable" error occurs during program scanning
▷ When system function SFC 46 STP is executed
▷ When requested by a communication function (stop request from the programming device or via communication function blocks from another CPU)

The CPU enters the reason for the STOP in the diagnostic buffer. In this mode, you can also read the CPU information with a programming device in order to localize the problem.

In STOP mode, the user program is not scanned. The CPU retrieves the settings – either the values which you entered in the hardware configuration data when you parameterized the CPU or the defaults – and sets the modules to the specified initial state.

In STOP mode, the CPU can receive global data via GD communication and carry out passive unilateral communication functions. The real-time clock keeps running.

You can parameterize the CPU in STOP mode, for instance you can also set the MPI address, transfer or modify the user program, and execute a CPU memory reset.

22.2.2 Memory Reset

A memory reset sets the CPU to the "initial state". You can initiate a memory reset with a programming device only in STOP mode or with the mode selector: hold the switch in the MRES position for at least 3 seconds then release, and after a maximum of 3 seconds hold it in the MRES position again for at least 3 seconds.

The CPU erases the entire user program both in work memory and in RAM load memory. System memory (for instance bit memory, timers and counters) is also erased, regardless of retentivity settings.

The CPU sets the parameters for all modules, including its own, to their default values. The MPI parameters are an exception. They are not changed so that a CPU whose memory has been reset can still be addressed on the MPI bus. A memory reset also does not affect the diagnostic buffer, the real-time clock, or the run-time meters.

If a memory card with Flash EPROM is inserted, the CPU copies the user program from the memory card to work memory. The CPU also copies any configuration data it finds on the memory card.

22.2.3 Retentivity

A memory area is retentive when its contents are retained even when the mains power is switched off as well as on a transition from STOP to RUN following power-up (at warm restart and cold restart).

Table 22.1 Start Information for the Restart OBs

Byte	Name	Data Type	Description
0	OB100_EV_CLASS	BYTE	Event class
1	OB100_STRTUP	BYTE	Restart request (see text)
2	OB100_PRIORITY	BYTE	Priority class
3	OB100_OB_NUMBR	BYTE	OB number
4	OB100_RESERVED_1	BYTE	Reserved
5	OB100_RESERVED_2	BYTE	Reserved
6..7	OB100_STOP	WORD	Number of the stop event
8..11	OB100_STRT_INFO	DWORD	Additional information on the current restart
12..19	OB100_DATE_TIME	DT	Date and time event occurred

Retentive memory areas may be those for bit memory, timers, counters and, on the S7-300, data areas. The number of data in which areas can be made retentive depends on the CPU. You can specify the retentive areas via the "Retentivity" tab when you parameterize the CPU.

The settings for retentivity are in the system data blocks (SDBs) in load memory, that is, on the memory card. If the memory card is a RAM card, you must operate the programmable controller with a backup battery to save the retentivity settings permanently.

When you use a backup battery, the signal states of the memory bits, timers and counters specified as being retentive are retained. The user program and the user data remain unchanged. It makes no difference whether the memory card is a RAM or a Flash EPROM card.

If the memory card is a Flash EPROM card and there is *no backup battery*, an S7-300 and an S7-400 respond differently. In S7-300 controllers, the signal states of the retentive memory bits, timers and counters are retained, in S7-400 controllers they are not.

The contents of retentive data blocks also remain unchanged in S7-300 controllers. Please note that on the S7-300, the contents of the retentive data areas are stored in the CPU and not on the memory card.

The remaining data blocks in an S7-300 controller and all data blocks in an S7-400 controller are copied from the memory card to work memory, as are the code blocks. The only data blocks whose contents are retained are those on the memory card. Data blocks generated with system function SFC 22 CREAT_DB are not retentive. After restart, the data blocks have the contents that are on the memory card, that is, the contents with which they were programmed.

22.2.4 Restart Parameterization

On the "Restart" tab of the CPUs, you can affect a restart with the following settings:

▷ Restart when the set configuration is not the same as the actual configuration
A restart is executed even if the parameterized hardware configuration does not agree with the actual configuration.

▷ Hardware test at complete restart (warm restart)
The S7-300 CPUs execute a hardware test on power up.

▷ Delete PIQ on warm restart
The S7-400 CPUs delete all process-image output tables at warm restart

▷ Disable warm restart at manual restart
Manual warm restart not permissible

▷ Restart following POWER UP
Definition of the type of restart following power up

▷ Monitoring time for ready signal of the modules
If the monitoring time for a module times out, the CPU remains at STOP. The result is

entered in the diagnostics buffer (important for switching on the power on expansion racks).

▷ Monitoring time for transferring the parameters to the modules
If the monitoring time runs out, the CPU remains at STOP. The event is entered in the diagnostics buffer. (In the event of this error, you can only parameterize the CPU with a higher monitoring time – without memory reset - if you transfer the system data of an "empty" project in which the new value of the monitoring time is entered, so that the module parameterization is completed within the "old" monitoring time.)

▷ Monitoring time for warm restart
If the time between power off and power on or the time between STOP and RUN is greater than the monitoring time, the CPU remains at STOP. The specification 0 ms switches the monitor off.

22.3 Types of Restart

22.3.1 START-UP Mode

The CPU executes a restart in the following cases:

▷ When the mains power is switched on

▷ When the mode selector is set from STOP to RUN or RUN-P

▷ On a request from a communication function (initiated from a programming device or via communication function blocks from another CPU)

A *manual* restart is initiated via the keyswitch or a communication function, an *automatic* restart by switching on the mains power.

The restart routine may be as long as required, and there is no time limit on its execution; the scan cycle monitor is not active.

During the execution of the restart routine, no interrupts will be serviced. Exceptions are errors that are handled as in RUN (call of the relevant error organization blocks).

In the restart routine, the CPU updates the timers, the run-time meters and the real-time clock.

During restart, the output modules are disabled, i.e., output signals cannot be transmitted. The output disable is only revoked at the end of the restart and prior to starting the cyclic program.

A restart routine can be aborted, for instance when the mode selector is actuated or when there is a power failure. The aborted restart routine is then executed from the beginning when the power is switched on. If a complete restart is aborted, it must be executed again. If a warm restart is aborted, all restart types are possible.

Figure 22.2 shows the activities carried out by the CPU during a restart.

22.3.2 Cold Restart

On a cold restart, the CPU sets both itself and the modules to the programmed initial state, deletes all data in the system memory (including the retentive data), calls OB 102, and then executes the main program in OB 1 from the beginning.

The current program and the current data in work memory are deleted and with them also the data blocks generated per SFC; the program from load memory is reloaded. (In contrast to memory reset, a RAM load memory is not deleted.)

Manual Cold Restart

A manual cold restart is initiated in the following instances:

▷ Via the mode selector on the CPU if the switch was held in the MRES position for at least 3 seconds on the transition from STOP to RUN or RUN-P.

▷ Via a communication function from a PG or with an SFB from another CPU; the mode selector must be in the RUN or RUN-P position.

A manual cold restart can always be initiated unless the CPU requests a memory reset.

Automatic cold restart

An automatic cold restart is initiated by switching on the mains power. The cold restart is executed if

▷ the CPU was not at STOP when the power was switched off

▷ the mode selector is at RUN or RUN-P

293

22 Restart Characteristics

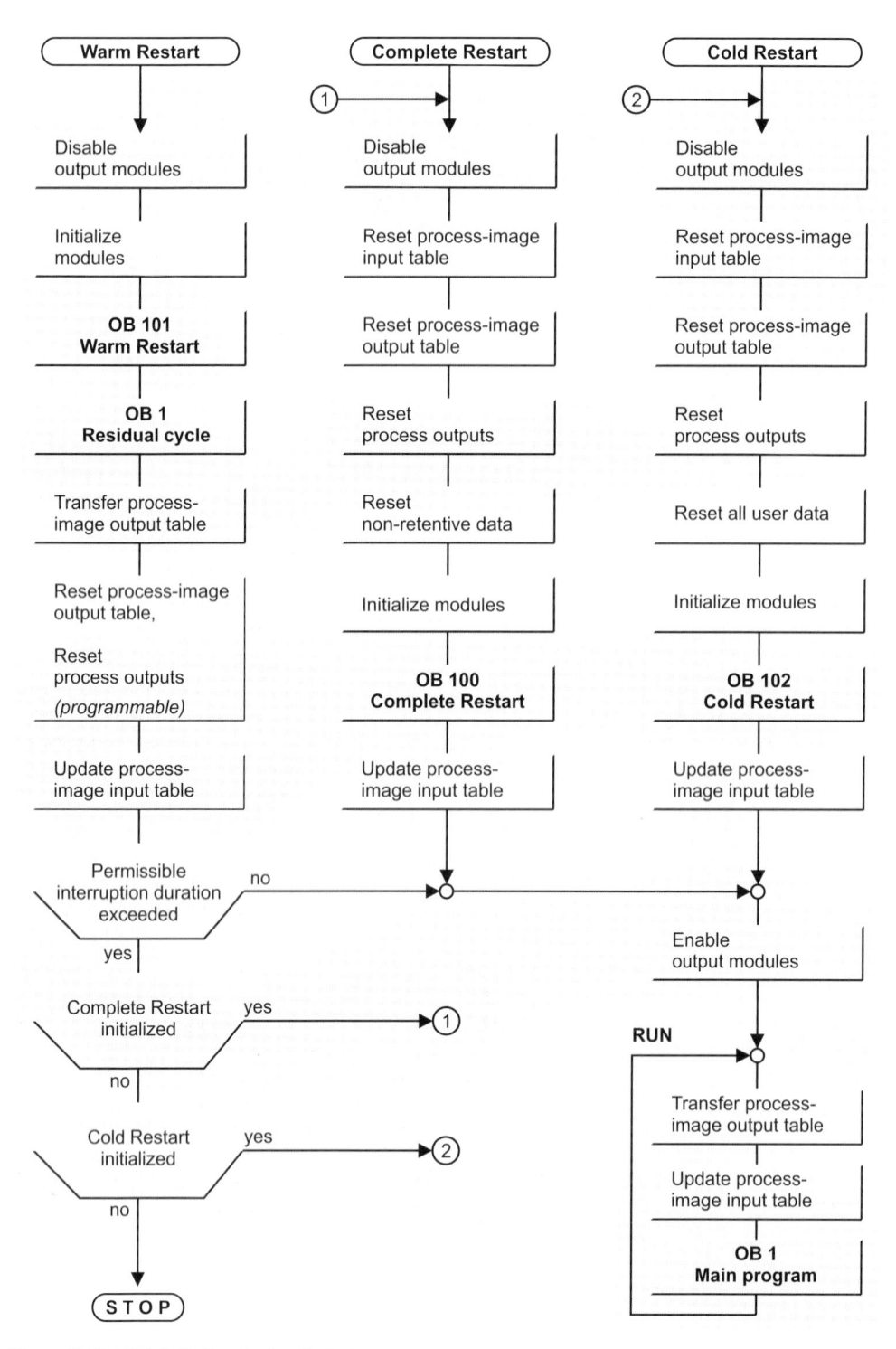

Figure 22.2 CPU Activities During Restart

▷ the CPU was interrupted by a power outage while executing a cold restart

▷ "Automatic cold restart on power up" is parameterized

When operated without a backup battery, the CPU executes an automatic non-retentive complete restart. The CPU starts the memory reset automatically, then copies the user program from the memory card to work memory. The memory card must be a Flash EPROM.

22.3.3 Complete Restart

On a complete restart, the CPU sets both itself and the modules to the programmed initial state, erases the non-retentive data in the system memory, calls OB 100, and then executes the main program in OB 1 from the beginning.

Complete manual restart

A complete manual restart is initiated in the following instances:

▷ Via the mode selector on the CPU on a transition from STOP to RUN or RUN-P (on S7-400 CPUs with restart type switch, this is in the CRST position.

▷ Via a communication function from a PG or with an SFB from another CPU; the mode selector must be in the RUN or RUN-P position.

A complete manual restart can always be initiated unless the CPU requests a memory reset.

Complete automatic restart

An automatic complete restart is initiated by switching on the mains power. The restart is executed if

▷ the CPU was not at STOP when the power was switched off

▷ the mode selector is at RUN or RUN-P

▷ the CPU was interrupted by a power outage while executing a complete restart

▷ "Automatic complete restart on power up" is parameterized

If there is a restart type switch, it remains without effect in the case of automatic complete restart.

When operated without a backup battery, the CPU executes an automatic non-retentive complete restart. The CPU starts the memory reset automatically, then copies the user program from the memory card to work memory. The memory card must be a Flash EPROM.

22.3.4 Warm Restart

A warm restart is possible only on an S7-400.

On a STOP or power outage, the CPU saves all interrupts as well as the internal CPU registers that are important to the processing of the user program. On a warm restart, it can therefore resume at the location in the program at which the interruption occurred. This may be the main program, or it may be an interrupt or error handling routine. All ("old") interrupts are saved and will be serviced.

The so-called "residual cycle", which extends from the point at which the CPU resumes the program following a warm restart to the end of the main program, counts as part of the restart. No (new) interrupts are serviced. The output modules are disabled, and are in their initial state.

A warm restart is permitted only when there have been no changes in the user program while the CPU was at STOP, such as modification of a block.

By parameterizing the CPU accordingly, you can specify how long the interruption may be for the CPU to still be able to execute a warm restart (from 100 milliseconds to 1 hour). If the interruption is longer, only a complete restart is allowed. The length of the interruption is the amount of time between exiting of the RUN mode (STOP or power-down) and reentry into the RUN mode (following execution of OB 101 and the residual cycle).

Manual warm restart

A manual warm restart is initiated

▷ If the mode selector was at RUN or RUN-P when the CPU was switched on By moving

22 Restart Characteristics

the mode selector from STOP to RUN or RUN-P when the restart switch is at WRST (only possible on CPUs with restart type switch)

▷ Via a communication function from a PG or with an SFB from another CPU; the mode selector must be at RUN or RUN-P.

A manual warm restart is possible only when the warm restart disable was revoked in the "Restart" tab when the CPU was parameterized. The cause of the STOP must have been a manual activity, either via the mode selector or through a communication function; only then can a manual warm restart be executed while the CPU is at STOP.

Automatic warm restart

An automatic warm restart is initiated by switching on the mains power. The CPU executes an automatic warm restart only in the following instances:

▷ If it was not at STOP when switched off

▷ If the mode selector was at RUN or RUN-P when the CPU was switched on

▷ If "Automatic warm restart on power up" is parameterized

▷ If the backup battery is inserted and in working order

The position of the restart switch is irrelevant to an automatic warm restart.

22.4 Ascertaining a Module Address

You can ascertain module addresses with the following SFCs:

▷ SFC 5 GADR_LGC
Ascertain logical address of a module channel

▷ SFC 50 RD_LGADR
Ascertain all logical addresses of a module

▷ SFC 49 LGC_GADR
Ascertain slot address of a module

Table 22.2 shows the parameters for these SFCs.

The SFCs have IOID and LADDR as common parameters for the logical address (= address in the I/O area). IOID is either B#16#54, which stands for the peripheral inputs (PIs) or B#16#55, which stands for the peripheral outputs (PQs). LADDR contains an I/O address in the PI or PQ area which corresponds to the specified channel. If the channel is 0, it is the module start address.

The hardware configuration data must specify an allocation between logical address (module start address) and slot address (location of the module in a rack or a station for distributed I/O) for the addresses ascertained with these SFCs.

SFC 5 GADR_LGC
Ascertain the logical address of a module channel

System function SFC 5 GADR_LGC returns the logical address of a channel when you specify the slot address ("geographic" address). Enter the number of the subnet in the SUBNETID parameter if the module belongs to the distributed I/O or B#16#00 if the module is plugged into a rack. The RACK parameter specifies the number of the rack or, in the case of distributed I/O, the number of the station. If the module has no submodule slot, enter B#16#00 in the SUBSLOT parameter. SUBADDR contains the address offset in the module's user data (W#16#0000, for example, stands for the module start address).

SFC 49 LGC_GADR
Ascertain the slot address of a module

SFC 49 LGC_GADR returns the slot address of a module when you specify an arbitrary logical module address. Subtracting the address offset (parameter SUBADDR) from the specified user data address gives you the module starting address. The value in the AREA parameter specifies the system in which the module is operated (Table 22.3).

SFC 50 RD_LGADR
Ascertain all logical addresses for a module

On the S7-400, you can assign addresses for a module's user data bytes which are not contiguous (under development).

296

22.4 Ascertaining a Module Address

Table 22.2 Parameters for the SFCs Used to Ascertain the Module Address

SFC	Parameter	Declaration	Data Type	Contents, Description
5	SUBNETID	INPUT	BYTE	Area identifier
	RACK	INPUT	WORD	Number of the rack
	SLOT	INPUT	WORD	Number of the slot
	SUBSLOT	INPUT	BYTE	Number of the submodule
	SUBADDR	INPUT	WORD	Offset in the module's user data address area
	RET_VAL	OUTPUT	INT	Error information
	IOID	OUTPUT	BYTE	Area identifier
	LADDR	OUTPUT	WORD	Logical address of the channel
50	IOID	INPUT	BYTE	Area identifier
	LADDR	INPUT	WORD	A logical module address
	RET_VAL	OUTPUT	INT	Error information
	PEADDR	OUTPUT	ANY	WORD field for the PI addresses
	PECOUNT	OUTPUT	INT	Number of PI addresses returned
	PAADDR	OUTPUT	ANY	WORD field for the PQ addresses
	PACOUNT	OUTPUT	INT	Number of PQ addresses returned
49	IOID	INPUT	BYTE	Area identifier
	LADDR	INPUT	WORD	A logical module address
	RET_VAL	OUTPUT	INT	Error information
	AREA	OUTPUT	BYTE	Area identifier
	RACK	OUTPUT	WORD	Number of the rack
	SLOT	OUTPUT	WORD	Number of the slot
	SUBADDR	OUTPUT	WORD	Offset in the module's user data address area

SFC 50 RD_LGADR returns all logical addresses for a module when you specify an arbitrary address from the user data area.

Use the PEADDR and PAADDR parameters to define an area of WORD components (a word-based ANY pointer, for example P#DBzD-BXy..x WORD nnn).

SFC 50 then shows you the number of entries returned in these areas in the RECOUNT and PACOUNT parameters.

Table 22.3 Description of SFC 49 LGC_GADR's Output Parameters

AREA	System	Meaning of RACK, SLOT and SUBADDR
0	S7-400	RACK = Number of the rack
1	S7-300	SLOT = Number of the slot SUBADDR = Address offset to the start address
2	Distributed I/O	RACK, SLOT and SUBADDR irrelevant
3	S5 P area	RACK = Number of the rack
4	S5 Q area	SLOT = Slot number of the adapter casing SUBADDR = Address in the S5 area
5	S5 IM3 area	
6	S5 IM4 area	

22.5 Parameterizing Modules

The following system functions are available for parameterizing modules:

▷ SFC 54 RD_DPARM
  Read predefined parameters

▷ SFC 55 WR_PARM
  Write dynamic parameters

▷ SFC 56 WR_DPARM
  Write predefined parameters

▷ SFC 57 PARM_MOD
  Parameterize module

▷ SFC 58 WR_REC
  Write data record

▷ SFC 59 RD_REC
  Read data record

The parameters for these and a number of other system functions are listed in Table 22.4.

The following data records can be transferred with the aforementioned system functions:

Record No.	Contents for Read	Contents for Write
0	Diagnostic data	Parameters
1	Diagnostic data	Parameters
2 to 127	User data	User data
128 to 255	Diagnostic data	Parameters

General remarks on parameterizing modules

Some S7 modules can be parameterized, that is to say, values may be set on the module which deviate from the default. To specify parameters, open the module in the hardware configuration and fill in the tabs in the dialog box. When you transfer the *System Data* object in the *Blocks* container to the PLC, you are also transferring the module parameters.

The CPU transfers the module parameters to the module automatically in the following cases

▷ On restart

▷ When a module has been plugged into a configured slot (S7-400)

▷ Following the "return" of a rack or a distributed I/O station.

The module parameters are subdivided into static parameters and dynamic parameters. You can set both parameter types offline in the Hardware Configuration. You can also modify the dynamic parameters at runtime using SFC calls. In the restart routine, the parameters set on the modules using SFCs are overwritten by the parameters set (and stored on the CPU) via the Hardware Configuration.

The parameters for the signal modules are in two data records: the static parameters in data record 0 and the dynamic parameters in data record 1. You can transfer both data records to the module with SFC 57 PARM_MOD, data record 0 or 1 with SFC 56 WR_DPARM, and only data record 1 with SFC 55 WR_PARM. The data records must be in the system data blocks on the CPU.

After parameterization of an S7-400 module, the specified values do not go into force until bit 2 ("Operating mode") in byte 2 of diagnostic data record 0 has assumed the value "RUN" (can be read with SFC 59 RD_REC).

As far as addressing for data transfer is concerned, use the *lowest* module start address (LADDR parameter) together with the identifier indicating whether you have defined this address as input or output (IOID parameter). If you assigned the same start address to both the input and output area, use the identifier for input. Use the I/O identifier regardless of whether you want to execute a Read or a Write operation.

Use the RECORD parameter with the data type ANY to define an area of BYTE components. This may be a variable of type ARRAY, STRUCT or UDT, or an ANY pointer of type BYTE (for example P#DBzDBX$y.x$ BYTE nnn). If you use a variable, it must be a "complete" variable; individual array or structure components are not permissible.

22.5 Parameterizing Modules

Table 22.4 Parameters for System Functions Used for Data Transfer

Parameter for SFC					Parameter	Declaration	Data Type	Contents, Description
55	56	57	58	59	REQ	INPUT	BOOL	"1" = Write request
55	56	57	58	59	IOID	INPUT	BYTE	B#16#54 = Peripheral inputs (PIs) B#16#55 = Peripheral outputs (PQs)
55	56	57	58	59	LADDR	INPUT	WORD	Module start address
55	56	-	58	59	RECNUM	INPUT	BYTE	Data record number
55	-	-	58	-	RECORD	INPUT	ANY	Data record
55	56	57	58	59	RET_VAL	OUTPUT	INT	Error information
55	56	57	58	59	BUSY	OUTPUT	BOOL	Transfer still in progress if "1"
-	-	-	-	59	RECORD	OUTPUT	ANY	Data record

SFC 54 RD_DPARM
Reading predefined parameters

System function SFC 54 RD_DPARM transfers the data record with the number specified in the RECNUM parameter from the relevant SDB system data block to the destination area specified at the RECORD parameter.

You can now, for example, modify this data record and write it to the module with SFC 58 WR_REC.

SFC 55 WR_PARM
Writing dynamic parameters

System function SFC 55 WR_PARM transfers the data record addressed by RECORD to the module specified by the IOID and LADDR parameters. Specify the number of the data record in the RECNUM parameter. The prerequisite is that the data record be in the correct SDB system data block and that it contain only dynamic parameters.

When the job is initiated, the SFC reads the entire data record; the transfer may be distributed over several program scan cycles. The BUSY parameter is "1" during the transfer.

SFC 56 WR_DPARM
Writing predefined parameters

System function SFC 56 WR_DPARM transfers the data record with the number specified in the RECNUM parameter from the relevant SDB system data block to the module identified by the IOID and LADDR parameters.

The transfer may be distributed over several program scan cycles; the BUSY parameter is "1" during the transfer.

SFC 57 PARM_MOD
Parameterizing a module

System function SFC 57 PARM_MOD transfers all the data records programmed when the module was parameterized via the Hardware Configuration.

The transfer may be distributed over several program scan cycles; the BUSY parameter is "1" during the transfer.

SFC 58 WR_REC
Writing a data record

SFC 58 WR_REC transfers the data record addressed by the RECORD parameter and the number RECNUM to the module defined by the IOID and LADDR parameters. A "1" in the REQ parameter starts the transfer. When the job is initiated, the SFC reads the complete data record.

The transfer may be distributed over several program cycles; the BUSY parameter is "1" during the transfer.

SFC 59 RD_REC
Reading a data record

When the REQ parameter is "1", SFC 59 RD_REC reads the data record addressed by the RECNUM parameter from the module and places it in destination area RECORD. The destination area must be longer than or at least as long as the data record. If the transfer is completed without error, the RET_VAL parameter contains the number of bytes transferred.

The transfer may be distributed over several program scan cycles; the BUSY parameter is "1" during the transfer.

S7-300s delivered prior to February 1997: the SFC reads as much data from the specified data record as the destination area can accommodate. The size of the destination area may not exceed that of the data record.

23 Error Handling

The CPU reports errors or faults detected by the modules or by the CPU itself in different ways:

- Errors in arithmetic operations (overflow, invalid REAL number) by setting status bits (status bit OV, for example, for a numerical overflow)
- Errors detected while executing the user program (synchronous errors) by calling organization blocks OB 121 and OB 122
- Errors in the programmable controller which do not relate to program scanning (asynchronous errors) by calling organization blocks OB 80 to OB 87.

The CPU signals the occurrence of an error or fault, and in some cases the cause, by setting error LEDs on the front panel. In the case of unrecoverable errors (such as invalid OP code), the CPU goes directly to STOP.

With the CPU in STOP mode, you can use a programming device and the CPU information functions to read out the contents of the block stack (B stack), the interrupt stack (I stack) and the local data stack (L stack) and then draw conclusions as to the cause of error.

The system diagnostics can detect errors/faults on the modules, and enters these errors in a diagnostic buffer. Information on CPU mode transitions (such as the reasons for a STOP) are also placed in the diagnostic buffer.

The contents of this buffer are retained on STOP, on a memory reset, and on power failure, and can be read out following power recovery and execution of a start-up routine using a programming device.

On the new CPUs, you can use CPU parameterization to set the number of entries the diagnostics buffer is to hold.

23.1 Synchronous Errors

The CPU's operating system generates a synchronous error when an error occurs in immediate conjunction with program scanning. If a synchronous error OB has not been programmed, the CPU goes to STOP. A distinction is made between two error types:

- **Programming error**, OB 121 is called and
- **Access error**, OB 122 is called.

Table 23.1 shows the start information for both synchronous error organization blocks.

A synchronous error OB has the same priority as the block in which the error occurred. It is for this reason that it is possible to access the registers of the interrupted block in the synchronous error OB, and it is also for this reason that the program in the synchronous error OB (under certain circumstances with modified content) can return the registers to the interrupted block.

Note that when a synchronous error OB is called, its 20 bytes of start information are also pushed onto the L stack for the priority class that caused the error, as are the other temporary local data for the synchronous error OB and for all blocks called in this OB.

In the case of S7-400, another synchronous error OB can be called in an error OB. The block nesting depth for a synchronous error OB is 3 for S7-400 CPUs and 4 for S7-300 CPUs.

You can disable and enable a synchronous error OB call with system functions SFC 36 MSK_FLT and SFC 37 DMSK_FLT.

23 Error Handling

Table 23.1 Start Information for the Synchronous Error OBs

Variable Name	Data Type	Description, Contents
OB12x_EV_CLASS	BYTE	B#16#25 = Call programming error OB 121 B#16#29 = Call access error OB 122
OB12x_SW_FLT	BYTE	Error code (see Section 23.2.1 "Error Filters")
OB12x_PRIORITY	BYTE	Priority class in which the error occurred
OB12x_OB_NUMBR	BYTE	OB number (B#16#79 or. B#16#80)
OB12x_BLK_TYPE	BYTE	Type of block interrupted (S7-400 only) OB: B#16#88, FB: B#16#8E, FC: B#16#8C

OB121_RESERVED_1
OB122_MEM_AREA
	BYTE	Byte assignments (B#15#xy):
		
		7... (x) ...4 3 ... (y) ...0
		
		1 Bit access
		   0 I/O area PI or PQ
		   1 Process-image input table I
		2 Byte access
		   2 Process-image output table Q
		   3 Memory bits M
		3 Word access
		   4 Global data block DB
		   5 Instance data block DI
		4 Doubleword access
		   6 Temporary local data L
		   7 Temporary local data of the predecessor block V

OB121_FLT_REG OB122_MEM_ADDR	WORD	OB 121: Error source Errored address (at read/write access) Errored area (in the case of area error) Incorrect number of the block, timer/counter function OB 122: Address at which the error occurred
OB12x_BLK_NUM	WORD	Number of the block in which the error occurred (S7-400 only)
OB12x_PRG_ADDR	WORD	Error address in the block that caused the error (S7-400 only)
OB12x_DATE_TIME	DT	Time at which programming error was detected

23.2 Synchronous Error Handling

The following system functions are provided for handling synchronous errors:

▷ SFC 36 MSK_FLT
 Mask synchronous errors (disable OB call)

▷ SFC 37 DMSK_FLT
 Unmask synchronous error (re-enable OB call)

▷ SFC 38 READ_ERR
 Read error register

The operating system enters the synchronous error in the diagnostic buffer without regard to the use of system functions SFC 36 to SFC 38.

The parameters for these system functions are listed in Table 23.2.

23.2.1 Error Filters

The error filters are used to control the system functions for synchronous error handling. In the programming error filter, one bit stands for each programming error detected; in the access error filter, one bit stands for each access error detected. When you define an error filter, you set the bit that stands for the synchronous error you want to mask, unmask or query. The error filters returned by the system functions show a "1" for synchronous errors that are still masked or which have occurred.

23.2 Synchronous Error Handling

Table 23.2 SFC Parameters for Synchronous Error Handling

SFC	Parameter Name	Declaration	Data Type	Contents, Description
36	PRGFLT_SET_MASK	INPUT	DWORD	New (additional) programming error filter
	ACCFLT_SET_MASK	INPUT	DWORD	New (additional) access error filter
	RET_VAL	OUTPUT	INT	W#16#0001 = The new filter overlaps the existing filter
	PRGFLT_MASKED	OUTPUT	DWORD	Complete programming error filter
	ACCFLT_MASKED	OUTPUT	DWORD	Complete access error filter
37	PRGFLT_RESET_MASK	INPUT	DWORD	Programming error filter to be reset
	ACCFLT_RESET_MASK	INPUT	DWORD	Access error filter to be reset
	RET_VAL	OUTPUT	INT	W#16#0001 = The new filter contains bits that are not set (in the current filter)
	PRGFLT_MASKED	OUTPUT	DWORD	Remaining programming error filter
	ACCFLT_MASKED	OUTPUT	DWORD	Remaining access error filter
38	PRGFLT_QUERY	INPUT	DWORD	Programming error filter to be queried
	ACCFLT_QUERY	INPUT	DWORD	Access error filter to be queried
	RET_VAL	OUTPUT	INT	W#16#0001 = The query filter contains bits that are not set (in the current filter)
	PRGFLT_CLR	OUTPUT	DWORD	Programming error filter with error messages
	ACCFLT_CLR	OUTPUT	DWORD	Access error filter with error messages

The access error filter is shown in Table 23.3; the Error Code column shows the contents of variable OB122_SW_FLT in the start information for OB 122.

The programming error filter is shown in Table 23.4, the Error Code column shows the contents of variable OB121_SW_FLT in the start information for OB 121.

Table 23.3 Access Error Filter

Bit	Error Code	Contents
3	B#16#42	I/O access error on read S7-300: Module does not exist or does not acknowledge S7-400: An existing module does not acknowledge after first access operation (time-out)
4	B#16#43	I/O access error on write S7-300: Module does not exist or does not acknowledge S7-400: An existing module does not acknowledge after first access operation (time-out)
5	B#16#44	S7-400 only: I/O access error on attempt to write to non-existent module (PZF) or on repeated access to modules which do not acknowledge
6	B#16#45	S7-400 only: I/O access error on attempt to write to non-existent module (PZF) or on repeated access to modules which do not acknowledge

23 Error Handling

Table 23.4 Programming Error Filter

Bit	Error Code	Contents
1	B#16#21	BCD conversion error (pseudo-tetrad detected during conversion)
2	B#16#22	Area length error on read (address not within area limits)
3	B#16#23	Area length error on write (address not within area limits)
4	B#16#24	Area length error on read (wrong area in area pointer)
5	B#16#25	Area length error on write (wrong area in area pointer)
6	B#16#26	Invalid timer number
7	B#16#27	Invalid counter number
8	B#16#28	Address error on read (bit address <>0 in conjunction with byte, word or double-word access and indirect addressing)
9	B#16#29	Address area on write (bit address <>0 in conjunction with byte, word or double-word access and indirect addressing)
16	B#16#30	Write error, global data block (write-protected block)
17	B#16#31	Write error, instance data block (write-protected block)
18	B#16#32	Invalid number of a global data block (DB register)
19	B#16#33	Invalid number of an instance data block (DI register)
20	B#16#34	Invalid number of a function (FC)
21	B#16#35	Invalid number of a function block (FB)
26	B#16#3A	Called data block (DB) does not exist
28	B#16#3C	Called function (FC) does not exist
30	B#16#3E	Called function block (FB) does not exist

The S7-400 CPUs distinguish between two types of access error: access to a non-existent module and invalid access attempt to an existing module. If a module fails during operation, that module is marked "non-existent" approximately 150 s after an access attempt, and an I/O access error (PZF) is reported on every subsequent attempt to access the module. The CPU also reports an I/O access error when an attempt is made to access a non-existent module, regardless of whether the attempt was direct (via the I/O area) or indirect (via the process image).

The error filter bits not listed in the tables are not relevant to the handling of synchronous errors.

23.2.2 Masking Synchronous Errors

System function **SFC 36 MSK_FLT** disables synchronous error OB calls via the error filters. A "1" in the error filters indicates the synchronous errors for which the OBs are not to be called (the synchronous errors are "masked").

The masking of synchronous errors in the error filters is in addition to the masking stored in the operating system's memory. SFC 36 returns a function value indicating whether a (stored) masking already exists on at least one bit for the masking specified at the input parameters (W#16#0001).

SFC 36 returns a "1" in the output parameters for all currently masked errors.

If a masked synchronous error event occurs, the respective OB is not called and the error is entered in the error register. The Disable applies to the current priority class (priority level). For example, if you were to disable a synchronous error OB call in the main program, the synchronous error OB would still be called if the error were to occur in an interrupt service routine.

23.2.3 Unmasking Synchronous Errors

System function **SFC 37 DMSK_FLT** enables the synchronous error OB calls via the error filters. You enter a "1" in the filters to indicate the synchronous errors for which the OBs are once again to be called (the synchronous errors are "unmasked"). The entries corresponding to the specified bits are deleted in the error register. SFC 37 returns W#16#0001 as function value if no (stored) masking already exists on at least one bit for the unmasking specified at the input parameters.

SFC 37 returns a "1" in the output parameters for all currently masked errors.

If an unmasked synchronous error occurs, the respective OB is called and the event entered in the error register. The Enable applies to the current priority class (priority level).

23.2.4 Reading the Error Register

System function **SFC 38 READ_ERR** reads the error register. You must enter a "1" in the error filters to indicate the synchronous errors whose entries you want to read. SFC 38 returns W#16#0001 as function value when the selection specified in the input parameters included at least one bit for which no (stored) masking exists.

SFC 38 returns a "1" in the output parameters for the selected errors when these errors occurred, and deletes these errors in the error register when they are queried. The synchronous errors that are reported are those in the current priority class (priority level).

23.2.5 Entering a Substitute Value

SFC 44 REPL_VAL allows you to enter a substitute value in accumulator 1 from within a synchronous error OB. Use SFC 44 when you can no longer read any values from a module (for instance when a module is defective). When you program SFC 44, OB 122 ("access error") is called every time an attempt is made to access the module in question. When you call SFC 44, you can load a substitute value into the accumulator; the program scan is then resumed with the substitute value. Table 23.5 lists the parameters for SFC 44.

You may call SFC 44 in only one synchronous error OB (OB 121 or OB 122).

23.3 Asynchronous Errors

Asynchronous errors are errors which can occur independently of the program scan. When an asynchronous error occurs, the operating system calls one of the organization blocks listed below:

OB 80   Timing error

OB 81   Power supply error

OB 82   Diagnostic interrupt

OB 83   Insert/remove module interrupt

OB 84   CPU hardware fault

OB 85   Program execution error

OB 86   Rack failure

OB 87   Communication error

The OB 82 call (diagnostic interrupt) is described in detail in section 23.4, "System Diagnostics".

On the S7-400H, there are two additional asynchronous error OBs:

OB 70 I/O redundancy errors

OB 72 CPU redundancy errors

The call of these asynchronous error organization blocks can be disabled and enabled with system functions SFC 39 DIS_IRT and SFC 40 EN_IRT, and delayed and enabled with system functions SFC 41 DIS_AIRT and SFC 42 EN_AIRT.

Table 23.5 Parameters for SFC 44 REPL_VAL

SFC	Parameter Name	Declaration	Data Type	Contents, Description
44	VAL	INPUT	DWORD	Substitute value
	RET_VAL	OUTPUT	INT	Error information

23 Error Handling

Timing errors

The operating system calls organization block OB 80 when one of the following errors occurs:

▷ Cycle monitoring time exceeded

▷ OB request error (the requested OB is still executing or an OB was requested too frequently within a given priority class)

▷ Time-of-day interrupt error (TOD interrupt time past because clock was set forward or after transition to RUN)

If no OB 80 is available and a timing error occurs, the CPU goes to STOP. The CPU also goes to STOP if the OB is called a second time in the same program scan cycle.

Power supply errors

The operating system calls organization block OB 81 if one of the following errors occurs:

▷ At least one backup battery in the central controller or in an expansion unit is empty

▷ No battery voltage in the central controller or in an expansion unit

▷ 24 V supply failed in central controller or in an expansion unit

OB 81 is called for incoming and outgoing events. If there is no OB 81, the CPU continues functioning when a power supply error occurs.

Insert/remove module interrupt

The operating system monitors the module configuration once per second. An entry is made in the diagnostic buffer and in the system status list each time a module is inserted or removed in RUN, STOP or START-UP mode.

In addition, the operating system calls organization block OB 83 if the CPU is in RUN mode. If there is no OB 83, the CPU goes to STOP on an insert/remove module interrupt.

As much as a second can pass before the insert/remove module interrupt is generated. As a result, it is possible that an access error or an error relating to the updating of the process image could be reported in the interim between removal of a module and generation of the interrupt.

If a suitable module is inserted into a configured slot, the CPU automatically parameterizes that module, using data records already stored on that CPU. Only then is OB 83 called in order to signal that the connected module is ready for operation.

CPU hardware faults

The operating system calls organization block OB 84 when an interface error (MPI network, PROFIBUS DP) occurs or disappears. If there is no OB 84, the CPU goes to STOP on a CPU hardware fault.

Program execution errors

The operating system calls organization block OB 85 when one of the following errors occurs:

▷ Start request for an organization block which has not been loaded

▷ Error occurred while the operating system was accessing a block (for instance no instance data block when a system function block (SFB) was called)

▷ I/O access error while executing (automatic) updating of the process image on the system side

On the S7-400 CPUs and the CPU 318, OB 85 is called at every I/O access error (on the system side), i.e. when updating the process image in each cycle. The substitute value or zero is then entered in the relevant byte in the process-image input table at every update.

On the S7-300 CPUs (with the exception of the CPU 318), OB 85 is not called in the event of an I/O access error during automatic updating of the process image. At the first errored access, the substitute value or zero is entered in the relevant byte; it is then no longer updated.

With appropriately equipped CPUs, you can use CPU parameterization to influence the call mode of OB 85 in the event of an I/O access error on the system side:

▷ OB 85 is called every time. The affected input byte is overwritten with the substitute value or with zero each time.

▷ OB 85 is called in the event of the first error with the attribute "incoming". An affected input byte is only overwritten with the substitute value or with zero the first time; following this it is no longer updated. If the error is then corrected, OB 85 is called with the attribute "outgoing"; following this, it is updated "normally".

▷ OB 85 is not called in the event of an access error. Affected input bytes are overwritten once with the substitute value or zero, and then no longer updated.

If there is no OB 85, the CPU goes to STOP on a program execution error.

Rack failure

The operating system calls organization block OB 86 if it detects the failure of a rack (power failure, line break, defective IM), a subnet, or a distributed I/O station. OB 86 is called for both incoming and leaving errors.

In multiprocessor mode, OB 86 is called in all CPUs if a rack fails.

If there is no OB 86, the CPU goes to STOP if a rack failure occurs.

Communication error

The operating system calls organization block OB 87 when a communication error occurs. Some examples of communication errors are

▷ Invalid frame identification or frame length detected during global data communication

▷ Sending of diagnostic entries not possible

▷ Clock synchronization error

▷ GD status cannot be entered in a data block

If there is no OB 87, the CPU goes to STOP when a communication error occurs.

I/O redundancy error

The operating system of an H CPU calls organization block OB 70 if a redundancy loss occurs on PROFIBUS-DP, e.g. in the event of a bus failure on the active DP master or in the event of a fault in the interface of a DP slave.

If OB 70 does not exist, the CPU continues to operate in the event of an I/O redundancy error.

CPU redundancy error

The operating system of an H CPU calls organization block OB 72 if one of the following events occurs:

▷ Redundancy loss of the CPU

▷ Comparison error (e.g. in RAM, in the PIQ)

▷ Standby-master changeover

▷ Synchronization error

▷ Error in a SYNC submodule

▷ Update abort

If OB 72 does not exist, the CPU continues to operate in the event of a CPU redundancy error.

23.4 System Diagnostics

23.4.1 Diagnostic Events and Diagnostic Buffer

System diagnostics is the detection, evaluation and reporting of errors occurring in programmable controllers. Examples are errors in the user program, module failures or wirebreaks on signaling modules. These *diagnostic events* may be:

▷ Diagnostic interrupts from modules with this capability

▷ System errors and CPU mode transitions

▷ User messages via system functions.

Modules with diagnostic capabilities distinguish between programmable and non-programmable diagnostic events. Programmable diagnostic events are reported only when you have set the parameters necessary to enable diagnostics. Non-programmable diagnostic events are always reported, regardless of whether or not diagnostics have been enabled. In the event of a reportable diagnostic event,

▷ The fault LED on the CPU goes on

▷ The diagnostic event is passed on to the CPU's operating system

▷ A diagnostic interrupt is generated if you have set the parameters enabling such interrupts (by default, diagnostic interrupts are disabled).

All diagnostic events reported to the CPU operating system are entered in a *diagnostic buffer* in the order in which they occurred, and with date and time stamp. The diagnostic buffer is a battery-backed memory area on the CPU which retains its contents even in the event of a memory reset. The diagnostic buffer is a ring buffer whose size depends on the CPU. When the diagnostic buffer is full, the oldest entry is overwritten by the newest.

You can read out the diagnostic buffer with a programming device at any time. In the CPU's *System Diagnostics* parameter block you can specify whether you want expanded diagnostic entries (all OB calls). You may also specify whether the last diagnostic entry made before the CPU goes to STOP should be sent to a specific node on the MPI bus.

23.4.2 Writing User Entries in the Diagnostic Buffer

System function **SFC 52 WR_USMSG** writes an entry in the diagnostic buffer which may be sent to all nodes on the MPI bus. Table 23.6 lists the parameters for SFC 52.

The entry in the diagnostic buffer corresponds in format to that of a system event, for instance the start information for an organization block. Within the permissible boundaries, you may choose your own event ID (EVENTN parameter) and additional information (INFO1 and INFO2 parameters).

The event ID is identical to the first two bytes of the buffer entry (Figure 23.1). Permissible for a user entry are the event classes 8 (diagnostic entries for signal modules), 9 (standard user events), A and B (arbitrary user events).

Additional information (INFO1) corresponds to bytes 7 and 8 of the buffer entry (one word) and additional information 2 (INFO2) to bytes 9 to 12 (one doubleword). The contents of both variables may be of the user's own choice.

Set SEND to "1" to send the diagnostic entry to the relevant node. Even if sending is not possible (because no node is logged in or because the Send buffer is full, for example), the entry is still made in the diagnostic buffer (when bit 9 of the event ID is set).

23.4.3 Evaluating Diagnostic Interrupts

When a diagnostic interrupt is incoming or outgoing, the operating system interrupts scanning of the user program and calls organization block OB 82. If OB 82 has not been programmed, the CPU goes to STOP on a diagnostic interrupt. You can disable or enable OB 82 with system function SFC 39 DIS_IRT or SFC 40 EN_IRT, and delay or enable it with system function SFC 41 DIS_AIRT or SFC 42 EN_AIRT.

In the first byte of the start information, B#16#39 stands for an incoming diagnostic interrupt and B#16#38 for a leaving diagnostic interrupt. The sixth byte gives the address identifier (B#16#54 stands for an input, B#16#55 for an output); the subsequent INT variable contains the address of the module that generated the diagnostic interrupt. The next four bytes contain the diagnostic information provided by that module.

You can use system function SFC 59 RD_REC (read data record) in OB 82 to obtain detailed error information. The diagnostic information are consistent until OB 82 is exited, that is, they remain "frozen". Exiting of OB 82 acknowledges the diagnostic interrupt on the module.

Table 23.6 Parameters for SFC 52 WR_USMSG

SFC	Parameter Name	Declaration	Data Type	Contents, Description
52	SEND	INPUT	BOOL	If "1": Send is enabling
	EVENTN	INPUT	WORD	Event ID
	INFO1	INPUT	ANY	Additional information 1 (one word)
	INFO2	INPUT	ANY	Additional information 2 (one doubleword)
	RET_VAL	OUTPUT	INT	Error information

Event ID

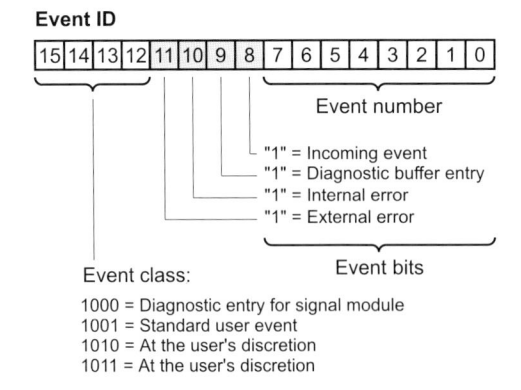

```
|15|14|13|12|11|10|9|8|7|6|5|4|3|2|1|0|
```
         Event number
      "1" = Incoming event
      "1" = Diagnostic buffer entry
      "1" = Internal error
      "1" = External error

Event class: Event bits
1000 = Diagnostic entry for signal module
1001 = Standard user event
1010 = At the user's discretion
1011 = At the user's discretion

Figure 23.1 Event ID for Diagnostic Buffer Entries

A module's diagnostic data are in data records DS 0 and DS 1. Data record DS 0 contains four bytes of diagnostic data describing the current status of the module. The contents of these four bytes are identical to the contents of bytes 8 to 11 of the OB 82 start information.

Data record DS 1 contains the four bytes from data record DS 0 and, in addition, the module-specific diagnostic data.

23.4.4 Reading the System Status List

The system status list (SZL) describes the current status of the programmable controller. Using information functions, the list can be read but not modified. you can read part of the list (that is, a sublist) with system function **SFC 51 RDSYSST**. Sublists are virtual lists, which means that they are made available by the CPU operating system only on request. The parameters for SFC 51 are listed in Table 23.7.

REQ = "1" initiates the read operation, and BUSY = "0" tells you when it has been completed. The operating system can execute several asynchronous read operations quasi simultaneously; how many depends on the CPU being used. If SFC 51 reports a lack of resources via the function value (W#16#8085), you must resubmit your read request.

The contents of parameters SZL_ID and INDEX are CPU-dependent. The SZL_HEADER parameter is of data type STRUCT, with variables LENGTHDR (data type WORD) and N_DR (WORD) as components. LENGTHDR contains the length of a data record, N_DR the number of data records read.

Use the DR parameter to specify the variable or data area in which SFC 51 is to enter the data records. For example, P#DB200.DBX0.0 WORD 256 would provide an area of 256 data words in data block DB 200, beginning with DBB 0. If the area provided is of insufficient capacity, as many data records as possible will be entered. Only complete data records are transferred. The specified area must be able to accommodate at least one data record.

Table 23.7 Parameters for SFC 51 RDSYSST

SFC	Parameter Name	Declaration	Data Type	Contents, Description
51	REQ	INPUT	BOOL	If "1": Submit request
	SZL_ID	INPUT	WORD	Sublist ID
	INDEX	INPUT	WORD	Type or number of the sublist object
	RET_VAL	OUTPUT	INT	Error information
	BUSY	OUTPUT	BOOL	If "1": Read not yet completed
	SZL_HEADER	OUTPUT	STRUCT	Length and number of data records read
	DR	OUTPUT	ANY	Field for data records read

Appendix

This section of the book contains useful supplements to the LAD and FBD programming languages, an overview of the contents of the STEP 7 Block Libraries, and a function overview of all LAD and FBD elements.

▷ You can also provide blocks with LAD/FBD program with **block protection.** For this purpose, you use the source-oriented Editor in the STL programming language.

▷ You can use a further function of STL, **indirect addressing**, to transfer data areas in the LAD and FBD programming languages; the addresses of these data areas are then not calculated until runtime. The "LAD_Book" and "FBD_Book" libraries on the accompanying diskette each contain a "Sample Message Frame" showing how to set up and transfer data areas.

▷ The standard STEP 7 package includes **block libraries** with loadable functions and function blocks and with block headers and interface descriptions for system blocks (SFCs and SFBs).

▷ A **function overview** of all LAD and FBD functions completes the book.

On the diskette accompanying the book you will find the archive libraries "LAD_Book" and "FBD_Book". You retrieve these libraries under the SIMATIC Manager with FILE → RETRIEVE. Select the archive (the diskette) from the dialog field displayed. You define the destination directory in the next dialog field. In general, libraries are located under ...\STEP7\S7LIBS; but you can choose any other directory, for example ...\STEP7\S7PROJ, which normally contains the projects. In the final dialog field "Retrieve - Options" you deactivate the option "Restore full path".

The "LAD_Book" and "FBD_BOOK" libraries each contain eight programs which are essentially illustrative examples of LAD resp. FBD representation. Two extensive examples show the programming of functions, function blocks and local instances (Conveyor Example) and the handling of data (Message Frame Example). The memory requirements are approximately 1.93 Mbytes.

To try out an example, set up a project that corresponds to your hardware configuration and copy the program, including the symbol table, from the library to the project. Now you can test the example online.

24 Supplements to Graphic Programming

Block Protection; Indirect Addressing; Message Frame Example

25 Block Libraries

Organization Blocks, System Function Blocks, IEC Function Blocks, S5-S7 Converting Blocks, TI-S7 Converting Blocks, PID Control Blocks, Communication Blocks

26 LAD Function Overview

All LAD functions

27 FBD Function Overview

All FBD functions

24 Supplements to Graphic Programming

24.1 Block Protection

The keyword KNOW_HOW_PROTECT represents block protection. You cannot view, print or modify a block with this attribute. The Editor only displays the block header and the declaration table with the block parameters. In source-oriented input, you can protect any block yourself with KNOW_HOW_PROTECT. This means that no one, including yourself, can view the compiled block (keep the source file in a safe place!).

You can enter the KNOW_HOW_PROTECT block protection with STL, and it must be source-oriented. To do so, proceed as follows:

1) Create the block under LAD or FBD in the usual way. Later, this block will be overwritten in the user program *Blocks* by the block with the keyword. If you want to retain the (original) block (strongly recommended when entering block protection), you can store the block in, for example, a (user-created) library before entering the keyword. You can also store your entire user program in this way.

2) Create a source container. If there is no object *Source Files* under the *S7 program* (on the same level as the user program *Blocks*), you must create one: Select the *S7 program* and insert the object *Source Files* with INSERT → S7 SOFTWARE → SOURCE DIRECTORY.

3) Generate an STL source from the block. Change to the Editor (via the taskbar, for example, or open any block in *Blocks* and then close it again) and select the menu item FILE → GENERATE SOURCE FILE. In the dialog form displayed, set your project, select the object *Source Files*, and assign a name for the source file under "Object Name". Confirm with "OK". The next dialog form shows you all blocks in the *Blocks* container; select the block(s) from which you want to create a source file. Confirm with "OK".

4) Open the source file (for example by double-clicking on the source file symbol in the SIMATIC Manager or with the Editor and FILE -> OPEN). You now see the ASCII source of your LAD/FBD block. If you have previously selected several blocks, these blocks will be arranged in order in the source file.

The entries for a code block are in the following order:

▷ Keyword for the block type (FUNCTION, FUNCTION_BLOCK, ORGANIZATION_BLOCK) with specification of the address. This may be followed by the block title (starting with TITLE=...) and the block comment (starting with //...).

▷ Block attributes (depending on whether you have filled in fields in the Properties page for the block and, if so, how many).

▷ Variable declaration (several sections with the keywords VAR_xxx, ..., END_VAR); depending on whether you have declared block-local variables and, if so, which ones.

▷ Program, starts with BEGIN and ends with the keyword for the end of the block (for example END_FUNCTION_BLOCK).

The entries for a data block are in the following order:

▷ Keyword for the block type (DATA_BLOCK) with specification of the address.

▷ Block attributes (depending on whether you have filled in fields in the Properties page for the block and, if so, how many).

▷ Variable declaration (starting with STRUCT and ending with END_STRUCT) or, in the case of instance data blocks, the address of the associated function block.

▷ Variable initialization, starting with BEGIN and ending with END_DATA_BLOCK.

5) You enter the keyword KNOW_HOW_PROTECT in the source file, in its own line in each case, following the block attributes and before the variable declaration. If you have created a source from several blocks, enter the keywords in all selected blocks. Finally, store the source file.

6) Compile the source file with FILE → COMPILE. The compiler creates a block with the specified block attributes (in the creation language STL in the case of code blocks; the creation language is not significant here since KNOW_HOW_PROTECT means the block can no longer be viewed or printed out). The compiled (new) block is in user program *Blocks* and replaces the (old) block with the same number.

24.2 Indirect Addressing

With the programming language STL, you have a method of accessing operands whose addresses are not calculated until runtime. This is also possible to a limited degree in LAD or FBD: You can wait until runtime to define which data areas you want to copy with SFC 20 BLKMOV.

However, first some useful information on pointers.

24.2.1 Pointers: General Remarks

For indirect addressing, you require a data format that contains the bit address as well as the byte address and, if applicable, the operand area. This data format is a *pointer*. A pointer is also used to point to an operand. There are three types of pointers:

▷ Area pointers; these are 32 bits long and contain a specific operand or its address

▷ DB pointers; these are 48 bits long and in addition to the area pointer they also contain the number of the data block

▷ ANY pointers; these are 80 bits long and, in addition to the DB pointer, they contain further information such as the data type of the operand

24.2.2 Area Pointer

The area pointer contains the operand address and, if applicable, the operand area. Without operand area, it is an area-internal pointer; if the pointer also contains an operand area, it is referred to as an area-crossing pointer.

The notation for constant representation is as follows:

P#y.x   for an area-internal pointer
         e.g. P#22.0

P#Zy.x for an area-crossing pointer
         e.g. P#M22.0

where x = bit address, y = byte address and Z = area. You specify the operand ID as the area. The contents of bit 31 differentiates the two pointer types.

Figure 24.1 shows all pointer types and their contents as provided by STEP 7.

The area pointer has, in principle, a bit address that must always be specified even with digital operands; in the case of digital operands, specify 0 as the bit address. Example: You can use area pointer P#M22.0 to address memory bit M 22.0, but also memory byte MB 22, memory word MW 22 or memory doubleword MD 22.

24.2.3 DB Pointer

In addition to the area pointer, a DB pointer also contains a data block number in the form of a positive INT number. It specifies the data block if the area pointer points to the global data or instance data area. In all other cases, the first two bytes contain zero.

The notation of the pointer is familiar to you from full addressing of data operands. Here also, the data block and the data operand are specified, separated by a period:
P#DataBlock.DataOperand

Example: P#DB 10.DBX 20.5

You can apply this pointer to a block parameter of parameter type POINTER in order to point to a data operand. The Editor uses this pointer type internally to transfer actual parameters.

24.2 Indirect Addressing

Area-internal pointer

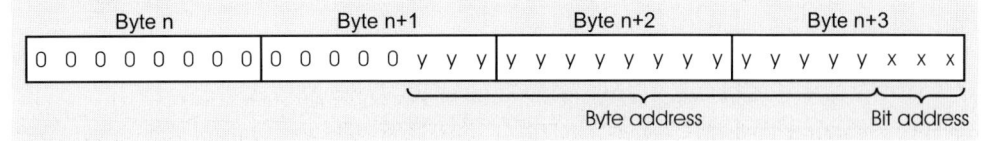

Byte n	Byte n+1	Byte n+2	Byte n+3
0 0 0 0 0 0 0 0	0 0 0 0 0 y y y	y y y y y y y y	y y y y y x x x

Byte address (n+1 low bits and n+2 and n+3 high bits) — Bit address (n+3 low 3 bits)

Area-crossing pointer

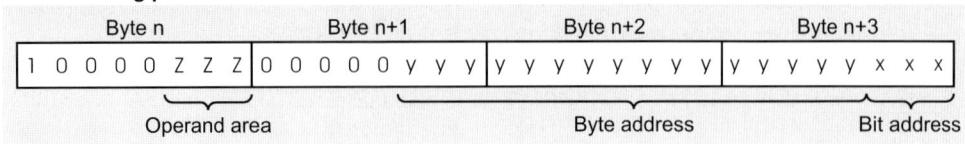

Byte n	Byte n+1	Byte n+2	Byte n+3
1 0 0 0 0 Z Z Z	0 0 0 0 0 y y y	y y y y y y y y	y y y y y x x x

Operand area — Byte address — Bit address

DB pointer

Byte n	Data block
Byte n+1	number
Byte n+2	
Byte n+3	Area
Byte n+4	pointer
Byte n+5	

ANY pointer for data types

Byte n	16#10
Byte n+1	Type
Byte n+2	Quantity
Byte n+3	
Byte n+4	Data block
Byte n+5	number
Byte n+6	
Byte n+7	Area
Byte n+8	pointer
Byte n+9	

ANY pointer for timers/counters

	16#10
	Type
	Quantity
	16#0000
	Type
	16#00
	Number

ANY pointer for blocks

	16#10
	Type
	Quantity
	16#0000
	16#0000
	Number

Address area:

0	0	0	Peripheral I/O (P)
0	0	1	Inputs (I)
0	1	0	Outputs (Q)
0	1	1	Memory bits (M)
1	0	0	Global data (DBX)
1	0	1	Instance data (DIX)
1	1	0	Temporary local data (L) [1]
1	1	1	Temporary local data of the predecessor block (V) [2]

[1] Not with area-crossing addressing
[2] Only with block parameter transfer

Type in the ANY pointer:

Elementary data types

01	BOOL
02	BYTE
03	CHAR
04	WORD
05	INT
06	DWORD
07	DINT
08	REAL
09	DATE
0A	TOD
0B	TIME
0C	S5TIME

Complex data types

0E	DT
13	STRING

Parameter types

17	BLOCK_FB
18	BLOCK_FC
19	BLOCK_DB
1A	BLOCK_SDB
1C	COUNTER
1D	TIMER

Zero pointer

00	NIL

Figure 24.1 Structure of the Pointers in STEP 7

24.2.4 ANY Pointer

In addition to the DB pointer, the ANY pointer also contains the data type and a repetition factor. This makes it possible to point to a data area.

The ANY pointer is available in two variants: For variables with data types and for variables with parameter types. If you point to a variable with a data type, the ANY pointer contains a DB pointer, the type, and a repetition factor. If the ANY pointer points to a variable with a parameter type, it contains only the number instead of the DB pointer in addition to the type. In the case of a timer or counter function, the type is repeated in byte (n+6); byte (n+7) contains B#16#00. In all other cases, these two bytes contain the value W#16#0000.

The first byte of the ANY pointer contains the syntax ID; in STEP 7 it is always 10hex. The type specifies the data type of the variables for which the ANY pointer applies. Variables of elementary data type, DT and STRING receive the type shown and the quantity 1.

If you apply a variable of data type ARRAY or STRUCT (also UDT) at an ANY parameter, the Editor generates an ANY pointer to the field or the structure. This ANY pointer contains the ID for BYTE (02hex) as the type, and the byte length of the variable as the quantity. The data type of the individual field or structure components is not significant here. Thus, an ANY pointer points with double the number of bytes to a WORD field. Exception: A pointer to a field consisting of components of data type CHAR is also created with CHAR type (03_{hex}).

You can apply an ANY pointer at a block parameter of parameter type ANY if you want to point to a variable or an operand area. The constant representation for data types is as follows: P#[DataBlock.]Operand Type Quantity

Examples:

▷ P#DB 11.DBX 30.0 INT 12
 Area with 12 words in DB 11 beginning DBB 30

▷ P#M 16.0 BYTE 8
 Area with 8 bytes beginning MB 16

▷ P#E 18.0 WORD 1
 Input word IW 18

▷ P#E 1.0 BOOL 1
 Input I 1.0

In the case of parameter types, you write the pointer as follows: L#Number Type Quantity

Examples:

▷ L#10 TIMER 1
 Timer function T 10

▷ L#2 COUNTER 1
 Counter function C 2

The Editor then applies an ANY pointer that agrees in type and quantity with the specifications in the constant representation. Please note that the operand address in the ANY pointer must also be a bit address for data types.

Specification of a constant ANY pointer makes sense if you want to access a data area for which you have not declared any variables. In principle, you can also apply variables or operands at an ANY parameter. For example, the representation "P#I 1.0 BOOL 1" is identical with "I 1.0" or the corresponding symbolic address.

If you do not specify any defaults when declaring an ANY parameter at a function block, the Editor assigns 10hex to the syntax ID and 00hex to the remaining bytes. The Editor then represents this (empty) ANY pointer (in the data view) as follows: P#P0.0 VOID 0.

24.2.5 "Variable" ANY Pointer

When copying with SFC 20, you specify in the source and destination parameters either an absolute-addressed area (for example, P#DB127.DBX0.0 BYTE 32) or a variable. In both cases, the source area and the destination area are fixed during programming (variable indexing is not possible even for array components). The following method is available for runtime modification of a data area created at a block parameter of type ANY:

You create a variable of data type ANY in the temporary local data and use this to initialize an ANY parameter. The Program Editor then does not generate an ANY pointer (as it would if you created a different variable), but uses the ANY variable in the temporary local data as an ANY pointer to a source or destination area. The ANY variable in the temporary local data is structured in the same way as an ANY pointer;

you can now modify the individual entries at runtime.

This procedure functions not only in the case of SFC 20 BLKMOV but also with block parameters of type ANY in other blocks.

The "LAD_BOOK" and "FBD_Book" libraries on the diskette accompanying the book each contain a "Message Frame Example" program, which in turn both contain an example of the "variable" ANY pointer.

24.3 Brief Description of the "Message Frame Example"

This example deals primarily with how data is handled, and is broken down as follows:

▷ Message frame data, shows how to handle data structures

▷ Time-of-day check, shows how to handle system blocks and standard blocks

▷ Editing the message frame, shows the use of SFC 20 BLKMOV with fixed addresses

▷ Indirect copying of the data area, shows an "indirect copy" function using "variable" ANY pointers

▷ Save message frame, shows the use of "indirect copying"

Figure 24.2 shows the program and data structure for this example.

Message frame data

The example shows how you can define frequently occurring data structures as your own data type and how you use this data type when declaring variables and parameters.

We construct a data store for incoming and outgoing message frames: A Send mailbox with the structure of a message frame, a Receive mailbox with the same structure, and a (receive) ring buffer that is to provide intermediate storage for incoming message frames. Since the da-

Sample message frame

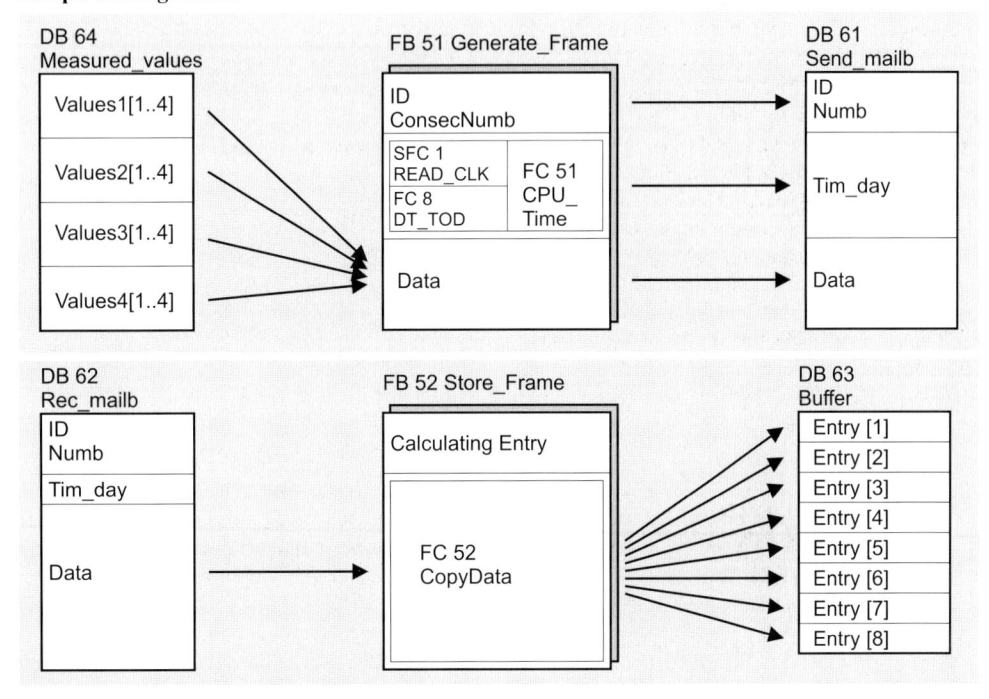

Figure 24.2 Data Structure for the Message Frame Data Example

ta structure of the message frame occurs frequently, we will define it as a user-defined data type (UDT) frame. The message frame contains a frame header whose structure we also want to give a name to. The Send mailbox and the Receive mailbox are to be data blocks that each contain a variable with the structure of the *frame*. Finally, there is also a ring buffer, a data block with an array consisting of eight components that also have the same data structure as the *frame*.

Time-of-day check

The example shows how to handle system and standard blocks (evaluating errors, copying from the library, renaming).

The time-of-day check function is to output the time-of-day in the integral CPU real-time clock as a function value. For this purpose, we require the system function SFC 1 READ_CLK, which reads the date and the time-of-day from the real-time clock in the DATE_AND_TIME or DT data format. Since we only want to read the time-of-day, we also require the IEC function FC 8 DT_TOD. This function fetches the time-of-day in the TIME_OF_DAY or TOD format from the DT data format.

Error evaluation

The system functions signal an error via binary result parameter BR and via function value RET_VAL. An error occurred if binary result BR = "0"; the function value is then also negative (bit 15 is set). The IEC standard functions signal an error only via the binary result. Both types of error evaluation are shown in the example. If an error was encountered, an invalid value is output for the time-of-day. In addition, the binary result is affected. After the time-of-day check function has been called, you can therefore also use the binary result to see if an error has occurred.

Offline programming of system functions

Before saving the input block, system function SFC 1 and standard function FC 8 must be included in the offline user program. Both functions are included in the STEP 7 standard package. You will find these functions in the block libraries provided. (For the system functions integrated into the CPU, the library contains only an interface description, not the actual system functions program. The function can be called offline via this interface description; the interface description is not transferred to the CPU. Loadable functions such as the IEC functions are available in the library as executable programs.)

Select the library *Standard Library* with FILE → OPEN → LIBRARY under the SIMATIC Manager and open the *System Function Blocks* library. Under *Blocks, you* will find all interface descriptions for the system functions. If you still have the project window of your project open, you can display the two windows side-by-side with WINDOW → ARRANGE → VERTICALLY and "drag" the selected system functions into your program using the mouse (select the SFC with the mouse, hold the mouse key down, drag to *Blocks* or into its open window, and "drop"). Copy standard function FC 8 in the same way. You will find it on the *IEC Function Blocks* library. FC 8 is a loadable function; it therefore reserves user memory, in contrast to SFC 1.

If a standard function block is called under "Libraries" in the Program Element Catalog using the Editor, it is automatically copied into *Blocks* and entered in the Symbol Table.

Renaming standard functions

You can rename a loadable standard function. Select the standard function (for example FC 8) in the project window and click once (again) on the designator. The name appears in a frame and you can specify a new address (for example, FC 98). If you press the F1 key while the standard function (renamed to FC 98) is still selected, you will nevertheless receive the online help for the original standard function (FC 8).

If an identically addressed block exists when copying, a dialog box appears where you can choose between overwriting and renaming.

Symbolic address

You can assign names to the system functions and the standard functions in the Symbol Table so that you can also access these functions symbolically. You can assign these names freely within the permissible limits applicable to block names. In the example, the block name in each case is selected as a symbolic name (for better identification).

24.3 Brief Description of the "Message Frame Example"

Editing the message frame

The data block *Send_Mailb is* to be filled with the data for a message frame. We use a function block that has the ID and the consecutive number stored in its instance data block. The net data are stored finally in a global data block; they are copied into the Send mailbox with system function BLKMOV. We use the time-of-day check function to take the time-of-day from the CPU's real-time clock.

The first network in the function block FB *Generate_Frame* transfers the ID stored in the instance data block to the frame header. The consecutive number is incremented by +1 and is also transferred to the frame header.

The second network contains the *READ_CLK* function call that takes the time-of-day from the real-time clock and enters it in the frame header in TIME_OF_DAY format.

In the subsequent networks, you will see a method of copying selected variables at runtime with the system function SFC 20 BLKMOV and without using indirect addressing. It is therefore also not necessary to know the absolute address or the structure of the variables. The principle is extremely simple: The desired copy function is selected using comparison functions. The numbers 1 to 14 are permissible as selection criteria.

FB *Generate_Frame is* programmed in such a way that it is called via a signal edge to generate a message frame.

Indirect copying of a data area

The example shows the editing and use of a "variable" ANY pointer with graphical program elements.

The *I_Copy* function copies a data area whose address and length you can set as required via block parameters. The individual block parameters correspond to the individual elements of an ANY pointer (see Section 24.2.4, "ANY Pointer"). The specifications in the block parameters must be valid values; they are not checked (SFC 20 BLKMOV signals a copy error in its function value parameter, which is transferred to the *I_Copy* function's function value parameter).

Essential elements are the two temporary variables *SoPointer* and *DesPointer*, which are *of data type ANY.* They contain the ANY pointers for the system function SFC 20 BLKMOV. *SoPointer* points to the source area of the data to be transferred and *DesPointer* points to the destination area. Figure 24.3 shows the structure of the *SoPointer* variable; *DesPointer* has the same structure. The individual bytes, words and doublewords of the ANY variables are accessed via their absolute addresses.

Save message frame

The example shows the use of the *i_Copy* function (copying a data area with programmable address).

A message frame in the data block *Rec_Mailb is* to be written to the next location in data block *Buffer*. The block-local variable *Entry* determines the location in the ring buffer; the value of this location is used to calculate the address in the ring buffer.

The *Entry* variable has a value range of from 0 to 7. In the first network, a comparator determines whether *Entry is* less than 7. If this is the case, *Entry is* incremented by 1 in the next network, otherwise it is set to zero. *Entry* multiplied by 16 gives the absolute byte address of the next entry in the ring buffer (the data structure *Message frame* consists of 16 bytes).

The *I_Copy* function, which copies the message frame from the Receive mailbox (data block DB 62) to the ring buffer (data block DB 63), is called in Network 3.

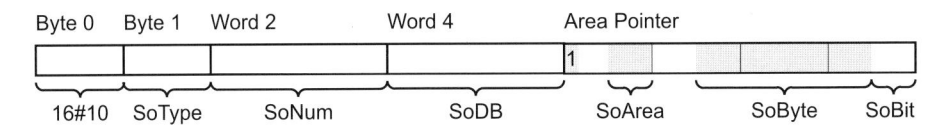

Byte 0	Byte 1	Word 2	Word 4	Area Pointer			
16#10	SoType	SoNum	SoDB	1 SoArea	SoByte	SoBit

Figure 24.3 Structure of the Variable *SoPointer*

25 Block Libraries

The STEP 7 Basic software includes the *Standard Library* which contains the following library programs:

▷ Organization Blocks
▷ System Function Blocks
▷ IEC Function Blocks
▷ S5-S7 Converting Blocks
▷ TI-S7 Converting Blocks
▷ PID Control Blocks
▷ Communication Blocks

You can copy blocks or interface descriptions into Version 3 projects from the library programs described. If you want to transfer blocks or interface descriptions into a Version 2 project, you must use the library *stdlibs*.

25.1 Organization Blocks

(Prio = Default priority class)

OB	Prio	Designation
1	1	Main program
10	2	Time-of-day interrupt 0
11	2	Time-of-day interrupt 1
12	2	Time-of-day interrupt 2
13	2	Time-of-day interrupt 3
14	2	Time-of-day interrupt 4
15	2	Time-of-day interrupt 5
16	2	Time-of-day interrupt 6
17	2	Time-of-day interrupt 7
20	3	Time-delay interrupt 0
21	4	Time-delay interrupt 1
22	5	Time-delay interrupt 2
23	6	Time-delay interrupt 3

OB	Prio	Designation
30	7	Watchdog interrupt 0 (5 s)
31	8	Watchdog interrupt 1 (2 s)
32	9	Watchdog interrupt 2 (1 s)
33	10	Watchdog interrupt 3 (500 ms)
34	11	Watchdog interrupt 4 (200 ms)
35	12	Watchdog interrupt 5 (100 ms)
36	13	Watchdog interrupt 6 (50 ms)
37	14	Watchdog interrupt 7 (20 ms)
38	15	Watchdog interrupt 8 (10 ms)
40	16	Hardware interrupt 0
41	17	Hardware interrupt 1
42	18	Hardware interrupt 2
43	19	Hardware interrupt 3
44	20	Hardware interrupt 4
45	21	Hardware interrupt 5
46	22	Hardware interrupt 6
47	23	Hardware interrupt 7
60	25	Multiprocessor interrupt
70	25	I/O redundancy error [1]
72	28	CPU redundancy error
80	26	Time error [1]
81	26	Power supply fault [1]
82	26	Diagnostics interrupt [1]
83	26	Insert/remove-module interrupt [1]
84	26	CPU hardware fault [1]
85	26	Priority class error [1]
86	26	DP error [1]
87	26	Communications error [1]
90	29	Background processing
100	27	Complete restart
101	27	Restart
102	27	Cold Restart
121	-	Programming error
122	-	I/O access error

[1] Prio = 28 at restart

25.2 System Function Blocks

IEC timers and IEC counters

SFB	Name	Designation
0	CTU	Up counter
1	CTD	Down counter
2	CTUD	Up/down counter
3	TP	Pulse
4	TON	On delay
5	TOF	Off delay

Communications via configured connections

SFB	Name	Designation
8	USEND	Uncoordinated send
9	URVC	Uncoordinated receive
12	BSEND	Block-oriented send
13	BRCV	Block-oriented receive
14	GET	Read data from partner
15	PUT	Write data to partner
16	PRINT	Write data to printer
19	START	Initiate complete restart in the partner
20	STOP	Set partner to STOP
21	RESUME	Initiate restart in the partner
22	STATUS	Check status of partner
23	USTATUS	Receive status of partner

SFC	Name	Designation
62	CONTROL	Check communications status

Integrated functions CPU 312/314/614

SFB	Name	Designation
29	HS_COUNT	High-speed counter
30	FREQ_MES	Frequency meter
38	HSC_A_B	Control "Counter A/B"
39	POS	Control "Positioning"
41	CONT_C	Continuous closed-loop control
42	CONT_S	Step-action control
43	PULSEGEN	Generate pulse

SFC	Name	Designation
63	AB_CALL	Call assembler block

System diagnostics

SFC	Name	Designation
6	RD_SINFO	Read start information
51	RDSYSST	Read SYS ST sublist
52	WR_USMSG	Entry in the diagnostics buffer

Create block-related messages

SFB	Name	Designation
33	ALARM	Messages with acknowledgment display
34	ALARM_8	Messages without accompanying values
35	ALARM_8P	Messages with accompanying values
36	NOTIFY	Messages without acknowledgment display
37	AR_SEND	Send archive data

SFC	Name	Designation
9	EN_MSG	Enable messages
10	DIS_MSG	Disable messages
17	ALARM_SQ	Messages that can be acknowledged
18	ALARM_S	Messages that are always acknowledged
19	ALARM_SC	Determine acknowledgment status

CPU clock and run-time meter

SFC	Name	Designation
0	SET_CLK	Set clock
1	READ_CLK	Read clock
2	SET_RTM	Set run-time meter
3	CTRL_RTM	Modify run-time meter
4	READ_RTM	Read run-time meter
48	SNC_RTCB	Synchronize slave clocks
64	TIME_TCK	Read system time

Drum

SFC	Name	Designation
32	DRUM	Drum

25 Block Libraries

Copy and block functions

SFC	Name	Designation
20	BLKMOV	Copy data area
21	FILL	Pre-assign data area
22	CREAT_DB	Generate data block
23	DEL_DB	Delete data block
24	TEST_DB	Test data block
25	COMPRESS	Compress memory
44	REPL_VAL	Enter substitute value
81	UBLKMOV	Copy Data area without gaps

Address modules

SFC	Name	Designation
5	GADR_LGC	Determine logical address
49	LGC_GADR	Determine slot
50	RD_LGADR	Determine all logical addresses

Distributed I/O

SFC	Name	Designation
7	DP_PRAL	Initiate hardware interrupt
11	DPSYN_FR	SYNC/FREEZE
12	D_ACT_DP	Deactivate or activate DP slave
13	DPNRM_DG	Read diagnostics data
14	DPRD_DAT	Read slave data
15	DPWR_DAT	Write slave data

Program control

SFC	Name	Designation
43	RE_TRIGR	Retrigger cycle time monitor
46	STP	Change to STOP state
47	WAIT	Wait for delay time

Data record transfer

SFC	Name	Designation
54	RD_DPARM	Read predefined parameter
55	WR_PARM	Write dynamic parameter
56	WR_DPARM	Write predefined parameter
57	PARM_MOD	Parameterize module
58	WR_REC	Write data record
59	RD_REC	Read data record

Process image updating

SFC	Name	Designation
26	UPDAT_PI	Update process-image input table
27	UPDAT_PO	Update process-image output table
79	SET	Set I/O bit field
80	RSET	Reset I/O bit field

Interrupt events

SFC	Name	Designation
28	SET_TINT	Set time-of-day interrupt
29	CAN_TINT	Cancel time-of-day interrupt
30	ACT_TINT	Activate time-of-day interrupt
31	QRY_TINT	Query time-of-day interrupt
32	SRT_DINT	Start time-delay interrupt
33	CAN_DINT	Cancel time-delay interrupt
34	QRY_DINT	Query time-delay interrupt
35	MP_ALM	Trigger multiprocessor alarm
36	MSK_FLT	Mask synchronous errors
37	DMSK_FLT	Unmask synchronous errors
38	READ_ERR	Read event status register
39	DIR_IRT	Disable asynchronous errors
40	EN_IRT	Enable asynchronous errors
41	DIS_AIRT	Delay asynchronous errors
42	EN_AIRT	Enable asynchronous errors

Communications via unconfigured connections

SFC	Name	Designation
65	X_SEND	Send data externally
66	X_RCV	Receive data externally
67	X_GET	Read data externally
68	X_PUT	Write data externally
69	X_ABORT	Abort external connection
72	I_GET	Read data internally
73	I_PUT	Write data internally
74	I_ABORT	Abort internal connection

Global data communications

SFC	Name	Designation
60	GD_SND	Send GD packet
61	GD_RCV	Receive GD packet

H-CPU

SFC	Name	Designation
90	H_CTRL	Control Operationg Modes on H-CPU

25.3 IEC Function Blocks

Comparisons

FC	Name	Designation
9	EQ_DT	Compare DT for equal to
28	NE_DT	Compare DT for not equal to
14	GT_DT	Compare DT for greater than
12	GE_DT	Compare DT for greater than or equal to
23	LT_DT	Compare DT for less than
18	LE_DT	Compare DT for less than or equal to
10	EQ_STRNG	Compare STRING for equal to
29	NE_STRNG	Compare STRING for not equal to
15	GT_STRNG	Compare STRING for greater than
13	GE_STRNG	Compare STRING for greater than or equal to
24	LT_STRNG	Compare STRING for less than
19	LE_STRNG	Compare STRING for less than or equal to

Date and time functions

FC	Name	Designation
3	D_TOD_DT	Combine DATE and TOD to DT
6	DT_DATE	Extract DATE from DT
7	DT_DAY	Extract day-of-the-week from DT
8	DT_TOD	Extract TOD from DT
33	S5TI_TIM	Convert S5TIME to TIME
40	TIM_S5TI	Convert TIME to S5TIME

1	AD_DT_TM	Add TIME to DT
35	SB_DT_TM	Subtract TIME from DT
34	SB_DT_DT	Subtract DT from DT

Math functions

FC	Name	Designation
22	LIMIT	Limiter
25	MAX	Maximum selection
27	MIN	Minimum selection
26	SEL	Binary selection

String functions

FC	Name	Designation
21	LEN	Length of a STRING
20	LEFT	Left section of a STRING
32	RIGHT	Right section of a STRING
26	MID	Middle section of a STRING
2	CONCAT	Concatenate STRINGs
17	INSERT	Insert STRING
4	DELETE	Delete STRING
31	REPLACE	Replace STRING
11	FIND	Find STRING
16	I_STRNG	Convert INT to STRING
5	DI_STRNG	Convert DINT to STRING
30	R_STRNG	Convert REAL to STRING
38	STRNG_I	Convert STRING to INT
37	STRNG_DI	Convert STRING to DINT
39	STRNG_R	Convert STRING to REAL

25.4 S5-S7 Converting Blocks

Floating-point arithmetic

FC	Name	Designation
61	GP_FPGP	Convert fixed-point to floating-point
62	GP_GPFP	Convert floating-point to fixed-point
63	GP_ADD	Add floating-point numbers
64	GP_SUB	Subtract floating-point numbers
65	GP_MUL	Multiply floating-point numbers
66	GP_DIV	Divide floating-point numbers
67	GP_VGL	Compare floating-point numbers
68	GP_RAD	Find the square root of a floating-point number

25 Block Libraries

Basic functions

FC	Name	Designation
85	ADD_32	32-bit fixed-point adder
86	SUB_32	32-bit fixed-point subtractor
87	MUL_32	32-bit fixed-point multiplier
88	DIV_32	32-bit fixed-point divider
89	RAD_16	16-bit fixed-point square root extractor
90	REG_SCHB	Bitwise shift register
91	REG_SCHW	Wordwise shift register
92	REG_FIFO	Buffer (FIFO)
93	REG_LIFO	Stack (LIFO)
94	DB_COPY1	Copy data area (direct)
95	DB_COPY2	Copy data area (indirect)
96	RETTEN	Save scratchpad memory (S5-155U)
97	LADEN	Load scratchpad memory (S5-155U)
98	COD_B8	BCD-binary conversion 8 decades
99	COD_32	Binary-BCD conversion 8 decades

Signal functions

FC	Name	Designation
69	MLD_TG	Clock pulse generator
70	MLD_TGZ	Clock pulse generator with timer function
71	MLD_EZW	Initial value single blinking wordwise
72	MLD_EDW	Initial value double blinking wordwise
73	MLD_SAMW	Group signal wordwise
74	MLD_SAM	Group signal
75	MLD_EZ	Initial value single blinking
76	MLD_ED	Initial value double blinking
77	MLD_EZWK	Initial value single blinking (wordwise) memory bit
78	MLD_EZDK	Initial value double blinking (wordwise) memory bit
79	MLD_EZK	Initial value single blinking memory bit
80	MLD_EDK	Initial value double blinking memory bit

Integrated functions

FC	Name	Designation
81	COD_B4	BCD-binary conversion 4 decades
82	COD_16	Binary-BCD conversion 4 decades
83	MUL_16	16-bit fixed-point multiplier
84	DIV_16	16-bit fixed-point divider

Analog functions

FC	Name	Designation
100	AE_460_1	Analog input module 460
101	AI_460_2	Analog input module 460
102	AI_463_1	Analog input module 463
103	AE_463_2	Analog input module 463
104	AE_464_1	Analog input module 464
105	AE_464_2	Analog input module 464
106	AE_466_1	Analog input module 466
107	AE_466_2	Analog input module 466
108	RLG_AA1	Analog output module
109	RLG_AA2	Analog output module
110	PER_ET1	ET 100 distributed I/O
111	PER_ET2	ET 100 distributed I/O

Math functions

FC	Name	Designation
112	SINUS	Sine
113	COSINUS	Cosine
114	TANGENS	Tangent
115	COTANG	Cotangent
116	ARCSIN	Arc sine
117	ARCCOS	Arc cosine
118	ARCTAN	Arc tangent
119	ARCCOT	Arc cotangent
120	LN_X	Natural logarithm
121	LG_X	Logarithm to base 10
122	B_LOG_X	Logarithm to any base
123	E_H_N	Exponential function with base e
124	ZEHN_H_N	Exponential function with base 10
125	A2_H_A1	Exponential function with any base

25.5 TI-S7 Converting Blocks

FB	Name	Designation
80	LEAD_LAG	Lead/lag algorithm
81	DCAT	Discrete control time interrupt
82	MCAT	Motor control time interrupt
83	IMC	Index matrix comparison
84	SMC	Matrix scanner
85	DRUM	Event maskable drum
86	PACK	Collect/distribute table data

FC	Name	Designation
80	TONR	Latching ON delay
81	IBLKMOV	Transfer data area indirectly
82	RSET	Reset process image bit by bit
83	SET	Set process image bit by bit
84	ATT	Enter value in table
85	FIFO	Output first value in table
86	TBL_FIND	Find value in table
87	LIFO	Output last value in table
88	TBL	Execute table operation
89	TBL_WRD	Copy value from the table
90	WSR	Save datum
91	WRD_TBL	Combine table element
92	SHRB	Shift bit in bit shift register
93	SEG	Bit pattern for 7-segment display
94	ATH	ASCII-hexadecimal conversion
95	HTA	Hexadecimal-ASCII conversion
96	ENCO	Least significant set bit
97	DECO	Set bit in word
98	BCDCPL	Generate ten's complement
99	BITSUM	Count set bits
100	RSETI	Reset PQ byte by byte
101	SETI	Set PQ byte by byte
102	DEV	Calculate standard deviation
103	CDT	Correlated data tables
104	TBL_TBL	Table combination
105	SCALE	Scale values
106	UNSCALE	Unscale values

25.6 PID Control Blocks

FB	Name	Designation
41	CONT_C	Continuous control
42	CONT_S	Step control
43	PULSGEN	Generate pulse

25.7 Communication Blocks

FC	Name	Designation
1	DP_SEND	Send data
2	DP_RECV	Receive data
3	DP_DIAG	Diagnostics
4	DP_CTRL	Control

26 Function Set LAD

26.1 Basic Functions

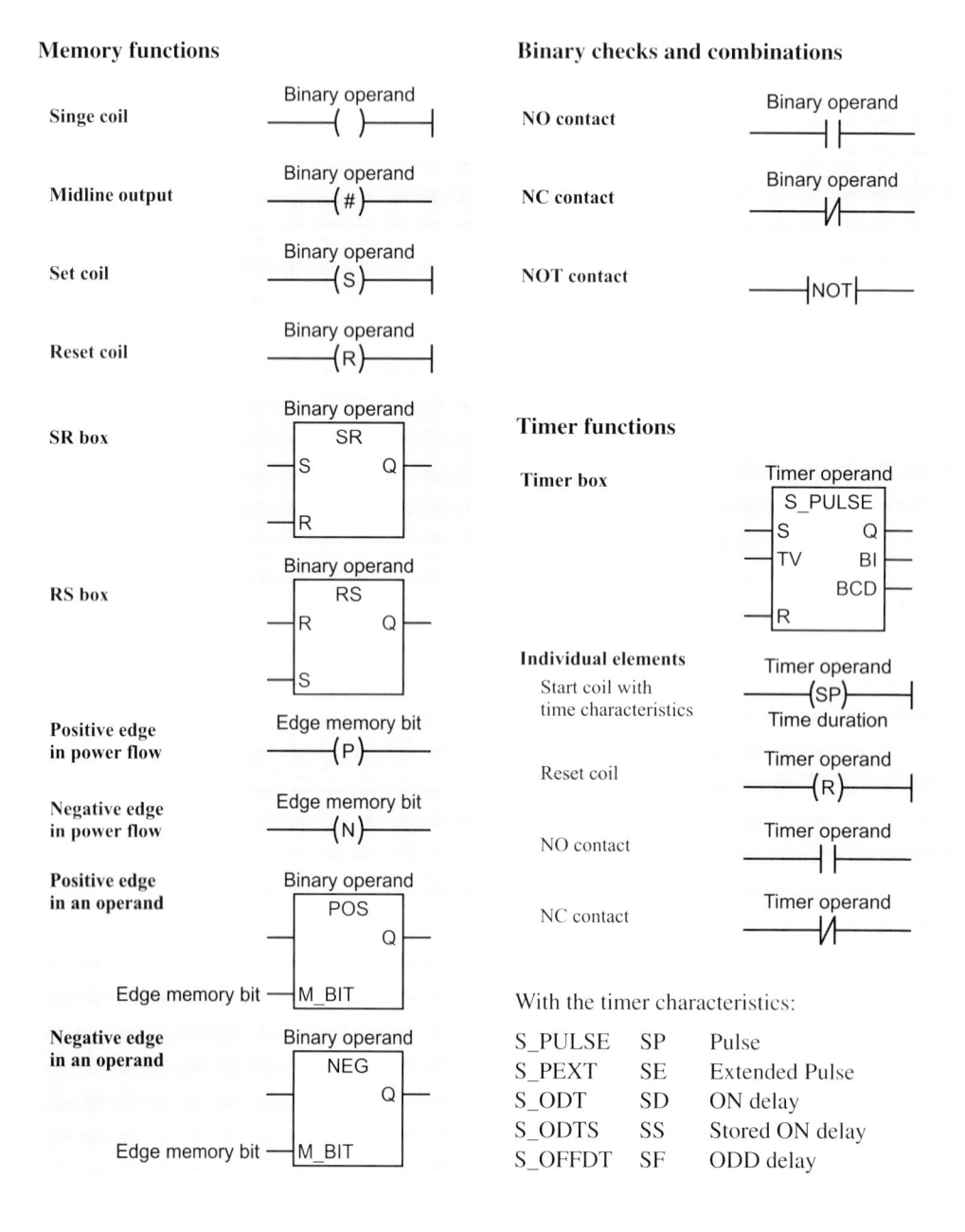

Memory functions

Singe coil — ( ) — Binary operand

Midline output — (#) — Binary operand

Set coil — (S) — Binary operand

Reset coil — (R) — Binary operand

SR box — Binary operand, SR: S, Q, R

RS box — Binary operand, RS: R, Q, S

Positive edge in power flow — (P) — Edge memory bit

Negative edge in power flow — (N) — Edge memory bit

Positive edge in an operand — Binary operand, POS: Q, M_BIT — Edge memory bit

Negative edge in an operand — Binary operand, NEG: Q, M_BIT — Edge memory bit

Binary checks and combinations

NO contact — | | — Binary operand

NC contact — |/| — Binary operand

NOT contact — |NOT| —

Timer functions

Timer box — Timer operand, S_PULSE: S, Q, TV, BI, BCD, R

Individual elements

Start coil with time characteristics — (SP) — Timer operand / Time duration

Reset coil — (R) — Timer operand

NO contact — | | — Timer operand

NC contact — |/| — Timer operand

With the timer characteristics:

S_PULSE	SP	Pulse
S_PEXT	SE	Extended Pulse
S_ODT	SD	ON delay
S_ODTS	SS	Stored ON delay
S_OFFDT	SF	ODD delay

324

26.2 Digital Functions

Transfer functions

MOVE box

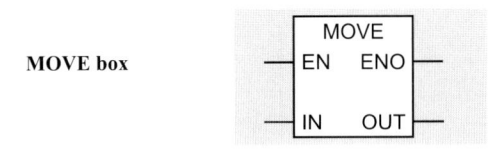

```
    MOVE
─┤EN   ENO├─
 ┤IN   OUT├─
```

Counter functions

Counter box

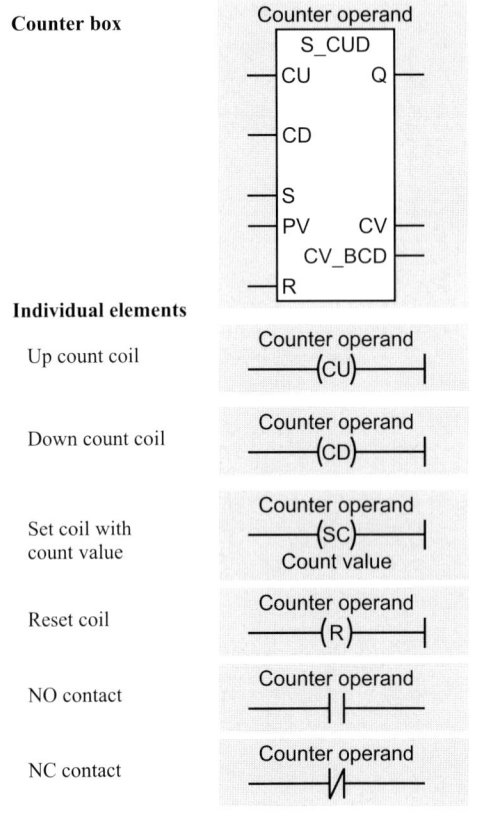

```
      Counter operand
        S_CUD
─┤CU       Q├─
─┤CD
─┤S
─┤PV      CV├─
         CV_BCD├─
─┤R
```

Individual elements

Up count coil

```
Counter operand
────(CU)────
```

Down count coil

```
Counter operand
────(CD)────
```

Set coil with count value

```
Counter operand
────(SC)────
  Count value
```

Reset coil

```
Counter operand
────(R)────
```

NO contact

```
Counter operand
────| |────
```

NC contact

```
Counter operand
────|/|────
```

With the counter characteristics:

S_CUD	Up/down counter
S_CU	Up counter
S_CD	Down counter

Comparison functions

Comparison box

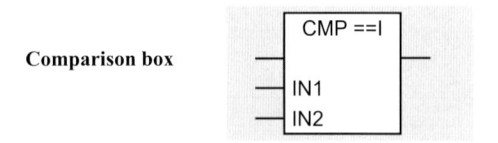

```
      CMP ==I
─┤IN1
─┤IN2
```

Compare for	according to		
	INT	DINT	REAL
equal to	==I	==D	==R
not equal to	<>I	<>D	<>R
greater than	>I	>D	>R
greater than or equal to	>=I	>=D	>=R
less than	<I	<D	<R
less than or equal to	<=I	<=D	<=R

Arithmetic functions

Arithmetic box

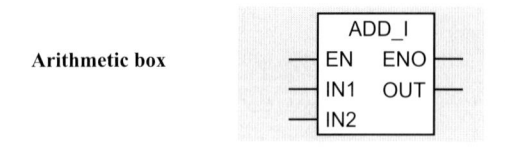

```
      ADD_I
─┤EN   ENO├─
─┤IN1  OUT├─
─┤IN2
```

Calculation	according to		
	INT	DINT	REAL
Addition	ADD_I	ADD_DI	ADD_R
Subtraction	SUB_I	SUB_DI	SUB_R
Multiplication	MUL_I	MUL_DI	MUL_R
Division	DIV_I	DIV_DI	DIV_R
Modulo	-	MOD_DI	-

Mathematical Functions

Math box

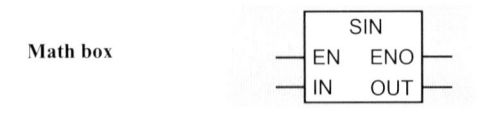

```
    ┌─────────┐
    │   SIN   │
   ─┤ EN  ENO ├─
   ─┤ IN  OUT ├─
    └─────────┘
```

SIN	Sine
COS	Cosine
TAN	Tangens
ASIN	Tangent
ACOS	Arc sine
ATAN	Arc tangent
SQR	Finding the square
SQRT	Finding the square root
EXP	Establishing the exponent
LN	Finding the logarithm

Conversion Functions

Conversion box

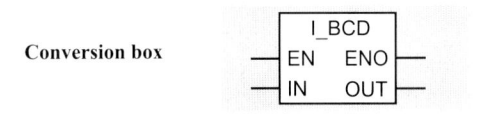

```
    ┌─────────┐
    │  I_BCD  │
   ─┤ EN  ENO ├─
   ─┤ IN  OUT ├─
    └─────────┘
```

I_DI	Conversion of INT to DINT
I_BCD	Conversion of INT to BCD
DI_BCD	Conversion of DINT to BCD
DI_R	Conversion of DINT to REAL
BCD_I	Conversion of BCD to INT
BCD_DI	Conversion of BCD to DINT
	Conversion of REAL to DINT with rounding
CEIL	to next higher number
FLOOR	to next lower number
ROUND	to next whole number
TRUNC	without rounding
INV_I	INT one's complement
INV_DI	DINT one's complement
NEG_I	INT negation
NEG_DI	DINT negation
NEG_R	REAL negation
ABS	REAL absolute-value generation

Shift Functions

Shift box

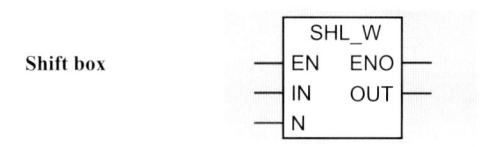

```
    ┌─────────┐
    │  SHL_W  │
   ─┤ EN  ENO ├─
   ─┤ IN  OUT ├─
   ─┤ N       │
    └─────────┘
```

SHL_W	Shift word left
SHL_DW	Shift doubleword left
SHR_W	Shift word right
SHR_DW	Shift doubleword right
SHR_I	Shift word with sign
SHR_DI	Shift doubleword with sign
ROL_DW	Rotate left
ROR_DW	Rotate right

Word Logic

Word logic box

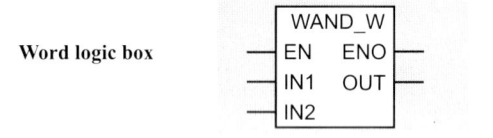

```
    ┌─────────┐
    │ WAND_W  │
   ─┤ EN  ENO ├─
   ─┤ IN1 OUT ├─
   ─┤ IN2     │
    └─────────┘
```

WAND_W	AND word
WOR_W	OR word
WXOR_W	Exclusive OR word
WAND_DW	AND doubleword
WOR_DW	OR doubleword
WXOR_DW	Exclusive OR doubleword

26.3 Program Flow Control

Status bits

Description	Symbol
Result greater than zero	—|>0|—
Result greater than or equal to zero	—|>=0|—
Result less than zero	—|<0|—
Result less than or equal to zero	—|<=0|—
Result not equal to zero	—|<>0|—
Result equal to zero	—|==0|—
Result invalid (unordered)	—|UO|—
Number range overflow	—|OV|—
Stored overflow	—|OS|—
Binary result	—|BR|—
SAVE coil	—(SAVE)—

Jump functions

Description	Symbol
Jump if RLO = "1"	Dest —(JMP)—
Jump if RLO = "0"	Dest —(JMPN)—
Jump destination	Dest

Master Control Relay

Description	Symbol
Activate MCR area	—(MCRA)—
Deactivate MCR area	—(MCRD)—
Open MCR zone	—(MCR<)—
Close MCR zone	—(MCR>)—

Block functions

Description	Symbol
Calling a function block with data block	DB x / FB x — EN ENO — IN1 OUT1 — IN2 OUT2
Calling a system function block with data block	DB x / SFB x — EN ENO — IN1 OUT1 — IN2 OUT2
Calling a function block or a system function block as local instance	#name — EN ENO — IN1 OUT1 — IN2 OUT2
Calling a function	FC x — EN ENO — IN1 OUT1 — IN2 OUT2
Calling a system function	SFC x — EN ENO — IN1 OUT1 — IN2 OUT2
Calling a parameter-free function	FC x —(CALL)—
Calling a parameter-free system function	SFC x —(CALL)—
RET coil, conditional block end	—(RET)—
Open data block	DB x —(OPN)—

27 Function Set FBD

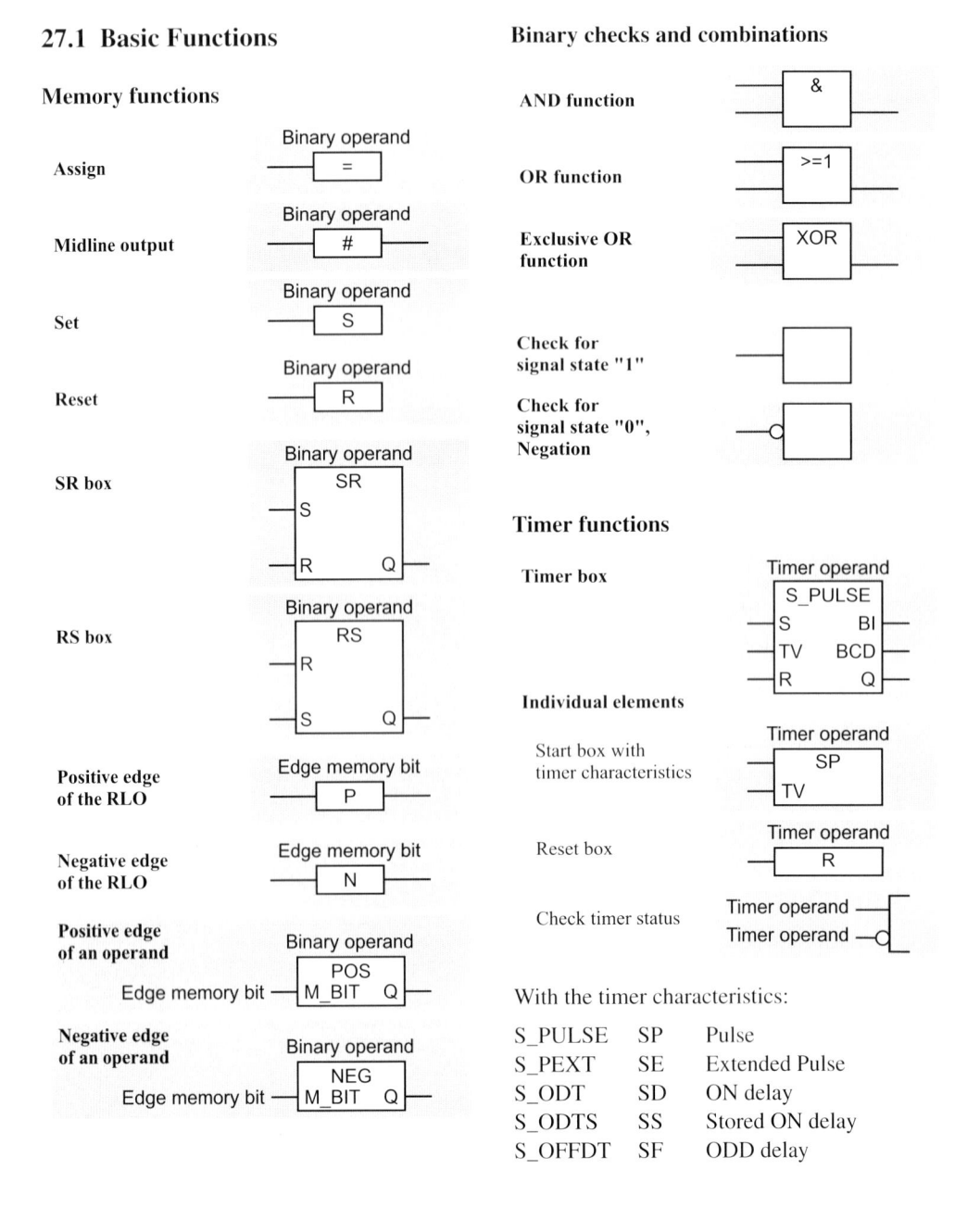

27.1 Basic Functions

Memory functions

Assign — Binary operand `=`

Midline output — Binary operand `#`

Set — Binary operand `S`

Reset — Binary operand `R`

SR box — Binary operand
```
   SR
—S
—R   Q—
```

RS box — Binary operand
```
   RS
—R
—S   Q—
```

Positive edge of the RLO — Edge memory bit `P`

Negative edge of the RLO — Edge memory bit `N`

Positive edge of an operand — Binary operand
```
      POS
—M_BIT  Q—
```
Edge memory bit

Negative edge of an operand — Binary operand
```
      NEG
—M_BIT  Q—
```
Edge memory bit

Binary checks and combinations

AND function — `&`

OR function — `>=1`

Exclusive OR function — `XOR`

Check for signal state "1"

Check for signal state "0", Negation

Timer functions

Timer box — Timer operand
```
   S_PULSE
—S        BI—
—TV      BCD—
—R        Q—
```

Individual elements

Start box with timer characteristics — Timer operand
```
   SP
—TV
```

Reset box — Timer operand `R`

Check timer status — Timer operand / Timer operand

With the timer characteristics:

S_PULSE	SP	Pulse
S_PEXT	SE	Extended Pulse
S_ODT	SD	ON delay
S_ODTS	SS	Stored ON delay
S_OFFDT	SF	ODD delay

27.2 Digital Functions

Transfer functions

MOVE box

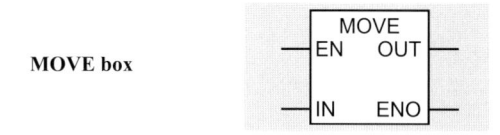

```
    ┌─────────┐
  ──┤EN  MOVE ├──
    │      OUT├──
  ──┤IN   ENO├──
    └─────────┘
```

Comparison functions

Comparison box

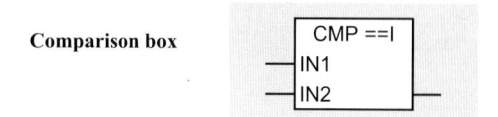

```
        ┌─────────┐
        │ CMP ==I │
      ──┤IN1      │
      ──┤IN2      │
        └─────────┘
```

Compare for	according to		
	INT	DINT	REAL
equal to	==I	==D	==R
not equal to	<>I	<>D	<>R
greater than	>I	>D	>R
greater than or equal to	>=I	>=D	>=R
less than	<I	<D	<R
less than or equal to	<=I	<=D	<=R

Counter functions

Counter box

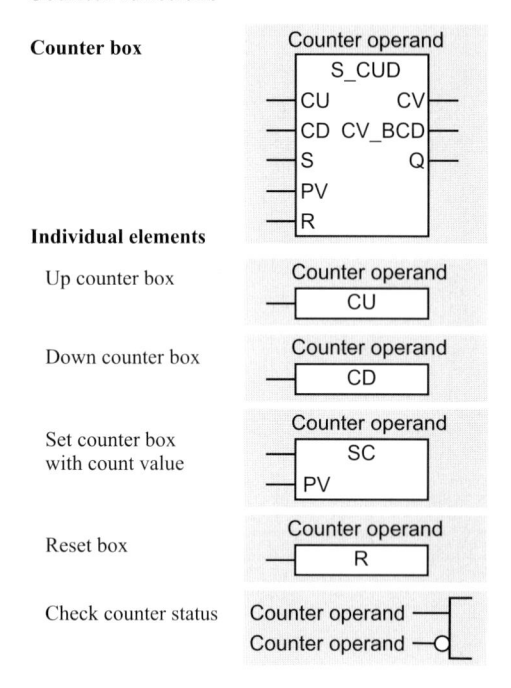

Counter operand
```
    ┌──────────┐
    │  S_CUD   │
  ──┤CU     CV ├──
  ──┤CD CV_BCD ├──
  ──┤S       Q ├──
  ──┤PV        │
  ──┤R         │
    └──────────┘
```

Arithmetic functions

Individual elements

Up counter box

Counter operand
```
    ┌────┐
  ──┤ CU │
    └────┘
```

Down counter box

Counter operand
```
    ┌────┐
  ──┤ CD │
    └────┘
```

Set counter box with count value

Counter operand
```
    ┌────┐
  ──┤ SC │
  ──┤ PV │
    └────┘
```

Arithmetic box

```
        ┌─────────┐
        │  ADD_I  │
      ──┤EN       │
      ──┤IN1   OUT├──
      ──┤IN2   ENO├──
        └─────────┘
```

Reset box

Counter operand
```
    ┌────┐
  ──┤ R  │
    └────┘
```

Check counter status

Counter operand ──┐
Counter operand ──o─

With the counter characteristics:

S_CUD	Up/down counter
S_CU	Up counter
S_CD	Down counter

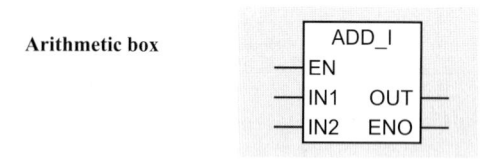

Calculation	according to		
	INT	DINT	REAL
Addition	ADD_I	ADD_DI	ADD_R
Subtraction	SUB_I	SUB_DI	SUB_R
Multiplication	MUL_I	MUL_DI	MUL_R
Division	DIV_I	DIV_DI	DIV_R
Modulo	-	MOD_DI	-

329

27 Function Set FBD

Mathematical Functions

Mathematics box

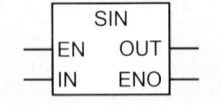

```
    SIN
─┤EN  OUT├─
─┤IN  ENO├─
```

SIN	Sine
COS	Cosine
TAN	Tangens
ASIN	Tangent
ACOS	Arc sine
ATAN	Arc tangent
SQR	Finding the square
SQRT	Finding the square root
EXP	Establishing the exponent
LN	Finding the logarithm

Shift Functions

Shift box

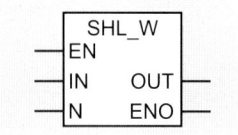

```
   SHL_W
─┤EN     ├─
─┤IN  OUT├─
─┤N   ENO├─
```

SHL_W	Shift word left
SHL_DW	Shift doubleword left
SHR_W	Shift word right
SHR_DW	Shift doubleword right
SHR_I	Shift word with sign
SHR_DI	Shift doubleword with sign
ROL_DW	Rotate left
ROR_DW	Rotate right

Conversion Functions

Conversion box

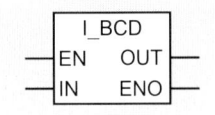

```
    I_BCD
─┤EN  OUT├─
─┤IN  ENO├─
```

Word Logic

Word logic box

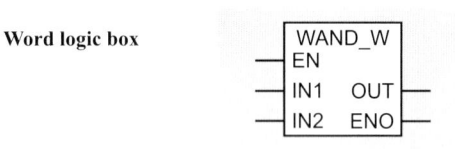

```
   WAND_W
─┤EN     ├─
─┤IN1 OUT├─
─┤IN2 ENO├─
```

I_DI	Conversion of INT to DINT
I_BCD	Conversion of INT to BCD
DI_BCD	Conversion of DINT to BCD
DI_R	Conversion of DINT to REAL
BCD_I	Conversion of BCD to INT
BCD_DI	Conversion of BCD to DINT
	Conversion of REAL to DINT with rounding
CEIL	to next higher number
FLOOR	to next lower number
ROUND	to next whole number
TRUNC	without rounding
INV_I	INT one's complement
INV_DI	DINT one's complement
NEG_I	INT negation
NEG_DI	DINT negation
NEG_R	REAL negation
ABS	REAL absolute-value generation

WAND_W	AND word
WOR_W	OR word
WXOR_W	Exclusive OR word
WAND_DW	AND doubleword
WOR_DW	OR doubleword
WXOR_DW	Exclusive OR doubleword

27.3 Program Flow Control

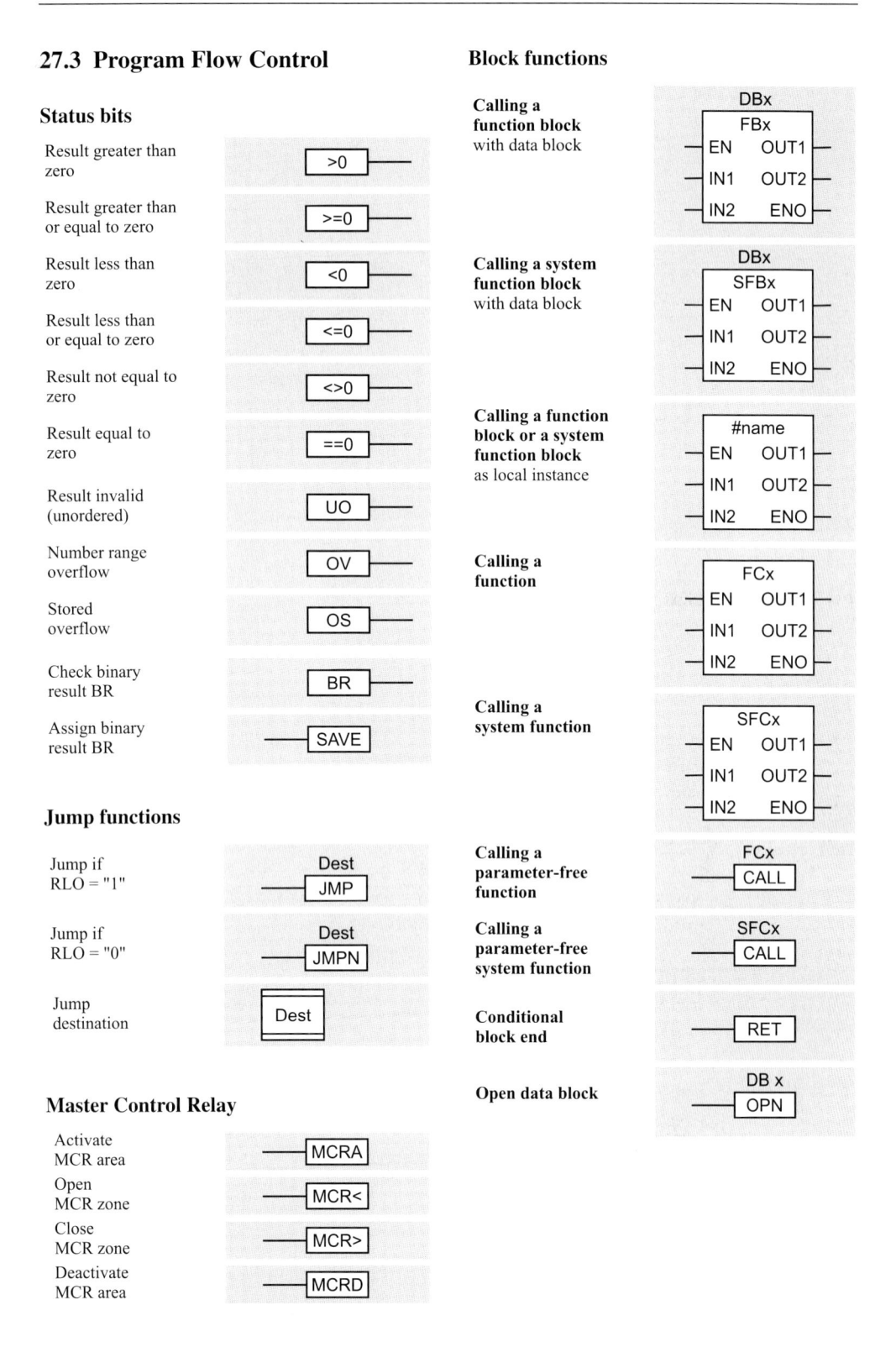

Status bits

Description	Symbol
Result greater than zero	—|>0|—
Result greater than or equal to zero	—|>=0|—
Result less than zero	—|<0|—
Result less than or equal to zero	—|<=0|—
Result not equal to zero	—|<>0|—
Result equal to zero	—|==0|—
Result invalid (unordered)	—|UO|—
Number range overflow	—|OV|—
Stored overflow	—|OS|—
Check binary result BR	—|BR|—
Assign binary result BR	—(SAVE)

Jump functions

Description	Symbol
Jump if RLO = "1"	Dest / —(JMP)
Jump if RLO = "0"	Dest / —(JMPN)
Jump destination	Dest

Master Control Relay

Description	Symbol
Activate MCR area	—(MCRA)
Open MCR zone	—(MCR<)
Close MCR zone	—(MCR>)
Deactivate MCR area	—(MCRD)

Block functions

Description	Symbol
Calling a function block with data block	DBx / FBx (EN, IN1, IN2, OUT1, OUT2, ENO)
Calling a system function block with data block	DBx / SFBx (EN, IN1, IN2, OUT1, OUT2, ENO)
Calling a function block or a system function block as local instance	#name (EN, IN1, IN2, OUT1, OUT2, ENO)
Calling a function	FCx (EN, IN1, IN2, OUT1, OUT2, ENO)
Calling a system function	SFCx (EN, IN1, IN2, OUT1, OUT2, ENO)
Calling a parameter-free function	FCx —(CALL)
Calling a parameter-free system function	SFCx —(CALL)
Conditional block end	—(RET)
Open data block	DB x —(OPN)

Index

A

Absolute-value generation   180
Actual parameters   222
Adapting block calls   81
Address priority   60
Addressing
   absolute   88
   indirect   312
   symbolic   89
AND function   107
ANY Pointer
   description   314
   introduction   136
Arc functions   172
Archiving projetcs   45
Area pointer   312
Arithmetic functions   165
ARRAY (data type)   97
Ascertaining a module address   296
Assign   117
Asynchronous errors   305
Authorization   39

B

Background Scanning OB 90   238
BCD numbers   92
Binary flags   189
Binary logic operations
   FBD   106
   LAD   102
Binary result   190
Bit memory   38
Block
   calling   205
   KNOW_HOW_PROTECT   311
   modifying online   64
   parameters   218
   programming description   78
   programming FBD   84
   programming general   57
   programming LAD   82
   properties   76
   structure   75
   transferring   64
   types   74
Block end function   207
Block functions   203
Block libraries   318
BOOL (Data type)   91
Box
   assign box   117
   call box   205
   memory box FBD   118
   memory box LAD   115
   reset box   118
   set box   118
BYTE (data type)   91

C

Call Box
   complex   205
Call box
   simple   206
Call coil   206
CHAR (data types)   91
Clock memory   38
Coil
   call coil   206
   Reset coil   113
   set coil   113
   single coil   113
Cold restart   293
Communication error OB 87   307
Communications
   distributed I/O   245
   global data   259
   indruduction   28
   SFB communication   269
   SFC communication   263

Index

Comparison functions   161
Complete restart   295
Compressing
    SFC 25 COMPRESS   243
    user program   65
Configuration table   47
Configuring stations   46
Connecting a PLC   62
Connection table   52
Constant representation   92
Contacts   102
Controlling I/O bits   202
Conversion functions   175
Counter functions
    IEC counters   154
    SIMATIC counters   150
CPU hardware faults OB 84   306
CPU information   63
Creating projetcs   43

D

Data block
    offline/online   65
    Open   214
    programming   86
Data block functions   212
Data block registers   212
Data operands
    absolute addressing   212
    fully-addressed access   213
    symbolic addressing   213
Data types
    complex   96
    elementary   91
    subdivision   90
    user-defined   99
DATE (data type)   95
DATE_AND_TIME (data type)   96
DB pointer   312
Diagnosing hardware   66
Diagnostic interrupt OB 82   308
Diagnostics address   35
Digital flags   189
DINT (data type)   93
DINT calculation   168

Disable output modules   290
Distributed I/O
    adressing   245
    configuring   248
    description   25
    system functions   256
Down counting   153
DP master system
    configuring   249
    description   25
    in network configuration   51
DWORD (data type)   91

E

Edge evaluation
    description   123
    FBD   125
    LAD   125
Editing projects   43
Editor
    FBD elements   84
    LAD elements   82
    program   57
    symbolic   56
EN/ENO Mechanism   193
Enabling peripheral outputs   69
Error handling   301
Example
    Binary scaler   126
    Conveyor belt   225
    Conveyor control   128
    Counter control   155
    Feed   228
    Parts counter   226
Exclusive OR function   108
Exponentiation   172
Extented pulse timer   144

F

First check   189
Forcing variables   68
Formal parameters   220
FREEZE   254
Fully-addressed data operands   213
Function value   220

Index

G

Global data communication 259
Global data table 261

H

Hardware catalog 47
Hardware configuration 46
Hardware interrupts 278

I

IEC counter functions 154
IEC functions library 321
IEC timer functions 147
Inputs 37
Insert/remove module interrupt OB 83 306
INT (data type) 93
INT calculation 167
Interrupt handling 277

J

Jump functions 195

L

Libraries
    communication blocks 323
    IEC function blocks 321
    organization blocks 318
    overview 318
    PID control blocks 323
    S5-S7 converting blocks 321
    system function blocks 319
    TI-S7 converting blocks 323
Library
    creating 44
    general 41
Load memory 23
Local data
    static 209
    temporary 207
Local instances 211
Logarithm 173

M

Main program OB1 234
Master Control Relay MCR 198
Mathematical functions 170
Memory bits 38
Memory box
    FBD 118
    LAD 115
Memory card 24
Memory functions
    FBD 117
    LAD 113
Memory reset 291
Midline outputs
    FBD 123
    general 120
    LAD 121
Minimum Scan Cycle Time 238
Modifying variables 68
Module start address 35
Monitoring variables 67
MOVE box 133
Move functions 132
Multiple instances, see Local instances
Multiprocessing mode 244
Multiprocessor interrupt 286

N

NC contact 102
Negation
    conversion function 180
    NOT contact 104
    scanning for "0" 110
Network configuration 49
Network templates 80
NO contact 102
Number range overflow 190

O

Off-dely timer 147
On-dely timer 144
One's complement 180
Online help 43
Open data block 214

Index

Operating modes
    HOLD   290
    intruduction   289
    RESTART   293
    RUN (main program)   234
    STOP   291
OR function   107
Organization blocks
    asynchronous errors OB 80 to OB 87   305
    background scanning OB 90   238
    interrupts OB 10 to OB 60   278
    main program OB 1   234
    restart OB 100 to OB 102   290
    synchronous errors OB 121, OB 122   301
Outputs   37
Overflow   190

P

Parallel circuits   103
Parameter types   99
Parameterizing modules   298
Peripheral inputs   36
Peripheral outputs   36
Pointer
    ANY pointer   314
    area pointer   312
    DB pointer   312
    general remarks   312
Power supply errors OB 81   306
Priority classes   72
Process image
    description   37
    subprocess images   236
    updating   236
Program editor
    FBD elements   84
    general   57
    LAD elements   82
Program elements catalog   81
Program execution errors OB 85   306
Program length   78
Program organization   235
Program processing methods   71
Program status   69
Program structure   234
Programming code blocks   78
Programming data blocks   212

Project
    archiving   45
    creating   43
    general   41
    object hierarchy   42
Project versions   45
Pulse timer   143

R

Rack failure OB 86   307
REAL (data type)   93
REAL calculation   168
Real-time clock   242
Reference data   61
Reset box   118
Reset coil   113
Response time   239
Restart types   293
Result of the logic operation   189
Retentive on-delay timer   146
Retentivity   291
Rewiring   60
Rotate functions   184
Rounding   178
Run-time meter   242

S

S5/S7 conversion library   321
S5TIME (data type)   95
Scan cycle monitoring time   237
Scan cycle statistics   238
Sensor type   110
Series circuits   103
Set box   118
Set coil   113
SFB 0 CTU   154
SFB 1 CTD   155
SFB 12 BSEND   272
SFB 13 BRCV   272
SFB 14 GET   273
SFB 15 PUT   273
SFB 16 PRINT   273
SFB 19 START   274

Index

SFB 2 CTUD   155
SFB 20 STOP   274
SFB 21 RESUME   274
SFB 22 STATUS   275
SFB 23 USTATUS   275
SFB 3 TP   148
SFB 4 TON   148
SFB 5 TOF   149
SFB 8 USEND   271
SFB 9 URCV   271
SFB communication   269
SFC 0 SET_CLK   242
SFC 1 READ_CLK   242
SFC 11 DPSYN_FR   256
SFC 12 D_ACT_DP   258
SFC 13 DPNRM_DG   258
SFC 14 DPRD_DAT   258
SFC 15 DPWR_DAT   258
SFC 2 SET_RTM   243
SFC 20 BLKMOV   136
SFC 21 FILL   137
SFC 22 CREAT_DB   216
SFC 23 DEL_DB   216
SFC 24 TEST_DB   216
SFC 25 COMPRESS   243
SFC 26 UPDAT_PI   237
SFC 27 UPDAT_PO   237
SFC 28 SET_TINT   284
SFC 29 CAN_TINT   284
SFC 3 CTRL_RTM   243
SFC 30 ACT_TINT   284
SFC 31 QRY_TINT   284
SFC 32 SRT_DINT   286
SFC 33 CAN_DINT   286
SFC 34 QRY_DINT   286
SFC 35 MP_ALM   287
SFC 36 MSK_FLT   304
SFC 37 DMSK_FLT   304
SFC 38 READ_ERR   305
SFC 39 DIS_IRT   287
SFC 4 READ_RTM   243
SFC 40 EN_IRT   287
SFC 41 DIS_AIRT   288
SFC 42 EN_AIRT   288
SFC 43 RE_TRIGR   238

SFC 44 REPL_VAL   305
SFC 46 STP   244
SFC 47 WAIT   243
SFC 48 SNC_RTCB   242
SFC 49 LGC_GADR   296
SFC 5 GADR_LGC   296
SFC 50 RD_LGADR   296
SFC 51 RDSYSST   309
SFC 52 WR_USMSG   308
SFC 54 RD_DPARM   299
SFC 55 WR_PARM   299
SFC 56 WR_DPARM   299
SFC 57 PARM_MOD   299
SFC 58 WR_REC   299
SFC 59 RD_REC   300
SFC 6 RD_SINFO   241
SFC 60 GD_SND   263
SFC 61 GD_RCV   263
SFC 62 CONTROL   276
SFC 64 TIME_TCK   242
SFC 65 X_SEND   267
SFC 66 X_RCV   268
SFC 67 X_GET   268
SFC 68 X_PUT   269
SFC 69 X_ABORT   269
SFC 7 DP_PRAL   256
SFC 72 I_GET   265
SFC 73 I_PUT   265
SFC 74 I_ABORT   266
SFC 79 SET   202
SFC 80 RSET   202
SFC 81 UBLKMOV   137
SFC Communication   263
Shift functions   181
SIMATIC counter functions   150
SIMATIC Manager   40
SIMATIC timer functions   139
Single coil   113
Slot address   34
Square-root extraction   172
Squaring   172
Start information
    interrupt handling   278
    OB 1   240
    restart   292

Index

Start-up characteristics   289
Static local data   209
Status bits
    Binary result BR   190
    Condition code bits CC0 and CC1   190
    description   189
    evaluating   192
    first check   189
    OR   190
    Overflow OV   190
    RLO   189
    Status   189
    Stored overflow OS   190
Stored Overflow   190
STRING (data type)   97
STRUCT (data type)   98
Subnets   31
Subprocess images   236
Symbol table   56
SYNC   254
Synchronous errors   301
System blocks
    calling   206
    description   75
    library   319
System diagnostics   307
System memory   24
System time   242

T

Temporary local data   207
TIME (data type)   95
TIME_OF_DAY (data type)   96
Time-delay interrupts   284
Time-of-day interrupts   282
Timer characteristics
    Extended pulse timer   144
    Off-delay timer   147
    On-delay timer   144
    Pulse timer   143
    Retentive on-delay timer   146
Timer functions
    IEC timers   147
    SIMATIC timers   139
Timing errors OB 80   306
Trigonometric functions   172
Two's complement   180

U

UDT (data type)   99
Up counting   153
User blocks   74
User data area   36
User data types   99
User memory   23
User program
    compressing   65
    load   64
    protection   63
    testing   65

V

Variable declaration table   79
Variable table   66

W

Warm restart   295
Watchdog interrupts   280
WORD (Data type)   91
Word logic operations   185
Work memory   23
Work memory   23

Abbreviations

AI	Analog input		FM	Function module
AO	Analog output		IM	Interface module
AS	Automation system		LAD	Ladder diagram
AS-I	Actuator-sensor interface		MCR	Master Control Relay
BR	Binary result		MPI	Multipoint interface
CFC	Continuous function chart		OB	Organization block
CP	Communications processor		OP	Operator panel
CPU	Central processing unit		PG	Programming device
DB	Data block		PLC	Programmable controller
DI	Digital input (module)		PS	Power supply
DO	Digital output (module)		RAM	Random access memory
DP	Distributed I/O		RLO	Result of logic operation
DR	Data record		SCL	Structured control language
DS	Data record		SDB	System data block
EPROM	Erasable programmable read-only memory		SFB	System function block
			SFC	System function call
FB	Function block		SM	Signal module
FBD	Function block diagram		SSL	System status list
FC	Function call		STL	Statement list
FEPROM	Flash erasable programmable read-only memory		UDT	User-defined data type
			VAT	Variable table